Complete ICT for Cambridge IGCSE®

A practical approach for the updated syllabus

Second edition

Stephen Doyle

Oxford excellence for Cambridge IGCSE®

OXFORD

Great Clarendon Street, Oxford, OX2 6DP, United Kingdom

Oxford University Press is a department of the University of Oxford. It furthers the University's objective of excellence in research, scholarship, and education by publishing worldwide. Oxford is a registered trade mark of Oxford University Press in the UK and in certain other countries

First published in 2015

British Library Cataloguing in Publication Data
Data available

978-0-19-839947-6

10 9 8 7 6 5

Paper used in the production of this book is a natural, recyclable product made from wood grown in sustainable forests. The manufacturing process conforms to the environmental regulations of the country of origin.

Printed in Great Britain by Bell & Bain Ltd., Glasgow

Acknowledgements
The publishers would like to thank the following for permissions to use their photographs:

Cover image: ALFRED PASIEKA/SCIENCE PHOTO LIBRARY; p1l: Fotolia; p1tr: Fotolia; p1m: Fotolia; p4: Shutterstock; p5: Alexandru Cristian Ciobanu/Shutterstock; p7: Fotolia; p8: Fotolia; p9br: Tatiana Popova/Shutterstock; p9l: Shutterstock; p9tr: Stavchansky Yakov/Shutterstock; p9mr: Shutterstock; p13tl: Shutterstock; p13bl: Fotolia; p14: Patryk Kosmider/Shutterstock; p16: Syda Productions/Shutterstock; p17: Hamik/Shutterstock; p18: Shutterstock; p19l: Fotolia; p19r: Kurt De Bruyn/Fotolia; p22: Fotolia; 23tl: Dmitry Goygel-Sokol/Fotolia; p23bl: Fotolia; p23r: Fotolia; p28: Fotolia; p29l: Nikon'as/Fotolia; p29r: Claudio Divizia/Fotolia; p30tl: Yong Hian Lim/Fotolia; p30bl: JAKRAPHONG/Shutterstock; p31: Shutterstock; p34tr: Sashkin/ Shutterstock; p34br: Fotolia; p35: Oleksandr Chub/Shutterstock; p36: Fotolia; p38: Paul Fleet/ Shutterstock; p40l: Martina Taylor/Fotolia; p41r: Gvictoria/Fotolia; p41br: Fotolia; p44: Corbis/ Image Library; p48: Rainer Plendl/Fotolia; p56tr: Redshinestudio/Fotolia; p56br: Redshinestudio/ Fotolia; p57tl: Talex/Fotolia;p57bl: Talex/Fotolia; p55: Shutterstock; p61: Fotolia; p64r: Fotolia; p68: WavebreakmediaMicro/Fotolia; p69: Onidji/Fotolia; p70: Fotolia; p71bl: Fotolia; p71tr: RosaIreneBetancourt 5 / Alamy; p74tr: Fotolia; p74br: Ron Chapple/Getty Images; p80: Kenishirotie/Shutterstock; p89: Barnaby Chambers/Shutterstock; p92: Fotolia; p105: Fotolia; p106: alexmillos/Fotolia.

Artwork by Six Red Marbles and OUP.

The author and publisher would like to thank Michael Gatens for his work as a consultant on this project.

Contents

Access your support website at: www.oxfordsecondary.com/9780198399476

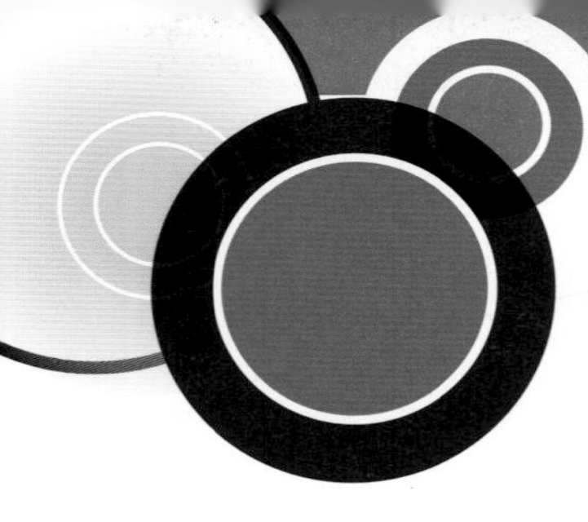

Introduction to Complete ICT for Cambridge IGCSE® 2nd Edition

What the book covers

This book supports the Cambridge IGCSE Information and Communication Technology syllabus. It has been written by an experienced author to ensure the content and features make learning as interesting and effective as possible. On the support website you will find interactive and revision tests, glossary and further past paper questions taken from the previous syllabus.

How you will be assessed

The assessment for the IGCSE consists of the following three papers:

Paper 1

This is a written paper of 2 hours' duration which tests the content of Sections 1 to 21 of the syllabus content. The questions are all compulsory and many of them consist of multiple choice or short answer questions. There are others which require longer answers.

The marks for paper 1 are 40% of the total.

Paper 2

This is a practical test that tests the practical skills to use the applications in Sections 17, 18 and19 of the syllabus content.

The marks for paper 2 are 30% of the total.

Paper 3

This is a practical test that tests the practical skills to use the applications in Sections 20 and 21 of the syllabus content.

The marks for paper 2 are 30% of the total. Throughout this book, you will find exam-style questions. Some of these are taken from past examination papers; these have been taken from the previous syllabus so may be different to what you encounter in your examination.

Sections for the content of the syllabus content

The content is divided into the following interrelated sections:

1 Types and components of computer systems
2 Input and output devices
3 Storage devices and media
4 Networks and the effects of using them
5 The effects of using IT
6 ICT applications
7 The systems lifecycle
8 Safety and security
9 Audience
10 Communication

The practical tests

For the practical tests you will also need to show knowledge and understanding of Sections 1–10, and demonstrate the practical skills relevant to Sections 11–21.

Practical tests assess the practical skills developed for applications in the following list:

11 File management
12 Images
13 Layout
14 Styles
15 Proofing
16 Graphs and charts
17 Document production
18 Data manipulation
19 Presentations
20 Data analysis
21 Website authoring

When your work is marked you will have to meet a series of learning outcomes in each of the sections outlined above.

For access to all the files used in the activities go to **www.oxfordsecondary.co.uk/9780198399476**, or the *Complete ICT for IGCSE Teacher Kit* CD-ROM.

1 Types and components of computer systems

Computer systems consist of two parts: hardware and software, and both are essential for computers to carry out tasks. This chapter looks at different types of computer and the hardware and software needed for them to do a useful job.

The key concepts covered in this chapter are:

- Hardware and software
- The main components of computer systems
- Operating systems
- Types of computer
- Impact of emerging technologies

Hardware and software

All computer systems consist of two main parts: hardware and software. Here are the definitions of hardware and software.

Hardware are the physical components of a computer system such as motherboards, memory, sound cards, screens/monitors, keyboards, printers, etc.

Software are the programs for controlling the operation of a computer or for the processing of electronic data.

KEY WORDS

Hardware the physical components of a computer system such as motherboards, memory, sound cards, screens/monitors, keyboards, printers, etc.
Software programs for controlling the operation of a computer or for the processing of electronic data.

Internal hardware devices

These are situated inside the computer casing and include:

- Motherboard – the main printed circuit board found in computers and contains electrical components such as the central processing unit and memory and also connectors for connecting external hardware devices such as keyboards, mice, speakers, etc.
- Random access memory (RAM) – these are the memory chips which lose their contents when the power is removed.
- Read-only memory (ROM) – memory chips where the contents can be read but not written to and where the contents are retained when the power is removed.
- Video card – an expansion card that generates the signals so that a video output device can display computer data such as text and graphics.
- Sound card – an expansion card that allows a computer to send audio signals to audio devices such as speakers and headphones. Most computers now have the sound card integrated into the motherboard.
- Internal hard disk drive – a rigid magnetic material coated disk onto which programs and data can be stored and situated inside the casing of the computer.

Motherboard

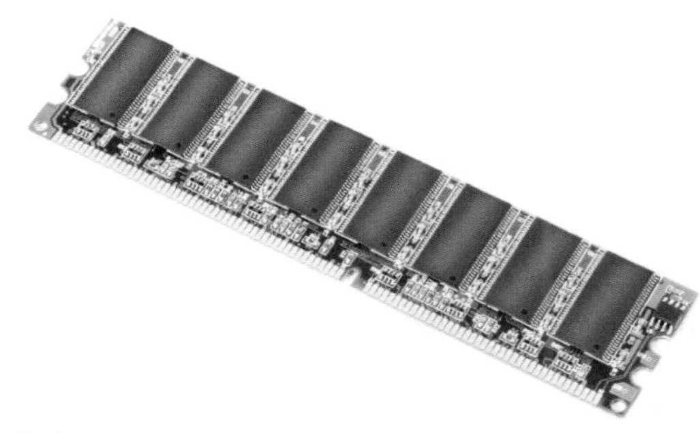

ROM

Sound card

External hardware devices

These are outside the computer casing and include:

- Monitors/screens
- Keyboards and mice
- External storage devices (e.g. external hard disk drive, memory sticks/pen drives)
- Printers
- Scanners
- Speakers and microphones

Software

Software are programs for controlling the operation of a computer or for processing electronic data.

Software consists of sets of instructions that tell the computer hardware what to do. Software is written in a computer language and there are quite a few different ones. Computer hardware is useless without software. Software is of two main types: applications software and system software.

Applications software

Applications software are programs that are designed to carry out certain operations for a particular application. The formatting of text and the organisation of page layout can be performed using word-processing software, which is an example of applications software.

Applications software cannot run on its own as it is dependent on system software, which is the other type of software.

Applications software includes:

- Word processing – software that allows the composition, editing, re-formatting, storing and printing of documents.
- Spreadsheets – software that stores data in cells formed by the intersection of rows and columns that can be used to help sort data, arrange data and calculate numerical values. It can calculate values using mathematical formulae and the data contained in the cells.
- Database management system – software that provides facilities for the organisation and management of a body of information required for a particular application.
- Control software – software used to supply instructions in order to tell devices such as washing machines, automatic cookers how to operate.
- Measuring software – software used to issue instructions to sensors to take measurements of physical quantities such as temperature, pressure etc. and input the readings into the computer for processing.
- Applets and apps – are little programs (i.e. applications) which range from games to social applications and are usually run using smartphones or tablets.
- Photo-editing software – used to improve/enhance photographs using a computer. This would include re-sizing, cropping (i.e. using only part of the photo) and colour/brightness/contrast adjustment.
- Video-editing software – allows a user to take sections of video recordings called clips and trim, splice, cut or arrange them across a timeline. Special visual effects can be included.
- Graphics manipulation software – used to alter the appearance of images. This typically involves altering size, altering orientation, cropping, adjusting colours etc.

System software

System software are programs that control the computer hardware directly by giving the step-by-step instructions that tell the computer hardware what to do. System software therefore operates and controls the computer hardware.

System software includes:

- Operating systems
- Device drivers
- Compilers
- Linkers
- Utilities

Operating systems

An operating system is a program or a suite of programs that controls the system resources and the processes using these resources on a computer. You will learn about the different tasks performed by an operating system later in this chapter.

Compilers

Compilers are programs that change the instructions written in programming languages, such as BASIC, into binary numbers (0's and 1's). A compiler converts the whole of a program written in the programming language into binary in one go. The program is then stored in binary and when it is run it runs much faster because it is now in binary so no time is wasted translating it.

Device drivers

Device drivers are software used to supply instructions to the hardware on how to operate equipment that may be connected to the computer such as printers, scanners, keyboards, mice, external hard disk drives, etc.

Most operating systems are able to recognise when hardware such as a pen drive/memory stick, camera, external hard disk drive, printer, scanner, etc., has been attached to the computer and automatically load the driver software needed to control it. Sometimes, when a new device is attached to a computer, you need to install device driver software included with the device on the computer so that the computer is able to recognise and operate the device.

Linkers

Linkers are programs that are usually part of the compiler. Linkers take care of the linking between the code that the programmer writes and other resources and libraries that make up the whole program file that can be executed (run).

Utilities

Utility programs are provided as part of the system software and they help the user with everyday tasks such as:

- File maintenance tasks such as creating new folders, copying files, renaming files, deleting files
- Compressing files so that they take up less storage space or can be transferred quickly over the internet
- Installing and uninstalling software
- Compacting files on the hard drive so they can be found faster
- Checking for and removing viruses
- Formatting a disk ready for use
- Burning CDs and DVDs (saving data onto them)

QUESTIONS A

1 a Explain the main difference between computer hardware and software. *(2 marks)*
  b Give **two** examples of computer hardware and **two** examples of computer software. *(4 marks)*

2 System software includes the following:
  Operating system
  Device drivers
  Utilities

  Write down the piece of software from the list above that performs each of the following tasks.
  a Switches the printer on/off and sends data to the printer for printing. *(1 mark)*
  b Compresses files so that they can be sent in less time over the internet. *(1 mark)*
  c Provides a user interface so that users can easily create folders, copy files, move files, etc. *(1 mark)*
  d Keeps track of where to store data on the hard disk drive. *(1 mark)*
  e Detects and removes viruses. *(1 mark)*
  f Used to supply instructions as to how to operate hardware attached to the computer such as printers and scanners. *(1 mark)*

3 a Hardware devices can be either internal hardware devices or external hardware devices. Explain the difference between the two. *(2 marks)*
  b From the following list, identify which are internal hardware devices and which are external hardware devices.

Mouse	Video card
Motherboard	Internal hard disk drive
Processor	RAM
ROM	Printer
Keyboard	Sound card *(10 marks)*

4 System software is a suite of programs that includes the operating system, device drivers and utilities.
  a Explain why all computers need an operating system. *(2 marks)*
  b i Is a device driver hardware or software? *(1 mark)*
   ii When a printer is attached or linked to a computer system, a device driver is loaded. Explain what the purpose of a device driver is. *(1 mark)*
  c Give **two** tasks that are performed by utility software. *(2 marks)*

KEY WORDS

Device driver a short, specially written program that understands the operation of the device it controls/operates. For example, driver software can operate/control a printer or scanner. Driver software is needed to allow the systems or applications software to control the device properly.

Operating system software that controls the hardware of a computer and is used to run the applications software. Operating systems control the handling of input, output, interrupts, etc.

System software any computer software that manages and controls the hardware thus allowing the applications software to do a useful job. System software consists of a group of programs.

Utility software part of the system software that performs a specific task such as creating a new folder, copying files, etc.

The main components of computer systems

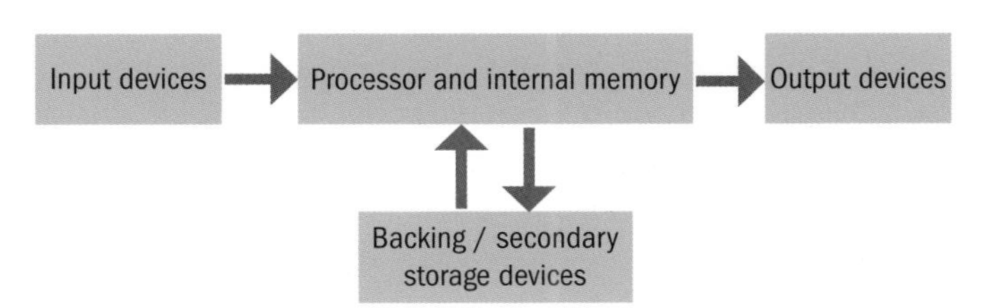

The main components of a computer are:

- Input devices (keyboard, mouse, etc.)
- Processor and internal memory
- Output devices (printer, monitor, speakers, etc.)
- Backing / secondary storage devices (DVD-R/W drive, internal hard disk drive, etc.)

The processor and internal memory include the central processing unit (CPU), read only memory (ROM), random access memory (RAM) and the internal hard disk drive.

The central processing unit and its role

The central processing unit (CPU), often called the processor, is the brain of the computer and it consists of millions of tiny circuits on a silicon chip. The central processing unit does a number of tasks: it controls the step-by-step running of the computer system, it performs all the calculations and the logical operations, and deals with the storage of data and programs in memory.

Internal memory: read only memory (ROM) and random access memory (RAM)

There are two types of memory called ROM and RAM. Both of these two types of memory are stored on chips and are available immediately to the central processing unit.

Computers also have a hard disk drive as internal memory and it is here that the application software is stored along with the user's files.

ROM (read only memory)	RAM (random access memory)
Data and program instructions are stored permanently.	Data and program instructions are stored temporarily.
The computer can only read the contents.	Can read contents as well as write new contents.
Non-volatile, meaning the contents of memory are retained when there is no power.	Volatile, meaning the contents are lost when there is no power.
Used to store the BIOS program used to boot the computer up when the power is turned on.	

KEY WORDS

BIOS (basic input/output system) stored in ROM and holds instructions used to 'boot' (start) the computer up when first switched on.
RAM random access memory – fast temporary memory which loses its contents when the power is turned off.
ROM read only memory – memory stored on a chip which does not lose data when the power is turned off.

Input devices

These are used to get raw data into the computer ready for processing by the CPU. Some input devices, such as a mouse, keyboard, touch screen, microphone, etc., are manual and need to be operated by a human. Others are automatic and once they are set up they can be left to input the data on their own. These include optical mark readers, optical character readers, etc.

Output devices

Once the raw data has been processed it becomes information and this information needs to be output from the computer using an output device. Output devices include monitors/screens, printers, speakers, plotters, etc.

Secondary/backing storage

Secondary/backing storage is used for the storage of programs and data that are not needed instantly by the computer. It is also used for long-term storage of programs and data as well as for copies in case the original data is lost.

Secondary/backing storage media includes internal hard disks, memory sticks, flash memory cards, optical disks (such as CDs and DVDs) and magnetic tape.

See Chapter 2 for more information on input, output and storage devices.

QUESTIONS B

1 The diagram represents a computer's main components:

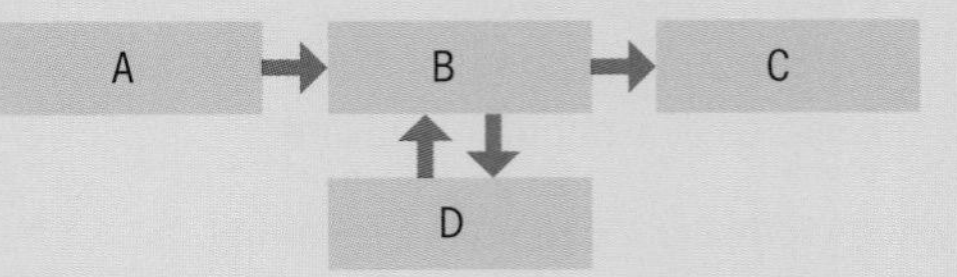

For each of these components, match it with a letter in the diagram above.

- Processor and internal memory devices
- Secondary storage devices
- Output devices
- Input devices

(4 marks)

2 ROM and RAM are both types of internal computer memory.

a **i** What do the letters ROM stand for? *(1 mark)*

ii What do the letters RAM stand for? *(1 mark)*

b Tick **one** box next to each item below to show which statements apply to ROM and which to RAM. *(4 marks)*

	ROM	RAM
Contents are lost when the computer is turned off		
Contents are not lost when the computer is turned off		
Stores the programs needed to start up the computer		
Stores application programs and data currently being used		

3 **a** Explain why it is important that a computer has a large amount of RAM and ROM. *(2 marks)*

b Computers contain RAM and ROM.
Explain what ROM would be used for in a computer.
Explain what RAM would be used for in a computer.
(4 marks)

Operating systems

Operating systems are programs that control the hardware directly by giving the step-by-step instructions that tell the computer hardware what to do. An operating system performs the following tasks:

- Handles inputs and outputs – selects and controls the operation of hardware devices such as keyboards, mice, scanners, printers, etc.
- Recognises hardware – the operating system will recognise that a hardware device such as a pen drive, camera, portable hard drive, etc., has been attached to the computer. The operating system loads the software it needs to control the device automatically.

- Supervises the running of other programs – it provides a way for applications software (i.e. the software that is used to complete a task such as word-processing, spreadsheet, stock control, etc.) to work with the hardware.
- Handles the storage of data – it keeps track of all the files and directories/folders on the disk.
- Maximises the use of computer memory – the operating system decides where in the memory the program instructions are placed. For example, some instructions are needed over and over again, whereas others are needed only now and again. It ensures that the parts of the program needed frequently are put in the fastest part of the memory.
- Handles interrupts and decides what action to take – when something happens such as the printer cannot print because the paper is jammed or it has run out of paper, it will stop the printer and alert the user.

The types of interface used with operating systems

An operating system needs a way of interacting with the user. The way the operating system communicates with a user is called the interface and there are two common interfaces in use and these are:

- Graphical user interface (GUI) – these are very easy to use and have features such as windows, icons, menus, pointers, etc. Examples of operating systems which make use of a GUI are:
  - Windows
  - Mac OS
  - Ubuntu
- Command line/driven interface (CLI) – here you have to type in a series of commands which must be precisely worded, and it can be hard to remember how to do this, so this type of interface is harder to use. An example of an operating system making use of a command line/driven interface is MSDOS (Microsoft Disk Operating System).

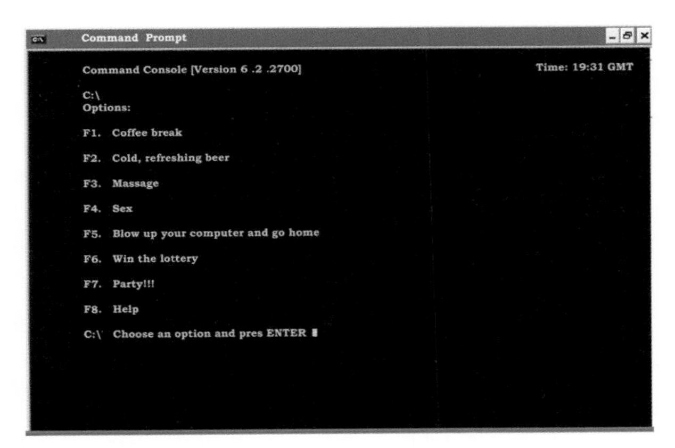

Example of a command line interface where the commands are entered at the prompt.

Graphical user interface

Graphical user interfaces (GUIs) are very popular because they are easy to use. Instead of typing in commands, you enter them by pointing at and clicking objects on the screen. Microsoft Windows and Macintosh operating systems use graphical user interfaces. The main features of a GUI include:

Windows – the screen is divided into areas called windows. Windows are useful if you need to work on several tasks.

Icons – these are small pictures used to represent commands, files or windows. By moving the pointer and clicking, you can carry out a command or open a window. You can also position any icon anywhere on your desktop.

Menus – these allow a user to make selections from a list. Menus can be pop-up or drop-down and this means they do not clutter the desktop while not in use.

Pointer – this is the little arrow that appears when using Windows. The pointer changes shape in different applications – it changes to an 'I' shape when using word-processing software. A mouse can be used to move the pointer around the screen.

Notice that the first letter of each feature in the above list spells out the term WIMP (i.e. **W**indows, **I**cons, **M**enus, **P**ointers).

The benefits and drawbacks between an operating system which uses a GUI and those which use a CLI

Benefits and drawbacks of a GUI

Benefits

- A GUI is considered by most people to be more user friendly. It is more obvious to new users how to do simple tasks such as printing by clicking on a printer icon.
- There are no commands to remember. You simply click on icons or drop-down menus using the mouse to make selections.
- It is much easier to find programs that are running using a GUI. For example, you can have one window open displaying a spreadsheet while using word-processing software in a different window.
- They enable data to be passed easily from one software package to another using drag and drop or cut and paste.

Drawbacks

- A GUI takes up more hard disk space when being stored.
- A GUI requires more memory (ROM and RAM) when being used.
- A more experienced user might find that typing in commands is quicker than moving the mouse or clicking.
- A powerful processor is needed to run the latest version of a GUI.

Benefits and drawbacks of a CLI

Benefits

- If the user is experienced in using the CLI it can be faster to type commands rather than move the mouse and click on drop-down menus and icons to issue the same commands.
- A CLI does not use as much hard disk space when being stored.
- A CLI does not need as much memory (ROM and RAM) when being run.
- You do not need a powerful processor to run a CLI.

Drawbacks

- You need to learn and remember lots of commands.
- Commands must be issued in a specific order otherwise the commands will be rejected.

Important note

Windows is an interface which allows easy user interaction with a multitasking graphical user interface (GUI) operating system as well as other software. MS Windows is a branded operating system owned by the company Microsoft.

QUESTIONS C

1 Graphical user interfaces are very popular interfaces that are used with computers, mobile phones, and other portable devices.
   a Give **three** features of a graphical user interface. *(3 marks)*
   b Other than a graphical user interface (GUI) give the name of **one** other type of user interface. *(1 mark)*

2 All computers need an operating system.
   a Explain what an operating system is. *(2 marks)*
   b List **three** different functions of an operating system. *(3 marks)*

3 Users need a way of interacting with an operating system.
   a Give the names of the **two** different types of user interface used with computers.
   (Note that the use of brand names will gain you no marks in an examination.) *(2 marks)*
   b Give a benefit and a drawback for each type of interface. *(2 marks)*

Types of computer

The type of computer you choose to use depends on the tasks you intend to do with the computer and where you intend using the computer. If a computer is to be used away from a desk then it needs to be small and light but there are some tasks such as desktop publishing where it becomes difficult to perform using a small screen.

Personal computers (PCs) or desktop computers

A personal computer is the type of computer that you are most likely to encounter at home or at school.

Computers can be classed as stand alone or networked:

- Stand alone computers are not connected to a network so they do not share resources or files with other computers. They are used when all the programs data/files needed for the application, are stored locally (i.e. on one of the drives or memory in the computer itself).
- Networked computers are linked to a network so they are able to share files, an internet connection and devices such as printers and scanners with other computers connected to the network. Many programs are stored on the internet (i.e. cloud storage) rather than locally on the computer and can therefore only be accessed using networked computers.

Laptops

Laptop computers are designed to be portable and used while on the move. A touch pad is typically used instead of a mouse to move the cursor and make selections. Laptops are often used in public places so there is a greater likelihood of them being stolen.

Laptops make use of LCD (liquid crystal displays) which use less power and are light in weight. This is important because laptops use rechargeable batteries when used away from a power supply. If a laptop is not connected to a network then it is a stand alone laptop and all the programs and files needed for the application used are stored on the computer itself. Once the computer is connected to a network then it is able to share files and resources and it becomes a networked laptop.

Tablet computers

Tablet computers, called tablets for short, are mobile computers with a display, electronic circuitry and battery in a single thin unit. They are designed to be portable yet have a screen size that makes them easy to use for accessing content on the internet, playing games, listening to music, as well as using all the other applications for which you would use a computer. Tablets usually feature an on-screen keyboard which pops up on the screen to enable text to be input.

Tablet computers are usually equipped with some or all of the following:

- A camera which enables still or video images to be recorded and saved as well as photo-editing software.
- Speakers to output music, sound from video recordings, instructions from GPS (global positioning systems), etc.
- A microphone to record voice for the phone facility, the web cam, and for voice recognition to enable commands to be issued verbally and to enable speech to be converted to text for documents and emails.

- Handwriting recognition enables a user to write on the touchscreen using a stylus and then convert it to text which can be used in email, documents, etc.
- Bluetooth which allows communication between the tablet and peripherals such as external speakers and other devices.
- Ability to use Wi-Fi which allows access to the internet wherever there is a wireless access point (e.g. in homes, coffee shops, libraries etc.).
- The ability to use 3G/4G telephony to access the internet when Wi-Fi is not available, provided you have a mobile signal.

Because they are light in weight and have a touch-screen, tablet computers can be used while standing up.

Smartphones

Smartphones have most of the features of a tablet computer but their small size limits their use for certain applications.

The following diagram shows some of the features of a smartphone.

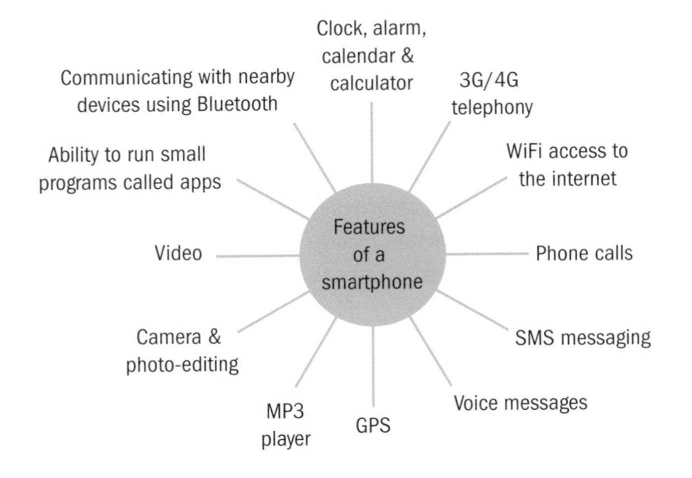

Desktop	Laptop	Tablet	Smartphone
Separate components means hard to dismantle/move.	All in one so easy to move.	Light and compact so easily transported.	Very light and pocket sized.
Used on a desk so less likelihood of RSI or backache.	Often used awkwardly so chance of backache or RSI.	Uses an on-screen keyboard which can be awkward to use.	On-screen keyboard is small and hard to use.
Needs a mains power supply.	Can be used away from the mains power.	Can be used away from the mains power supply.	Can be used away from the mains power.
Easy to upgrade and repair.	Harder to upgrade and repair.	Harder to upgrade and repair.	Harder to upgrade and repair.
Full size keyboard and mouse are easy to use.	Touch pad is more cumbersome to use.	Uses a touch screen which makes it easy to surf the net.	Small screen makes it more difficult to use.
Flat surface needed to move mouse on.	No flat surface is needed.	Can be used in most positions including standing up.	Easiest type to use when on the move.

Impact of emerging technologies

Emerging technologies are those that are likely to make a great impact on our everyday lives in the future and these are outlined here.

Artificial intelligence

Artificial intelligence is the science of getting computers to learn in a similar way to the way the human brain learns new things. The aim of this is to make computers more intelligent and make them able to work things out for themselves.

Artificial intelligence is a reasoning process performed by computers that allow the computer to:

- Draw deductions
- Produce new information
- Modify rules or write new rules.

KEY WORD

Artificial intelligence (AI) creating computer programs or computer systems that behave in a similar way to the human brain by learning from experience, etc. The computer, just like a human, is able to learn as it stores more and more data.

Biometrics

Biometrics uses a property of the human body to identify people. The three main properties used to identify a person using ICT are:

- Fingerprints
- The pattern on their retina (the pattern of blood vessels at the back of the eye).
- The pattern on their iris (the coloured, circular area around the pupil in the eye).

These have the following uses.

Fingerprint scanning:

- Used for recording attendance in school/college
- Used for access to buildings and rooms
- Used to restrict access to computers/smartphones
- Used to restrict access to countries at border control at airports, ports, etc.

Retinal scanning:

- Used for passport control.
- Used for access to buildings and rooms.

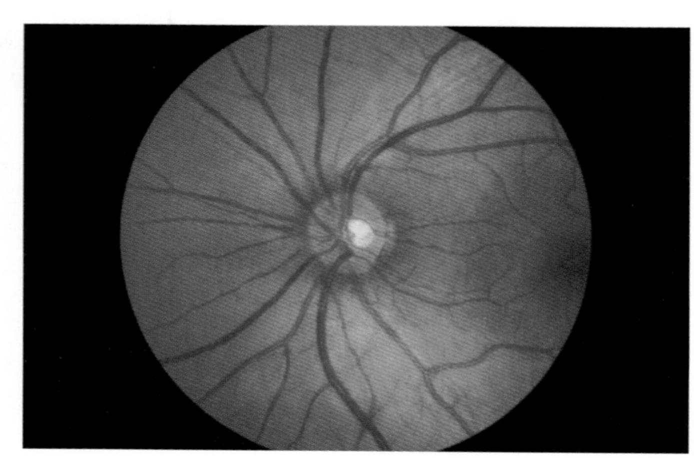

Retinal scanning makes use of the unique pattern of blood vessels in the retina to identify people.

There are some new developments in biometrics:

- Face recognition systems – where a person can be identified by their face. Faces are stored on a database and the image from a CCTV camera is used to find a match. Face recognition systems are used at passport/border control where a face is identified using a camera and a check is made to confirm it matches the photograph in the passport.
- Walk recognition systems – people have a unique way of walking and this can be used to identity people from behind. This is used by the security and police forces.

Computer-assisted translation

Computer-assisted translation is a type of translation from one language into another where the person doing the translation uses computer software to aid them with the translation.

All you need to do is enter text into a window in one language and then select the language you want it translating into and then all the text will appear in another window translated in the new language. There is even the option of hearing the spoken words in each language. In some cases it is not necessary to tell the computer the language being translated as it can recognise it automatically. One popular computer-assisted translation service is provided by Google Translate.

Uses for computer-assisted translation

- Changing the language on websites – many websites are used internationally and it is possible for you change the language used by the website to the language you use.
- Translation of documents – documents such as contracts, letters, brochures, notes etc may need to be translated into different languages. There is no need for translators to perform the translation as computer software can do it saving both time and money.

Quantum cryptography

Quantum cryptography uses a special branch of physics called quantum mechanics to produce a very secure method of encryption. It allows a sender and receiver to communicate in private using a special key. If someone is trying to eavesdrop on the communication, the system is able to detect this and the system stops the communication. This method is used by the military to send orders to troops and also for sending financial and banking details over networks.

Uses for quantum cryptography

- Used for the sending of credit/debit card or banking details – these are sent over the Internet without hackers being able to view and understand the data.
  Provides a completely secure method of sending secret information – (e.g. details of informants and spies) which improves the security of a country.
- Used for securely performing elections using the Internet – in the past, paper votes or even electronic votes could be tampered with.

Vision enhancement

Vision enhancement uses special sensors (which are usually special cameras) that detect information from images outside the visible spectrum. This information is then put together with the ordinary image to make it clearer.

Vision enhancement is used in some luxury cars to provide a screen at which a driver can glance when driving in poor visibility (e.g. fog, spray, lights from oncoming cars, rain, and at night). It can also be used by the military to enable troops to see when it is almost dark.

People who have low vision (i.e. they are almost blind) can use vision enhancement to enable them to see using their remaining sight. The system takes their remaining sight and magnifies objects up to 50 times allowing them to change the brightness and contrast of the image to enable them to see much more clearly.

Uses for vision enhancement

- Improving vision – pilots/drivers in poor light conditions can have their vision improved making it safer.
- Improves the vision of partially sighted people – thus improving their quality of life.

3D and holographic imaging

3D imaging is a technique that gives the illusion of depth in an image. It works by using two slightly offset images with one image sent to the right eye and the other image sent to the left eye. The brain then uses the 2D images on the screen to give the impression of depth.

Holographic imaging uses laser light to form an image of an object that is almost indistinguishable from the object itself. Research is currently taking place to use computer-generated holograms with special holographic displays to produce holographic television. The main problem at the moment is that producing a holographic image requires huge numbers of calculations to be performed very quickly and current processors are not fast enough. There is also the problem of transmitting holographic images, because the bandwidth needed is much higher than that currently available. As soon as these technical problems are solved, we will have true 3D images.

Virtual reality

Virtual reality is a simulation of the real world, or an imaginary world that is created using computers. Sometimes the image is displayed on a computer screen and sometimes it is displayed on a specially designed headset that you wear. In some cases users are able to interact with the virtual environment using a keyboard and mouse or by using a specially wired glove.

Robotics

Robots have been widely used in manufacturing for years, especially for painting and welding in car factories. Robots are also used for picking and packing goods in large warehouses. Robots have been developed that will do some of the tasks humans hate to do, such as mowing the lawn or vacuuming floors.

There are robots available for the home that will wash floors, clean gutters, and clean swimming pools. The robots that are available at the moment in the home are usually capable of performing one task. In the future you will probably buy a single multifunctional robot capable of carrying out a range of different tasks.

Mowing the lawn is a chore for many people, so this robot lawnmower is a useful device.

Advantages and disadvantages of robots

Compared to manual control, computer-based control systems or robots have the following advantages and disadvantages.

Advantages:

- Can operate continuously, 24 hours per day and 7 days per week.
- Less expensive to run as you don't have to pay wages.
- Can work in dangerous places (e.g. a robot to remove a bomb).
- Can easily change the way the device works by re-programming it.
- More accurate than humans.
- Can react quickly to changes in conditions.

Disadvantages:

- Initial cost of equipment is high.
- Equipment can go wrong.
- Fewer people needed so leads to unemployment.
- Need specialist to program, which is expensive.

This robot is used by the army to make bombs safe.

Computer control is used in manufacturing.

Robots will eventually be seen in all homes. This vacuuming robot is already in the shops.

REVISION QUESTIONS

1 Explain the difference between hardware and software. [2]

2 Desktop computers consist of a number of components.

Explain the purpose of each of the following components.

a Central processing unit (CPU) [1]

b Main memory [1]

c Backing storage [1]

3 Give the names of:

a **two** input devices [2]

b **two** output devices [2]

c **two** backing storage devices [2]

4 Computers come in all sizes and can be used in different situations.

a Describe what is meant by a desktop computer and explain how it differs from a laptop computer. [3]

b Tablet computers are very popular. Give **two** ways in which a tablet computer differs from a laptop computer. [1]

5 A salesperson travels to newsagents taking orders for sweets using a tablet.
Give **one** advantage of using a tablet rather than a laptop computer. [1]

6 Computers need memory and backing storage.

a Give **one** difference between memory and backing storage. [1]

b Give **one** example of what would be stored in memory. [1]

c Give **one** example of what would be stored in backing storage. [1]

7 a Explain how you could distinguish between an internal and an external hardware device. [1]

b Give **two** examples of internal hardware devices and **two** examples of external hardware devices. [4]

8 a Give the meanings of the abbreviations RAM and ROM. [2]

b Describe the differences between RAM and ROM. [4]

9 There are many emerging technologies that are set to change peoples' lives.
Give the name of **one** emerging technology and explain what the technology is and how it is likely to change peoples' lives. [5]

10 Many computer systems make use of biometrics.

a Explain the meaning of the term biometrics. [1]

b Give the names of **two** biometric measurements that are likely to be used by a computer system. [2]

c For **one** of the measurements in your answer to part **b** explain an application for the measurement. [2]

Test yourself

The following notes summarise this chapter, but they have missing words. Using the words below, copy out and complete sentences **A** to **M**. Each word may be used more than once.

software ROM information hardware desktop output CPU input RAM backing applications

A Computer systems consist of two main parts, hardware and ______.

B The physical components of a computer system are called ______.

C The ______ is a piece of hardware that is the brains of the computer and it turns data into ______.

D Computer hardware is useless without the ______ which is used to give it instructions as to what to do.

E There are two types of software called system software and ______ software.

F A computer which has a full-sized keyboard and full-sized screen and is normally used in one place is called a ______ computer.

G ______ devices such as the keyboard and mouse are used to input data into the computer for processing.

H After data has been processed, the results of processing are passed to a(n) ______ device.

I Storage which is not memory is called ______ storage.

J ______ is fast temporary memory where programs and data are stored only when the power is supplied.

K ______ is fast permanent memory used for holding instructions needed to start up the computer.

L ______ is held on a computer chip and is called non-volatile memory because it does not lose its contents when the power is turned off.

M ______ is held on a computer chip and is called volatile memory because the contents disappear when the power is turned off.

EXAM-STYLE QUESTIONS

1 Choose **three** tasks from the list below that are carried out by system software:

Renaming a file
Deciding where to store data on a hard disk drive
Underlining text in a word-processing package
Cropping a picture
Loading a file from the disk drive *[3]*

2 **a** Give three tasks (other than those in Q1) performed by system software. *[3]*

b Some printers come with device driver software. Give one purpose of device driver software. *[1]*

3 Write down the name of the piece of system software that carries out each of the following by matching the name to the correct description.

Utility
Device driver
Operating system

	Description	Name
(A)	Deals with errors that occur when the computer is working on tasks	
(B)	Software needed when a new piece of hardware is attached to the computer	
(C)	Scans the hard disk drive to detect and remove viruses	

[3]

4 Tick **True** or **False** next to each of these statements.

	True	False
Computer programs are examples of software.		
Creating a new folder would be performed by a device driver.		
A compiler is an example of an operating system.		
Word-processing software is an example of applications software.		

[4]

5 Which **three** of the following tasks are carried out by all operating systems?

a Transferring data to a printer
b Allocating storage space on a disk
c Positioning text in a word-processing document
d Finding a database record
e Accepting keyboard input
f Adding colour to a drawing on screen *[3]*

6 Give **three** functions of an operating system. *[3]*

7 There are two types of internal memory called ROM and RAM. Describe the differences between ROM and RAM. *[4]*

8 Complete each sentence using **one** item from the list:

Router Output Communication Software

Processor Microphone ROM RAM Input

a .... means the programs that supply the instructions to the hardware to tell it what to do.
b .... devices such as keyboards, mice, and scanners are used to supply data to the computer.
c .... devices are hardware such as printers, speakers, and screens.
d .... is memory which is used to hold the boot program needed to start the computer up when first switched on.
e .... is memory held on a chip that can have its contents changed by the user. *[5]*

9 Graphical user interfaces are very popular particularly with tablets and smartphones.

a Give **two** advantages of a tablet compared to a laptop computer. *[2]*
b Give **two** disadvantages of a tablet compared to a laptop computer. *[2]*
c Give **three** features of a graphical user interface. *[3]*

10 Give **two** examples of tasks that are completed by robots. *[2]*

11 Washing machines are assembled by robots in a factory.

Tick three advantages in using robots rather than humans to assemble washing machines.

Robots are more intelligent than humans.	
Robots are cheap to buy and maintain.	
Robots can work 24 hours a day.	
Robots always assemble the parts correctly.	
Robots need to be programmed to perform a task.	
It is cheaper as robots are not paid wages.	

[3]

12 Describe the differences between a **CLI** (Command Line Interface) and a **GUI** (Graphical User Interface). *[4]*

(Cambridge IGCSE Information and Communication Technology 0417/11 q13 Oct/Nov 2013)

13 Describe the differences between a stand-alone computer and a networked computer. *[4]*

2 Input and output devices

Input devices such as a keyboard and mouse/touch pad are available with nearly all computers. There are, however, other input devices which reduce the amount of work and improve accuracy when entering data into the computer for processing.

Once data has been processed, the results of the processing, called the output, will need to be produced. Various output devices such as printers, screens, speakers, etc., are used for output.

In this chapter you will be covering both input and output devices.

The key concepts covered in this chapter are:

- Input devices and their uses
- Direct data entry and associated devices
- Output devices and their uses

Input devices and their uses

Input devices are those hardware devices that are used for the entry of either instructions or data into the computer for processing. There are many different input devices; which one is chosen depends mainly on the task being performed. For the examination you will need to know how each of the following input devices is used and their relative advantages and disadvantages.

Keyboard

Keyboards are used to enter text into word-processed documents, numbers into spreadsheets, text into online forms, and so on.

The main advantages of using keyboard entry are:

- Text appears on the screen as you type so it can easily be checked for accuracy. This is called visual verification.
- They are ideal for applications such as word-processing or composing emails where you have to create original text.
- Can be used for other instructions (e.g. Ctrl+P to print).

The main disadvantages of using keyboard entry are:

- It is a slow method for entering large amounts of text compared to methods such as dictating or scanning text in.
- It can be inaccurate as it is easy to make typing mistakes.
- It can be frustrating if you do not have good typing skills.
- Typing for a long time can cause a painful condition called repetitive strain injury (RSI).

Numeric keypad

Numeric keypads are used where only numeric data is to be entered. For example, you may see them being used in ATMs (Automatic Teller Machines—cash dispensers) to enter the PIN (Personal Identification Number) and the amount to be withdrawn or deposited.

Numeric keypads are also used in chip and PIN machines where you insert a credit/debit card and then enter your PIN to pay for goods or services.

The main advantage of using a numeric keypad is:

- It is smaller than a full keyboard, making it easier to carry or move, e.g. at a shop counter.

The main disadvantages of using a numeric keypad are:

- Small keys might be difficult for some people to see.
- Keys might be arranged differently, e.g. on a phone versus on a chip and PIN pad.

Pointing devices

Pointing devices are needed to make selections from a graphical user interface. In order to move the pointer onto a button, icon, hyperlink, etc., it is necessary to use one of the following devices:

- *Mouse* – The cursor moves in response to the movement of the mouse. The left mouse button is used to make selections while the right button is used to display a drop-down menu. A scroll button or wheel is used to help move quickly through long documents.

  Advantages of using a mouse include:

  - Easier than using a keyboard to select or move items.
  - Enables rapid navigation through an application, such as a web browser.
  - Small, so it occupies only a small area on the desktop.

  Disadvantages of using a mouse include:

  - Hard to use if the user has limited wrist or finger movement.
  - Needs a flat surface so it can't be used 'on the move'.

- *Touch pad* – these are used on laptop computers when there is no flat surface to use a mouse, though many people find them more difficult to use. They also have buttons to act like mouse buttons.

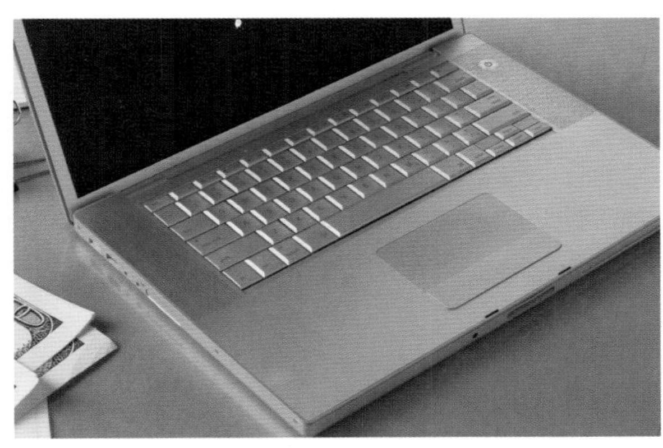

Touch pads are used when there is not enough room for a mouse.

Advantages of using a touch pad include:

- Can be used 'on the move' where there is no flat surface on which to use a mouse.
- There is no extra device to carry as the touch pad is built into the computer.
- It is faster to navigate through an application compared to using a keyboard.

Disadvantages of using a touch pad include:

- They are hard to use if the person using the touch pad has limited wrist or finger movement.
- Controlling the pointer is more difficult compared to using a mouse.
- They are hard to use for actions such as 'drag and drop'.
- They tend to be used in more cramped conditions so their long-term use could lead to health problems, such as RSI.

- *Trackerball* – these are a bit like an upside-down mouse and you move the pointer on the screen by rolling your hand over the ball. They are used by people with poor motor skills such as the very young or people with disabilities.

A trackerball is an alternative to a mouse.

Advantages of using a trackerball include:

- They are easier to use than a mouse if the user has wrist or finger problems.
- They allow faster navigation compared to a mouse.
- They are stationary as it is only the ball that moves.

Disadvantages of using a trackerball include:

- The cost, as trackerballs are not normally included with a computer system.
- Time is needed to get used to using them.

Remote control

Remote controls are input devices because they issue instructions to control devices such as TVs, video players/recorders, DVD players/recorders, satellite receivers, music systems, data projectors or multimedia projectors.

Advantages of using a remote control include:

- The user can operate a device wirelessly from a short line of sight distance where a device's controls are difficult to reach (e.g. a projector positioned on a ceiling).
- Can control devices safely from a distance (e.g. a robot used to investigate a bomb).
- Enables people with disabilities to operate devices at a distance.

Disadvantages of using a remote control include:

- Typically uses infra-red signals and these can be blocked by objects in their path, so the remote control may not operate.
- Needs batteries to operate and these need regular replacement.
- Small buttons can be pressed by mistake and this can alter the settings of the device and make it hard for the user to go back to the original state.

Joystick

Joysticks are used mainly for playing computer games. They can also be used in car-driving simulators and flight simulators. The stick moves on-screen in the same way as a mouse, and buttons are used to select items.

Advantages of joysticks:

- They are ideal for quick movement.
- They can be used by disabled people because they can be operated by foot, mouth, etc.

Disadvantages of joysticks:

- Entering text is very slow as you have to select individual letters (e.g. the name of a high scorer in a game).
- You have to purchase them separately from a computer system.

Touch screen

Touch screens are displays that can detect the presence and location of a touch to the screen. Usually a finger is used to make selections on the screen, although some systems allow several fingers to be used, for example re-sizing an image on the screen by moving two fingers closer together or further part. Touch screens can be found in information kiosks, tablet computers and smartphones.

Touch screens are popular in restaurants and shops for inputting orders/sale details because employees need little training to use them. They are also features of some point of sale terminals (i.e. computerised tills). They are ideal for tourist information kiosks, transport information, airport self check-in or ticket collection points.

Sat Navs use touch screens.

Advantages of touch screens:

- They are very simple to use.
- They are easier to use while standing, compared to keyboards.
- They are ideal where space is limited such as on a smartphone, tablet, etc.
- They are tamper-proof so other data cannot be entered which could corrupt the system.
- Enables faster selection of options than using a keyboard or mouse.

Disadvantages of touch screens:

- The screen can get quite dirty and this can make items on the screen hard to see.
- There is a danger that germs can be spread with many people touching the same screen.
- On small screens the icons can be hard to see or select.
- They often cost more than input devices such as a keyboard.

Scanners

Scanners can be used to scan in photographs and other images on paper to put into documents or web pages, thus providing a way of digitising hard copy documents. They can also be used for scanning text into word-processing or other packages thus saving having to re-type the text. Scanning and recognising text in this way is called OCR (optical character recognition).

Advantages of scanners:

- They can be used in conjunction with special optical character recognition software to scan and enter text into word-processing or other software without the need for re-typing.
- They can be used to digitise old documents such as maps, pictures and photographs for archive purposes.
- They can be used with software to repair/improve old photos.

Disadvantages of scanners:

- They take up a lot of space.
- It can take a long time to digitise all the pages in a long document.
- The quality of the scanned item might be poor depending on scanner resolution.

Copyright 2005 by Randy Glasbergen.
www.glasbergen.com
CHIPS
GLASBERGEN
"I HAVE TO STAY HOME TONIGHT AND HELP MY DAD WITH HIS NEW CAMERA PHONE. WE NEED TO DELETE 750 PICTURES OF HIS HAND."

Digital cameras

Digital cameras store the digital photographs on an internal memory card and display the picture on a screen. The digital image can then be saved on a computer or edited using photo-editing software and then saved. Digital cameras can often be connected directly to a photo printer (a printer designed specifically for printing photos) to print out the photographs without the need for a computer. Nearly all mobile phones also have a digital camera facility.

The dots making up a digital image are called pixels, and the number of pixels used is called the resolution. Generally, the higher the number of pixels the more detailed the photograph. This also means the file size will be larger and fill more space on the memory card. Digital video cameras are available which capture moving images and sound; these are very useful for adding to websites, presentations, and other multimedia materials.

Advantages of digital cameras:

- No film to develop, so quicker to produce a photo.
- No need to use chemicals in developing films.
- A large number of photographs can be taken of the same subject and the best one can be chosen.
- The images are in a form that can be placed in documents, in presentations and on websites.
- The digital images they produce are easily sent from place to place using phones, email attachments, etc.
- They can be placed on sharing sites for others to see.
- Can use software to improve photos (e.g. to remove red-eye).

Disadvantages of digital cameras:

- The photographs may not be of as high a quality as those produced using film. You are limited by the number of pixels (small dots) that make up the image.
- A photographer needs to have computer skills to use digital photographs properly.

Microphones

Microphones are input devices that allow sound signals to be converted into data. They are used to digitise sound (convert the sound wave into a series of pulses consisting of 0's and 1's) so that it can be stored and processed by the computer. Microphones are used to give instructions to the computer or enter data using special software called voice recognition software.

Here is what a microphone allows you to do:

- You can instruct without using hands (e.g. hands-free phone in a car).
- You can dictate letters and other documents directly into your word-processor or email package. This is called voice recognition.
- You can take part in video-conferencing.
- You can add narration (i.e. spoken words) to websites, presentations, and other multimedia products.
- You can input speech for VoIP (Voice over Internet Protocol, which is an inexpensive way of conducting phone calls using the internet) and video-conferencing systems.

Advantages of microphones:

- Can be used by disabled people to input data/instruct.
- Can improve safety for car phones.
- They are relatively inexpensive.
- Can be faster to speak instructions rather than type them.

Disadvantages of microphones:

- Background noises can cause problems with voice recognition systems.
- Voice recognition is not completely accurate so mistakes may occur.
- Sound files, when stored, take up a lot of disk space.

Sensors

Sensors are able to measure quantities such as temperature, pressure and amount of light. The signals picked up by the sensors can be sent to and then analysed by the computer. Sensors can:

- be found in lots of devices such as burglar alarms, central heating systems, washing machines, etc., and as well as sensing, they are also used to control the device in some way.
- be used to input data for the control of certain devices.

Advantages of using sensors:

- The readings are more accurate than those taken by humans.
- The readings can be taken more frequently than by a human.
- They work when a human is not present, so costs are less.
- They can work in dangerous environments (e.g. down deep mines or in nuclear reactors).

Disadvantages of using sensors:

- Purchase cost.
- Dirt and grease may affect performance.

Temperature sensors – used to monitor temperature. When the temperature data is sent to the computer/microprocessor it can be used to control heaters, coolers, windows, etc. Temperature sensors can be used to control automatic washing machines, central heating systems, air conditioning systems, greenhouses, and environmental monitoring systems.

Pressure sensors – pressure sensors measure liquid pressure (the pressure in a liquid increases with depth) and physical pressure (when something is pressing down on a pad). Some pressure sensors measure atmospheric pressure which is an important quantity for predicting the weather. Pressure sensors can be used in burglar alarm systems to detect the pressure exerted by an intruder. They are also used in washing machines to detect when the water has reached a certain depth.

Other uses for pressure sensors include robotics (so that robot arms can pick things up without squashing them), production line control, and environmental monitoring (e.g. flood warnings in rivers).

Light sensors – these are used to measure light intensity and can be used in security lights which come on when it gets dark and go off when it gets light. Light sensors can be used for burglar alarm systems, production line control, scientific experiments, and environmental monitoring (e.g. the conditions in a greenhouse).

Humidity/moisture sensors – these are used to measure the amount of moisture in the air or soil. They are used in greenhouses to ensure ideal growing conditions for plants.

Graphics tablet

A graphics tablet is a flat board (or tablet) which you use to draw or write on using a pen-like device called a stylus. Graphics tablets are ideal for inputting freehand drawings. They can also be used for retouching digital photographs.

Graphics tablets are used for specialist applications such as when designing a kitchen using computer-aided design (CAD), and have a range of specialist buttons for certain shapes and items. You can add these items by clicking on them with the stylus.

A graphics tablet.

Advantages of graphics tablets:

- Used in countries such as Japan and China where graphical characters are used instead of letters for words.
- It is more accurate to draw freehand on a tablet than to use a mouse to draw.
- The icons/buttons are on the graphics tablet rather than the screen, leaving more screen space for the design/drawing.

Disadvantages of graphics tablets:

- Specialist tablets are expensive.
- Requires more desk space.

Video camera

Most modern video cameras are digital, which means the files produced are digital and can be transferred to a computer and saved or edited and then saved. Like any digital file, it may be added to presentations, websites, etc. You can also add the video to sites such as YouTube so that the video can be shared with others.

Advantages of video cameras:

- Can capture both still and moving images.
- Easy to transfer video to the computer.
- Can store many images/videos for editing later.

Disadvantages of video cameras:

- Can erode privacy as CCTV cameras are present in many public places.
- It is very difficult to produce good video without training.
- Editing is usually required to make a good video and this can be complicated and time-consuming.
- Can be expensive to buy.
- The picture quality of the video is determined by the number of pixels used (the resolution) and this may be low.

Web cam

Web cams are digital cameras that can take both still and video images which can then be transferred to other computers or saved. They are used for conducting conversations over the internet where you can see the person you are talking to.

Web cams are now included with many computers and are easily added to computers not having one. The uses for web cams are many and include:

- To conduct simple video-conferencing.
- Allow parents to see their children in nurseries when they are at work.
- To record video for the inclusion on websites.
- Allow people to view the traffic in local road systems.
- Allow people to view the actual weather in a place they are going on their holidays.

Advantages of web cams:

- Can see the reactions of people as you are talking to them.
- Parents can see their children and grandchildren and speak to them if they do not live near.

Disadvantages of web cams:

- Limited extra features to improve image quality.
- The picture quality can be poor at low resolution.
- Generally have a fixed position so do not move around.

KEY WORDS

Sensors devices which measure physical quantities such as temperature, pressure, etc.
Touch screen a special type of screen that is sensitive to touch. A selection is made from a menu on the screen by touching part of it.
Voice recognition the ability of a computer to 'understand' spoken words by comparing them with stored data.

QUESTIONS A

1 a Which **six** of the following are input devices?

Graphics tablet	Digital still camera
Colour laser printer	Magnetic hard disk drive
Mouse	CD ROM drive
Microphone	Web cam
Speakers	Photocopier
LCD screen	Touch screen

(6 marks)

b Name **two** output devices given in the choices above. *(2 marks)*

2 a Explain the purpose of an input device. *(1 mark)*

b Give the names of **two** input devices that would be used by a desktop computer. *(2 marks)*

c Give the name of **one** input device that would be found on a laptop computer that you would not find on a desktop computer. *(1 mark)*

3 Laptop computers enable people to do work or keep in touch when they are travelling.

Name **two** input devices typically used with a laptop computer. *(2 marks)*

4 Name **two** items in this list which are used as input devices.

3D printer	inkjet printer	temperature sensor
motor	touch screen	dot matrix printer
web cam	laser printer	buzzer

(2 marks)

5 Laptop computers usually use a touch pad rather than a mouse for pointing and making selections.

Explain why laptops use a touch pad rather than a mouse. *(2 marks)*

6 Match each input device with the correct use.

Input device	Use
a Keyboard	**i** Used to enter a PIN when making a purchase
b Remote control	**ii** Used when checking the depth of water in a container
c Touch screen	**iii** For inputting a hand-drawn drawing into the computer
d Numeric keypad	**iv** Used to make selections on tablets and some smartphones
e Pressure sensor	**v** Used as the input for an air-conditioning unit to keep the temperature of a room constant
f Temperature sensor	**vi** Entering text when writing a book
g Scanner	**vii** Operating a television

(7 marks)

Direct data entry and associated devices

Direct data entry (DDE) is a method of data processing where the data is input directly into the computer system for immediate processing. The data are entered automatically by one of the devices outlined here.

Magnetic stripe readers

Magnetic stripe readers read data stored in the magnetic stripes on plastic cards such as loyalty cards. The stripe contains data such as account numbers and expiry date.

They are used for reading data off credit/debit cards where a chip and PIN reader is not available. They are also used in ID cards where the card is swiped through a reader to gain access to buildings and rooms. Other uses include pre-payment cards for using services such as the internet, photocopiers, etc.

Advantages of magnetic stripe readers:

- Faster input of data by swiping as opposed to typing.
- Can be used as an alternative method to chip and PIN for credit/debit cards.
- Avoids possible typing errors which could be introduced by keying in.
- Stripes are not affected by water so they are robust.

Disadvantages of magnetic stripe readers:

- Magnetic stripes can store only a small amount of data.
- Magnetic stripes can be easily damaged by magnetic fields.
- Magnetic stripes can be duplicated relatively easily (called card cloning) and this leads to card fraud.
- The cards cannot be read at a distance as you have to put them into the reader and swipe them.

Chip readers and PIN pads

Chip readers are the devices into which you place a credit/debit card to read the data which is encrypted in the chip on the card.

The chip on a credit card.

The PIN pad is the small numeric keypad where the personal identification number (PIN) is entered, and the holder of the card can be verified as the true owner of the card.

The main use of chip and PIN is to read card details when making purchases for goods or services where the cardholder is present to input the PIN.

Advantages of chip readers and PIN pads:

- They reduce fraud as the true cardholder has to input their PIN.
- Chips are harder to copy compared to a magnetic stripe.
- The storage capacity for data on a chip is much higher than that for a magnetic stripe.
- A chip is less likely to be damaged than a magnetic stripe.

Disadvantages of chip readers and PIN pads:

- Not all countries can use chip and PIN, so they have to use magnetic stripe readers.
- People might forget their PIN.
- Other people might furtively observe your PIN as you enter it (called 'shoulder surfing').

Chips are also found in ID cards, some passports, public transport tickets, and these are fed into readers to allow access, display information stored on the chip, etc.

RFID (radio frequency identification) readers

RFID (radio frequency identification) obtains data stored on a tag (a small chip) using radio signals, which means that the reading device and tag do not have to come into contact with each other; the data on the tag can be read from a distance. This is therefore a wireless system.

Applications of RFID tags:

- Used for stock control in factories.
- Cattle can be tagged so the milking system can identify which cow is being milked and information about yield, etc. can be stored.
- Used for season tickets to car parks.
- Used in libraries as a replacement for bar codes.
- Used for automated passport/border control gates.

Advantages of RFID tags:

- There is no need for the reader and tag to come into contact with each other so you could have a tag with you which could be read without getting it out of your pocket or bag.
- You can store more data on the tags compared to bar codes.

Disadvantages of RFID tags:

- The reader and the tags themselves are expensive.
- As tags can be read at a distance, some people are worried that the personal information contained in the tag can be hacked into. This information could then be used to commit identity theft.

Magnetic ink character recognition (MICR)

Cheques are still used and millions of them go through a process called cheque clearing each day. Check clearing uses input methods that use magnetic ink characters printed on the cheque.

⑆ ⑇ ⑈ ⑉ 0 1 2 3 4 5 6 7 8 9

Numbers are printed onto the cheque using a special magnetic ink, which can be read at very high speed by the magnetic ink character reader. Most of the data (cheque number, bank sort code, and account number) are pre-printed onto the cheque but the amount is not known until the person writes the cheque.

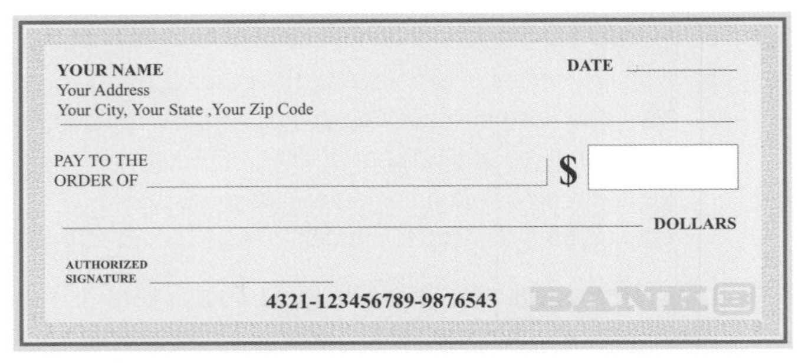
YOUR NAME
Your Address
Your City, Your State ,Your Zip Code
DATE
PAY TO THE ORDER OF $
DOLLARS
AUTHORIZED SIGNATURE
4321-123456789-9876543
BANK

Bank cheques are read at high speed during the clearing process.

When a cheque is presented for payment, the amount is then printed onto the cheque, again in magnetic ink. All cheques are batched together at a clearing centre operated by all the banks. All the cheques are read and processed in one go by the machine; this is an example of batch processing.

Advantages of magnetic ink character recognition:

- Accuracy – the documents (usually cheques) are read with 100% accuracy.
- Difficult to forge – because of the sophisticated magnetic ink technology used, it would be difficult to forge cheques.
- Can be read easily – cheques are often folded, crumpled up, etc. Methods such as OCR or OMR would not work with these. Magnetic ink character recognition uses a magnetic pattern, which is unaffected by crumpling.
- Speed – documents can be read at very high speed; this is particularly important for clearing cheques.

Disadvantage of magnetic ink character recognition:

- Expense – the high-speed magnetic ink character readers are very expensive.

Optical mark reader (OMR)

Optical mark readers use paper-based forms or cards with marks on them which are read automatically by the device. OMR readers can read marked sheets at typical speeds of 3000 sheets per hour. OMR is an ideal method for marking multiple choice question answer sheets for examinations. Students mark the bubbles or squares by shading them in and the reader can read and process the results at high speed.

Advantages of optical mark recognition include:

- Only need one computer and optical mark reader to read the marked sheets, thus less expensive.
- If the forms have been filled in correctly, then almost 100% accuracy can be achieved as there are no typing errors.
- The computer is fast at reading the forms and analysing the results. This is particularly important if a large volume of data needs to be input and processed in a short space of time.

Disadvantages of optical mark recognition include:

- Only suitable for capturing certain data - data needs to be in a form where there are multiple-choice style answers.
- If you have not given precise instructions, users may fill in the forms incorrectly, which will lead to high rejection rates.
- If the form is creased or folded it may be rejected or jam the machine.

Suitable applications for optical mark recognition include:

- Voting forms
- Lottery tickets
- Tests/assessments
- School/college attendance registers.

Optical character reader (OCR)

An optical character reader works by scanning an image of the text and then using special recognition software to recognise each individual character. Once this is done the text can be used with software such as a word-processor, desktop publishing or presentation software.

OCR is used for reading account numbers and details on utility (gas, electricity, water or telephone) bills. It is also used where there is a large amount of text that needs inputting, such as where the text of a book needs to be digitised or for forms used in passport applications.

OCR is used in conjunction with CCTV cameras to recognise car registration plates so that the vehicle can be checked to see if it is taxed and has valid insurance.

Advantages of OCR:

- A faster way of inputting text compared to typing.
- Avoids having to type the text in, which reduces the risk of RSI (repetitive strain injury).
- Can recognise handwriting so can be used to handwrite notes on a tablet computer and convert them to word-processed text.
- Reduces typing errors.

Disadvantages of OCR:

- Text needs to be clearly typed or written (e.g. handwriting is poorly read).
- The forms may be rejected if they are incorrectly filled in.

Bar code reader

Bar codes are a series of parallel bars of differing widths. They are used to input the number using a bar code reader, which appears below the bar code; this is then used to look up the item details in a database.

The bar code is used to input the number below the bar code without having to type it in.

Using the code, the system can determine from a product database:

- The country of origin
- The manufacturer
- The name of the product
- The price
- Other information about the product.

Suitable applications for bar code recognition include:

- Producing itemised customer receipts in supermarkets/stores
- Warehouse stock control systems
- Tracking the progress of parcels during delivery
- Recording books loaned to members in a library
- Luggage labelling at airports.

Advantages of bar code input:

- Faster than typing in numbers.
- Greater accuracy compared to typing in long codes manually.
- Can be read from a distance using a hand-held scanner – useful for wholesalers where the goods are often too heavy to be removed from the trolley.
- Can record bar code and time/date at the same time, which is useful for tracking parcels or bags at an airport.
- Prices can be changed by altering the data in the shop's database so you do not need price stickers on each item.

Disadvantages of bar code input:

- Bar codes can sometimes be damaged – this means having to type in the long number underneath manually.
- Expensive – the laser scanners in supermarkets are expensive, although hand-held scanners are relatively cheap.
- Only a limited amount of data can be stored in a bar code – for example, sell-by dates cannot be stored.
- Computers and network infrastructure must be at each point where the bar code may be read, which adds to the expense.

QUESTIONS B

1 Here is a list of direct data entry devices:

Optical mark reader
Magnetic stripe reader
Chip and PIN reader
Magnetic ink character reader
Optical character reader

Write down the name of the direct data entry device from the above list that is most suited for each of these tasks:

a Reading the numbers on a large number of lottery tickets at high speed.

b Reading the number-plate of a car at an airport where a parking place has been pre-booked and paid for so that the barrier can be raised.

c For inputting loyalty card details when a customer makes a purchase in a supermarket.

d To verify that you are the true owner of a credit card when paying for a meal at a restaurant.

e For reading the details on cheques as part of the process called cheque clearing. *(5 marks)*

2 Here are some input devices being used.

Give the name of the input device. *(4 marks)*

a

b

c

d

3 Ahmed is travelling overseas on holiday. His passport contains an RFID chip which will allow him to pass through automated passport control gates.

a Describe what an RFID chip consists of. *(3 marks)*

b Describe another application where RFID chips are used. *(1 mark)*

c Rather than using RFID chips, a bar code could be used. Describe **two** advantages of using RFID chips rather than bar codes for passport/border control. *(2 marks)*

5 Complete sentences **a** to **f** using one item from the options **i** to **vi**.

a A magnetic ink character reader ...
b An optical mark reader ...
c A magnetic stripe reader ...
d An optical character reader ...
e A bar code reader ...
f An RFID reader ...

i is used to read information about library books at a distance.
ii is used to input data from a school register.
iii is used by banks to read the information contained on a cheque.
iv is used to read the details contained in a magnetic stripe on a credit/debit card.
v is used to read number plates for the issuing of speeding tickets.
vi is used to read information about products in a supermarket. *(6 marks)*

Output devices and their uses

After data entered into the computer has been processed, the resulting information often needs to be output. There are many output devices, with a screen (monitor) and a printer being the most popular.

CRT monitors

CRT stands for cathode ray tube and these monitors are the older, fatter monitors that you still see being used. They take up more space than the more modern thinner TFT screens.

Advantages of CRT monitors:

- Can be used with a light pen to produce and edit drawings using CAD software.
- Sometimes they are used where several people need to view the screen at the same time, for example where several designers are viewing a prototype as they have a wide angle of viewing.

Disadvantages of CRT monitors:

- They are bulky and take up a large amount of space on the desk.
- They generate a lot of heat and can make rooms hot in the summer.
- Glare on the screen can be a problem.
- They are much heavier and present a safety problem when being moved, due to their weight.
- Flicker can cause headaches and eyesight problems.

TFT (thin film transistor/LCD) monitors

These are the thin, flat panel screens that are the more modern type of monitor. The screen is made up of thousands of tiny pixels. Each pixel has three transistors – red, green and blue in a liquid crystal display. The colour is generated by the intensity of each transistor that forms each pixel.

Advantages of TFT/LCD monitors:

- They are very light and this is why they are used in laptop computers.
- They are thin so they do not take up much desk space.
- They are cheaper to run as they use less power than CRT monitors.
- They can be easily positioned to suit the user because they are light.
- The radiation given off is much less than that by a CRT monitor.
- They do not get as hot as CRT monitors and energy is not wasted in the summer using air conditioning to cool offices down.
- They do not create glare and may be less likely to cause eye strain and headaches.
- They are ideal where only one person needs to view the screen.

Disadvantages of TFT/LCD monitors:

- They are not easy to repair, so if they go wrong it is usually cheaper to replace them, which is not very environmentally friendly.
- They have a narrow viewing angle so the image is not as clear as CRT when viewed from the side. They are less useful if many people are viewing the screen.

IPS monitor

IPS (in-plane switching) is a screen technology used with LCD (liquid crystal display) which improves the response time, thus images on the screen do not appear to ghost when they move. It also improves the image when viewed from an angle.

Advantages of IPS monitors:

- Wider viewing angle makes it easier for groups to watch.
- They are much thinner than CRT monitors and therefore take up less desk space.
- Less glare on the screen compared to CRT monitors and are therefore better when used in brightly lit rooms.
- Much more energy efficient compared to a CRT monitor.

Disadvantages of IPS monitors:

- Not as energy efficient as an LED monitor.
- Not as good as graduating shades of colour compared to an LED monitor.
- Not as good when used in dimly-lit rooms.

LED monitor

An LED monitor makes use of light emitting diodes (LEDs) to form the image. LED monitors have the advantage that they use very little power, thus very little heat is produced.

Advantages of LED monitors:

- They are a lot thinner than CRT monitors and about the same thickness as LCD monitors.
- They are much lighter than CRT monitors.
- Energy savings compared to LCD and CRT monitors.
- Better picture quality when watching sports or playing games.

Disadvantages of LED monitors:

- More expensive compared to LCD monitors.
- If you sit off-centre the visibility is poor.

Touch screens as an output device

We have already covered touch screens as input devices; touch screens as output devices can be found on smartphones, tablets, and some laptops and desktops, and the output is produced usually on a TFT/LCD screen from choices made by touching the screen.

Printers

Printers are used to provide users with output in hard copy form. This means the output is printed on paper.

Laser printers

Laser printers are the type of printer mainly used by businesses and organisations because of their high speed. For this reason they are usually chosen as printers used for networks. Most laser printers print in black and white. Although you can buy colour laser printers, they are relatively expensive.

Advantages of laser printers:

- Very quiet – important when used as a network printer in an office where phones are also being used.
- Supplies last longer – the toner cartridge lasts longer than inkjet cartridges.
- High printing speed – essential to have high speed if lots of people on a network are using one printer.
- Very reliable – fewer problems compared to inkjet printers.
- No wet pages that smudge – inkjet pages can smudge but there is no such problem with a laser printer.
- They are cheaper to run (i.e. they have a lower cost per page compared to inkjet printers).
- High quality printouts.

Disadvantages of laser printers:

- More expensive to buy – but they are cheaper to run.
- Colour lasers are very expensive to buy.
- Power consumption is high.
- Most laser printers are larger than inkjet printers.

Inkjet printers

Inkjet printers are popular with home users because they are relatively cheap to buy. However, they are more expensive to run, because of the high cost of the ink cartridges. They work by spraying ink onto the paper and can produce very good colour or black and white printouts. They are ideal printers for stand-alone computers but are less suitable for networked computers as they are too slow.

Inkjet cartridge refills can be expensive.

Advantages of inkjet printers:

- Relatively small compared to laser printers.
- They do not produce ozone or organic vapours which can cause health problems.
- High quality print – ideal for printing photographs, brochures, and illustrated text.
- Quietness of operation – this is important in an office as telephone calls or conversations with colleagues can be conducted during printing.
- Inkjet printers are usually cheaper than laser printers.

Disadvantages of inkjet printers:

- High cost of the ink cartridges – this is okay for low volume work but for large volume it is cheaper to use a colour laser printer.
- Ink smudges – when the printouts are removed the paper can get damp which tends to smudge the ink.
- Cartridges do not last long – this means they need to be replaced frequently.

Dot matrix printers

Dot matrix printers are used in offices and factories where multi-part printouts are needed and noise is not an issue. For example, one copy of an invoice could be kept by the sales office, one copy could be sent to the customer and another given to the factory. They are impact printers and work by hitting little pins against an inked ribbon to form the characters on the paper, which creates a lot of noise.

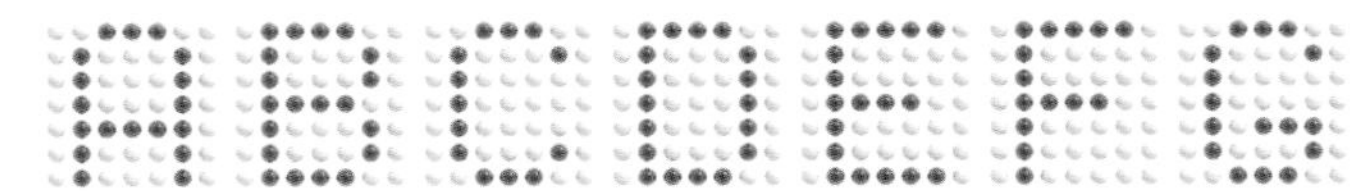

With a dot matrix printer, characters are formed by an arrangement of dots.

Dot matrix printers may seem out-of-date but they are ideal in some situations. You will see them being used to print out invoices in warehouses, car part stores, and garages. You might also see them being used in pharmacies for printing labels.

Advantages of dot matrix printers:

- Can be used to print multi-part stationery – this is useful where the printout needs to be sent to different places.
- Can be used with continuous stationery – this makes it ideal for stock lists, invoices, etc.
- Cheaper to run than inkjet printers.

Disadvantages of dot matrix printers:

- Very noisy – the pins make a lot of noise when they strike the inked ribbon.
- Low quality printouts. The text can look 'dotty'.
- Unsuitable for printing graphics.
- More expensive to buy than inkjet printers.

Multimedia projectors

Multimedia projectors are used to project what appears on the computer onto a much larger screen, enabling an audience to view it. They are used for training presentations, teaching, advertising presentations, etc. They can also be used to provide a larger picture from televisions and video/DVD players.

Advantages of multimedia projectors:

- Many people can view what is being presented.
- Can show video to a large audience who may be seated some distance from the screen.

Disadvantages of multimedia projectors:

- The image quality may not be as high as when seen on a computer screen.
- The cooling fans on the projector create noise which can be distracting.
- Requires a darkened room otherwise the image will appear dim.
- Requires the user to know how to set up and run the equipment.

Wide format printers

Wide format printers can handle printing on sheets of paper 24 inches or wider. Not many people use these printers regularly enough to purchase one, so instead they go to special print shops who will print out on wide format printers for a fee. Some businesses may have a need to produce posters, maps, plans and other large format documents and may decide to purchase one. Wide format printers use the same technology as smaller printers so it is possible to purchase ink-jet and laser wide format printers.

Advantages of wide format printers

- Can print on paper larger than A4 so is ideal for printing maps, scale drawings, posters etc.
- Do not need to use the services of professional print shops which can be expensive.
- Printouts larger than A4 size such as maps, banners, plans etc. can be printed.

Disadvantages of wide format printers

- Expensive especially if only used occasionally.
- Wide format printers are quite large which can be a problem in small offices where space is limited.

As the paper size is larger the ink cartridges will not last long and they are expensive to replace.

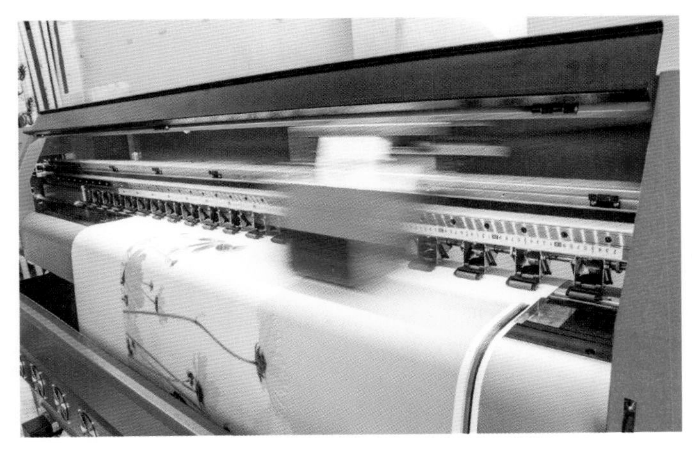

Printing a poster on vinyl rather than paper

3D printers

A digital model in three dimensions is created on the screen using design software. A model is then produced in 3D using a 3D printer. 3D printing uses an additive technique where successive layers of material such as plastic, glass, ceramic, or metal are laid down to produce the 3D model.

Some companies use 3D printing to manufacture parts.

3D printing is very popular in health care applications because it is possible to produce items that fit the patient exactly. Here are some health care applications of 3D printing:

Dentistry – scanning a patient's mouth and then using computer-aided design software along with computer-aided manufacturing software enables crowns, bridges, implants, etc., to be produced. The speed of manufacture using a 3D printer means that false-teeth can be produced quickly so patients are not left with missing teeth for long.

Hearing aids – these need to be comfortable to wear. Using scanning and then 3D printing ensures hearing aids fit perfectly.

Prosthetic limbs – a near-perfect bridge between a patient's body and a prosthetic can be manufactured using a 3D printer.

Advantages of 3D printers

- Can covert designs to prototypes a lot faster compared to traditional manufacturing of prototypes. This makes making small design changes and seeing the finished product in much less time.
- It is cheaper to use 3-D printing rather than traditional methods of manufacture despite 3-D printers being expensive.
- It is much easier to produce customised products as they can just be printed out from designs made using design software on the computer.

Disadvantages of 3D printers

Manufacturing of goods is much easier using 3D printing rather than by traditional methods so it is easier to produce counterfeit goods. Selling of counterfeit goods steals money copyright/trademark owners.

- The manufacture of goods using 3D printers is less skilled compared to the use of older techniques which takes work from skilled manufacturers.
- Dangerous items such as weapons can be manufactured using 3D printers.
- There is a limited number of materials that can be used with 3D printers which limits what can be manufactured.

A model tank is designed using computer-aided design (CAD) software and then output to the 3D printer where a 3D model of the tank is produced

Speakers

Any application which requires sound will need speakers or earphones to output the sound. Applications which use sound include multimedia presentations and websites. Many online encyclopedias make use of sound with explanations, famous speeches, music, etc.

People also use their computer for playing their CDs while working or maybe listening to music via the internet. Speakers are also needed to output the sound when you watch a film on the computer. With laptop computers the speakers are usually built in. Desktop computers may have basic speakers included but many people choose to upgrade these to better sound quality speakers.

Motors

Computers can issue control signals to turn electric motors on or off. Some special motors, called stepper motors, turn through only a certain angle depending on the signals they receive from the computer.

Here are some applications of motors being used in control systems:

- Motors used in automatic washing machines to turn valves on/off to allow the water in/out.
- Motors are used to rotate the washing in the drum of a washing machine.
- Motors can operate pumps in central heating systems to pump hot water around the radiators.
- Motors in computer-controlled greenhouses can turn watering systems on and off.
- Motors in computer-controlled greenhouses can open windows or close them to decrease or increase humidity.
- Motors in microwave ovens can rotate a turntable.
- Motors can control the movements of robots in factories.
- Motors can control the passage of items along a production line in a factory.

Buzzers

Buzzers often act as warning signs or indicate that a control process has been completed. For example, when a washing machine or drier has finished, the control system issues a buzz or other sound to inform the householder. Buzzers can also be found in automatic cookers and microwaves.

Heaters

Heaters can be switched on or off according to control signals issued by a computer. Some control devices with heaters are: automatic washing machines, automatic cookers, central heating controllers, and computer-controlled greenhouses.

Lights/lamps

Lights and lamps can be controlled using a control system that has a light sensor to sense the light level. In a computer-controlled greenhouse, optimum lighting conditions are maintained by switching on lights when the natural light level falls below a certain value. Security lights and outside lighting can be controlled to come on when it gets dark and go off when it gets light again. Some cars also turn lights on and off automatically depending on light levels.

QUESTIONS C

1 Write down the names of **five** items from the following list that are used as output devices:

Temperature sensor	Touch screen	Speakers
Multimedia projector	Joystick	Web cam
Pressure sensor	Buzzer	Mouse
Graphics tablet	Bar code reader	Motor

(5 marks)

2 Match each output device on the left with the correct use on the right.

Output device	Use
Multimedia projector	A large monitor used by older computers
Buzzer	Used when checking the depth of water in a container
Light	Used for outputting large posters, maps and plans
Wide format printer	A thin lightweight monitor
Dot matrix printer	Used to make presentations to a large audience
CRT monitor	Used for producing multi-part invoices
TFT/LCD monitor	Used as an alert on smartphones

(7 marks)

3 Describe **three** differences between a laser printer and an ink-jet printer. *(3 marks)*

4 Touch screens are both input and output devices.

Describe **one** application where a touch screen is used as an output device. *(1 mark)*

5 A home computer user is about to purchase a printer and they have the choice of the following printers:

Laser printer
Ink-jet printer
Dot matrix printer

a Describe the advantages and disadvantages of each of these types of printer. *(6 marks)*

b State which printer you think they should buy and give reasons for your choice. *(2 marks)*

REVISION QUESTIONS

1 Here is a list of devices that may be attached to a computer system:

LCD monitor	keyboard
portable hard drive	mouse
touch pad	laser printer
flash/pen drive	microphone
digital camera	speakers
web camera	CD ROM drive

a Write down the names of all the output devices in the above list. [3]

b List **two** other output devices **not** in the above list. [2]

c Give the name of **one** input device that is **not** included in the list above. [1]

2 a Give **two** uses for each of the following input devices in a personal computer: [6]

i mouse

ii microphone

iii digital camera

b Give the names of **four** output devices and give **one** use for each of them. [8]

3 Copy and complete the table below. [10]

Application	Most suitable output device
Alerting the user that an error has occurred by making a beep	
Printing a poster in colour	
Listening to a radio station using the internet	
Producing a large plan of a house	
Producing a hard copy of a spreadsheet	
Producing a colour picture on paper taken with a digital camera	
Producing a series of invoices with several copies that can be sent to different departments	
Producing a warning when a bar code is read incorrectly	
Listening to messages from a voicemail system	
Displaying the results of a quick search on the availability of a holiday	

4 Touch screens can often be seen at tourist information offices.

a Describe what a touch screen is and how it works. [2]

b Give **one** advantage of using a touch screen as an input device for use by the general public rather than using a monitor and a mouse. [1]

5 There are a number of different printers, each with their own advantages and disadvantages. The names of these printers are listed here:

Laser printer Inkjet printer 3D Printer

Identify the name of the printer being described for each one of the following:

a A printer which is used in offices for printing lots of documents in a short period of time. [1]

b An inexpensive printer that is ideal for the home which can print in colour as well as black and white. [1]

c A printer which sprays the ink onto the page. [1]

d The type of printer where you have to be careful not to smudge the damp printouts as they come out the printer. [1]

e The type of printer which uses a toner cartridge. [1]

f The printer which is cheap to buy but which has high running costs owing to the high cost of the ink cartridges. [1]

g A printer that dentists can use to print crowns, bridges, implants, and false teeth for their patients. [1]

Test yourself

The following notes summarise this chapter, but they have missing words. Using the words below, copy out and complete sentences **A** to **O**. Each word may be used more than once.

input **touch pad** **keyboard** **microphone**
scanner **optical character recognition** **laser**
joysticks **output** **sensors** **mouse**
inkjet **digital** **touch** **RFID**

A Devices used to get data from the outside world into the computer are called ______ devices.

B The most common input device, which comes with all computers, is the ______.

C A ______ is used to move a pointer or cursor around the screen and to make selections when a desktop computer is used.

D Where space is restricted, such as when a laptop is being used on your knee, a ______ is used instead of a mouse.

E ______ are used primarily with games software.

F In voice recognition systems a ______ is used as the input device.

G The device used to input text and images is called a ______.

H Special software can be used to recognise the individual letters in a scanned piece of text and this is called ______.

I Cameras that do not use film and can transfer an image to the computer are called ______ cameras.

J A(n) ______ reader reads the data contained in a chip at a distance and can be used at passport/border control and also at libraries.

K Quantities such as temperature and pressure can be detected and measured using ______.

L Printers and monitors are examples of ______ devices.

M The type of printer that is very fast and uses a toner cartridge is called a ______ printer.

N A cheaper printer, which squirts a jet of ink at the paper, is called a(n) ______ printer.

O ______ screens are sometimes used as an input device for multimedia systems and are popular for information points to be used by the general public.

EXAM-STYLE QUESTIONS

1 Ring **two** items which are output devices. *[2]*

Blu-ray disk | Wide format printer | Scanner
Optical mark reader | Multimedia projector | Web cam

2 Draw **four** lines on the diagram to match the input device to its most appropriate use. *[5]*

Input device	Use
Scanner	For recording narration to be used with a presentation
Touch screen	For digitizing an old photograph so it can be put on a website
Chip reader	For inputting selections when buying a train ticket
Microphone	Reading information on a credit/debit card

3 Complete each sentence using **one** item from the list. *[4]*

Laser | Dot matrix | Wide format printers | Projectors
Speakers | Inkjet | Motors | Touch screens

a .... printers are printers that use continuous stationery.

b .... are used when very large hard copy is required.

c .... are used with tablets and smartphones for both input and output.

d .... printers are used for the printing of high quality photographs.

4 A payroll office, which prints out large numbers of payslips every month, has decided to install a new printer. Discuss the advantages and disadvantages of using a laser printer, an inkjet printer or a dot matrix printer in this office. *[6]*

(Cambridge IGCSE Information and Communication Technology 0417/13 q20 Oct/Nov 2010)

5 Name the input devices **A**, **B**, **C** and **D** using the words from the list.

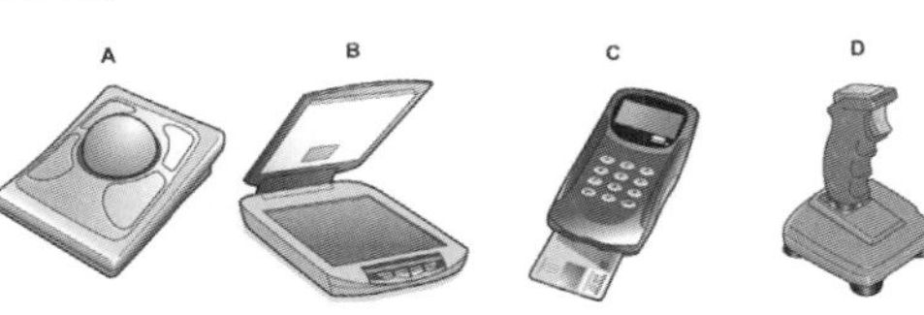

Chip reader | Digital camera | Joystick
Light pen | Microphone | Remote control
Scanner | Trackerball *[4]*

(Cambridge IGCSE Information and Communication Technology 0417/13 q1 Oct/Nov 2010)

6 Here is a list of devices used with computer systems:

Pressure sensor
Touch screen
Wide format printer
Mouse
RFID reader
Magnetic ink character reader (MICR)
Laser printer
Motor
Light
Bar code reader
Light sensor

a Write down the names of all the devices in the list which are **input** devices. *[7]*

b Write down the names of all the devices in the list which are **output** devices. *[5]*

c Write down the name of the device in the list most suited for performing each of the following tasks:

i Determining whether the liquid in a large vessel is at the correct level.

ii Printing out a large-scale plan of a house to be used by a builder.

iii For scanning goods in quickly at a supermarket checkout.

iv For giving the instruction to turn on the lights outside a house automatically when it goes dark.

v Used for raising a barrier in an airport car park after the number plate has been read and matched that a payment has been made. *[5]*

7 Identify **two** devices which are used in control systems to output data. *[2]*

(Cambridge IGCSE Information and Communication Technology 0417/11 q2 May/June 2014)

8 **a** State what is meant by OMR and OCR. *[2]*

b Compare and contrast the use of OMR, OCR and a keyboard as methods of data entry. *[6]*

(Cambridge IGCSE Information and Communication Technology 0417/11 q5 Oct/Nov 2012)

9 Describe **two** advantages and **two** disadvantages of wide format printers. *[4]*

10 **a** Explain what is meant by 3D-printing. *[2]*

b Describe two uses of 3D-printing. *[2]*

3 Storage devices and media

Backing storage is any storage which is not classed as ROM or RAM. It is used to hold programs and data. Backing storage devices include magnetic hard disk drives, optical drives (CD or DVD), flash/pen drives, etc. Storage devices are the pieces of equipment which record data onto the storage media (i.e. optical, magnetic, etc.) or read it from the storage media.

In this chapter you will be covering both storage devices and media.

The key concepts covered in this chapter are:

- Identification of storage devices, their associated media and their uses
- The advantages and disadvantages of the storage devices

Backing storage devices and media

Data needs a material on which to be stored (called the storage media) and the device recording the data onto the storage media is called the storage device.

Memory (ROM and RAM) is an expensive component of a computer and the RAM loses its contents when the power is switched off. Backing storage is used for the storage of programs and data that are not needed immediately by the computer. They are also used for storing backups of programs and data in case the originals are damaged or destroyed.

Backing storage is defined as storage other than ROM or RAM that is:

- Non-volatile, which means it holds its contents when the power is removed
- Used to hold software/files not being currently used.

Magnetic backing storage media

Magnetic backing storage media are those types of media including magnetic disk and magnetic tape where data is stored by writing a magnetic pattern onto the magnetisable surface of the media.

Fixed hard disks and drives

It should be noted that the magnetic hard disk is the media and a magnetic hard disk drive includes all the components such as the read/write heads which read the data off the disk and also record data onto the surface of the disk. The words magnetic hard disk and magnetic hard disk drive are often used interchangeably.

Fixed hard disks consist of a series of disks coated with a magnetic material and a series of read/write heads which record data onto the surface or read it off the surface. Fixed hard disks are used to store operating systems, applications software (i.e. the software you use to complete tasks such as word-processing software), and any files of data needed by the user. All of these require that the data can be both accessed and stored onto the media quickly.

Fixed hard disks can consist of one or more magnetically coated disks.

Fixed hard disks are, as the name suggests, fixed in the computer and are not designed to be portable. Portable hard disks are available and these will be looked at later.

Fixed hard disks in PCs and laptops are used for storing files that have been created using software and then saved, e.g. C:/My Documents/letter.doc.

Fixed hard disks are used for online processes. For example, when you create a personal website, it is stored on the fixed hard disk of the organisation you use to connect to the internet. When other people access your website, the data is obtained off this fixed hard disk. Computers connected to the internet which store web pages are called web servers.

Many fixed hard disks are used in file servers for networks. In most networks all the data and program files are stored centrally rather than on each individual computer. This central store of data uses fixed hard disks as the storage media. Because many users will want access to this store, the speed at which data is accessed and transferred needs to be very high.

Advantages of fixed hard disks:

- A very high access speed (the speed of finding and reading data off the disk).
- A very high transfer rate (the speed of storing data onto the disk).
- A very high storage capacity.

Disadvantages of fixed hard disks:

- Fixed hard disks cannot be transferred between computers, unlike a portable hard disk.
- The hard disk cannot be taken out of the computer and locked away for security purposes.

Portable hard disks and drives

It is possible to buy additional hard disks. These hard drives are called portable hard disks and may be removed each night and stored safely.

Portable hard disks are also used to store very large files which need to be transported from one computer to another. Generally portable hard disks are more expensive than other forms of removable media but their very large storage capacity, high access speed, and transfer rate are the reasons why they are chosen. Portable hard disks are used for reasons other than the back-up of data and programs, e.g. by a writer who works in two or three locations during the week.

Most portable hard disks connect to a USB port and need no installation. You simply connect them and the operating system knows how to work them.

Advantages of portable hard drives include:

- Their very large storage capacity means large files can be transferred between computers.
- They are very fast at accessing and storing files so transferring large files such as multimedia files takes little time.
- They can be attached to and used by any computer that has a USB port.

Disadvantages of portable hard drives include:

- Their size and portability means they are easily stolen.
- It makes it easy for employees to copy confidential data such as a medical database which is a security risk to companies/ organisations.
- Their use with lots of different computers can mean there is a danger of viruses being transferred.

Magnetic tapes

Magnetic tape stores data on a plastic tape coated with a magnetic layer. Here are the main features of magnetic tape:

- Magnetic tape has a huge storage capacity.
- Magnetic tape is used to back up the data stored on hard disks.
- Because it takes time to move the tape to the position where the data is stored, tape storage is much less common than disk storage.

Magnetic tape is used in any application where extremely large storage capacity is needed and the speed of access is not important.

Magnetic tape has a huge storage capacity but is used for only a few limited applications.

Magnetic tape provides serial access. What this means is that it is necessary to access each record in turn on the tape until the correct record is found. This takes a long time and is the reason why magnetic tape is being taken over by fixed hard disks. Magnetic tape is useful for when every record on the tape needs to be accessed or stored in turn on the tape. This is the reason why magnetic tape can be used for backups of file servers for computer networks. These servers can have huge storage capacities and can be backed up on a single tape rather than a series of hard disks.

Magnetic tape is also used in a variety of batch processing applications where the computer can just be left to process the data without any human intervention during the processing. Examples of this include reading bank cheques, payroll processing and general stock control.

Advantages of magnetic tape include:

- They have extremely large storage capacities and this makes them ideal for backup purposes where all data and programs need to be stored.
- They are less expensive compared to similar capacity magnetic hard disk drives.
- The data transfer rate is high (writing to tape).

These magnetic tape cartridges are used for daily backups of data.

Disadvantages of magnetic tape include:

- They are not suitable for an application that requires fast access to data because the access speed is low.
- In order to update details on a tape it is necessary to create a new tape containing some of the previous details along with the updated details.

Optical drives

Optical drives are storage devices that make use of lasers to read data off optical disks.

Optical disks are flat circular disks on which data is stored as a series of bumps. The way the bumps reflect laser beam light is used to read the data off the disk.

CDs are used to hold large files (<1 gigabyte) and are ideal for holding music and animation files. DVDs have a much larger capacity (4.7 to 8.5 gigabytes) and are used mainly for storing films/video. Both CD and DVD can be used to store computer data and can be used for backup purposes.

Optical disks include CDs and DVDs and are used to store digital data as a binary pattern on the disks.

CD ROM (compact disk read only memory)

CD ROMs are used mainly for the distribution of software and the distribution of music. Although most home computers are equipped with DVD drives, a lot more computers, especially those used in businesses, still only have CD drives. You can read a CD using a DVD drive but you cannot read a DVD with a CD drive. This is why software is still being sold on CD rather than DVD.

With CD ROM:

- Data is read only
- Data is stored as an optical pattern
- There is a large storage capacity (700 Mb)
- They can be used for the distribution of software.

Advantage:

- Once written, data cannot be erased, making it useful for the distribution of software/music or backing up.

Disadvantage:

- Data transfer rate and access rate are lower than for a hard disk.

DVD ROM (digital versatile disk read only memory)

DVD ROM is used for the distribution of movies where you can only read the data off the disk. A DVD ROM drive can also be used for reading data off a CD. DVD is mainly used for the distribution of films, computer games and multimedia encyclopaedias.

Advantage:

- High storage capacity means full-length movies can be stored.

Disadvantages:

- The user cannot store their files on the disk.
- Older computers may not have a drive capable of reading DVDs.

CD R (CD recordable)/DVD R (DVD recordable)

CD R allows data to be stored on a CD, but only once. DVD R allows data to be stored on a DVD once. Both these disks are ideal where there is a single 'burning' of data onto the disk. For example, music downloaded off the internet could be recorded onto a CD in case the original files were damaged or lost.

They can be used for archive versions of data. Storing archive data on the fixed hard disk would clutter up the disk so it is better to store it on removable media and keep it in a safe place. DVD R is ideal for storing TV programmes where you do not want to record over them.

CD RW (CD rewriteable)

A CD RW disk allows data to be stored on the disk over and over again – just like a hard disk. This is useful if the data stored on the disk needs to be updated. You can treat a CD RW like a hard drive but the transfer rate is less and the time taken to locate a file is greater. The media is not as robust as a hard drive.

Advantages:

- Re-writable, so they can be re-used.
- The data stored can be altered.

Disadvantages:

- The data transfer rate is lower than for a magnetic hard disk.
- Optical drives such as CD RW are more easily damaged than hard drives.

DVD RW (digital versatile disk read/write)

A DVD RW drive can be used to write to as well as read data from a DVD. DVD RW are sometimes called DVD burners because they are able to be written to and not just read from. Like CD RW, they are ideal for storing data that needs regularly updating.

Typical storage capacities are:

- 4.7 Gb for older DVD drives
- 8.5 Gb for latest DVD drives

DVD RAM (digital versatile disk random access memory)

DVD RAM has the same properties as DVD RW in that you can record data onto it many times but it is faster and it is easier to overwrite the data. The repeated storage and erasure of data acts in a similar way to RAM – hence the name. A typical storage capacity for DVD RAM is 10 Gb. DVD RAM is used for the storage of TV/film at the same time as watching another program.

Advantages:

- Can read and write at the same time
- Fast transfer rate
- Fast access rate.

Disadvantage:

- The devices are expensive compared to other devices.

Blu-ray

The Blu-ray disk is an optical disk that has a much higher storage capacity than a DVD. Blu-ray disks have capacities of 25 Gb, 50 Gb, and 100 Gb. These high capacity Blu-ray disks are used to store high definition video. They are used for storing films; a 25 Gb Blu-ray disk is able to store 2 hours of HDTV or 13 hours of standard definition TV. It is possible to play back video on a Blu-ray disk while simultaneously recording HD video. Newer computers now come with Blu-ray drives and eventually Blu-ray disks will become the norm for the storage of data on computers.

Solid state backing storage

Solid state backing storage is the smallest form of memory and is used as removable storage. Because there are no moving parts and no removable media to damage, this type of storage is very robust. The data stored on solid state backing storage is rewritable and does not need electricity to keep the data. Solid state backing storage includes the following:

- Memory sticks/pen drives
- Flash memory cards.

Revision Tip

When selecting backing storage media, ensure that you take into account the relative access speeds, whether the media needs to be re-used and the size of the files to be stored.

Memory sticks/pen drives

Memory sticks/pen drives are very popular storage media which offer large storage capacities and are ideal media for photographs, music, and other data files. Memory sticks are more expensive per Gb than CD/DVD/hard disks. They consist of printed circuit boards enclosed in a plastic case.

The main advantages are:

- Small and lightweight – easy to put on your key ring or in your pocket.
- Can be used in almost any computer with a USB drive.
- Large storage capacity (up to 256 Gb).
- No moving parts so they are very reliable.
- Not subject to scratches like optical media.

The main disadvantages are:

- Their small size means they are easily stolen.
- They are often left in the computer by mistake and lost.
- They have lower access speeds than a hard disk.

Memory sticks/pen drives are the most popular portable storage media. Their portability is their main advantage and you simply plug them into the USB port where they are recognised by the operating system automatically.

Memory sticks/pen drives are ideal for the transfer of relatively small amounts of data between computers.

Flash memory cards

Flash memory cards are the small thin rectangular or square removable cards that are used for storage of digital images by digital cameras. They can also be used in any situation where data needs to be stored and so are used with desktop computers, laptops, mobile phones, and MP3 players. You can see the card readers in supermarkets and other stores where you can take your cards containing photographs and get them printed out.

Backups and the reasons for taking them

A backup is a copy of data and program files kept for security reasons. Should the originals be destroyed or corrupted then the backups can be used. Using a file server and storing both programs and data on it means that backups can be taken in one place. Backups should be held on removable devices or media that are taken off site each day. The individual users do not need to take their own backups. The person in charge of the network (usually the network manager) will take the backups needed. Many systems now take backups automatically at a certain time of the day and send the data using the internet to a company that specialises in storing backups. Backups should always be removed off-site. This is in case of fire or if the building is destroyed.

QUESTIONS A

1 Which **three** of the following are backing storage devices? *(3 marks)*

RAM	Plotter
Hard drive	Speaker
CD RW drive	Keyboard
ROM	Pen drive

2 Give the meaning of the following abbreviations.
   a DVD R *(1 mark)*
   b DVD RW *(1 mark)*

3 Backups of programs and data should be taken on a regular basis.
   a Explain what is meant by a backup. *(1 mark)*
   b Give **one** reason why backups should be taken on a regular basis. *(1 mark)*
   c Give **one** example of backing storage suitable for the taking of backup copies and explain why it is suitable. *(3 marks)*

Activity 3.1

Everyone who uses a computer needs to store their data and programs somewhere.

For this activity you have to find out about the backing storage devices that are available for computer users.

Use the internet to find out:

- Types of storage devices and their storage capacity.
- The main advantages and disadvantages of each device.
- Type of access.
- Speed of access.
- The cost of the storage device.
- The media they need if applicable.

Online computer equipment stores are good places to look for information.

Test yourself

The following notes summarise this chapter, but they have missing words. Using the words below, copy out and complete sentences **A** to **G**. Each word may be used more than once.

cartridges **immediately** **DVD**
backing **hard** **programs** **backups**

A ______ are copies of data and program files kept for security reasons.

B Backing storage is used for the storage of programs and data that are not needed ______ by the computer.

C Flash/pen drives are popular ______ storage media because they are very small and cheap compared to the alternatives.

D Currently they are also used for storing backups of ______ and data in case the originals are damaged or destroyed.

E ______ drives consist of a series of disks with a magnetic coating and a series of read/write heads which put the data onto or record it off each surface.

F ______ usually use reels of magnetic tape in a hard plastic case.

G Optical media include CD and ______ for storing data.

REVISION QUESTIONS

1 **a** Give **two** uses for each of the following devices in a personal computer:
   i Magnetic hard disk drive [2]
   ii CD ROM drive [2]
   b Give **two** ways small, high capacity storage devices have influenced the development of portable equipment that can be used for work and play. [2]

2 Computers need memory and backing storage.
   a Give **one** difference between main internal memory and backing storage. [1]
   b Give **one** example of what would be stored in main internal memory. [1]
   c Give **one** example of what would be stored in backing storage. [1]

3 Data needs to be stored for future use. For each of these storage/media devices, explain a suitable use and explain clearly why it is suited to the application.
   a Memory card [3]
   b CD ROM [3]
   c Magnetic hard disk drive [3]
   d Flash/pen drive [3]

EXAM-STYLE QUESTIONS

1 Ring **two** items which are storage media.

Flash memory card **Graph plotter** **Magnetic disc**

OCR **OMR** **Touch pad** [2]

(Cambridge IGCSE Information and Communication Technology 0417/13 q2 Oct/Nov 2010)

2 A student wants to transfer work from a computer in school to their home computer.
They have the choice of using a CD or a pen drive.
Give **three** reasons why a student would choose a pen drive rather than a CD. [3]

3 Name the methods of storage **A**, **B**, **C** and **D** using the words from the list.

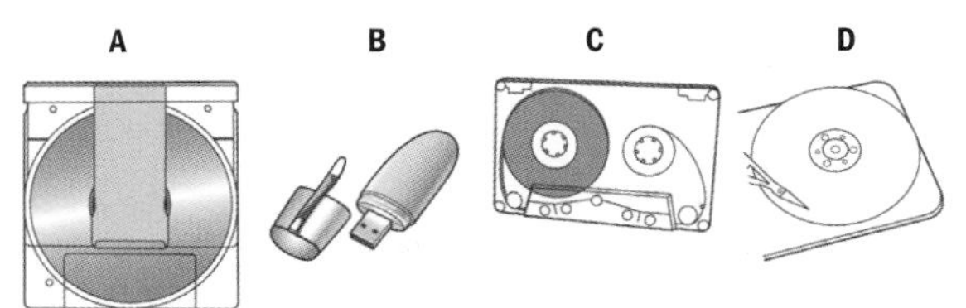

Bar code **Chip** **DVD RAM**

Graphics tablet **Light pen** **Magnetic disc**

Magnetic tape **Pen drive** [4]

(Cambridge IGCSE Information and Communication Technology 0417/11 q1 May/June 2010)

4 A website designer has the choice of backing their work up using either a pen drive or a portable hard disk.
Give **two** advantages of using the portable hard disk for backup. [2]

5 Give **two** advantages of a fixed magnetic hard disc compared to an optical disc to store data. [2]

(Cambridge IGCSE Information and Communication Technology 0417/11 q3 May/June 2014)

6 Draw **four** lines to match each storage medium to the **most appropriate** use in this list.

Storage medium	Use
Fixed hard disc	**To transfer files from one computer to another**
DVD ROM	**Batch processing applications**
Pen drive	**To store operating systems**
Magnetic tape	**Publishers distributing encyclopaedias**

[4]

(Cambridge IGCSE Information and Communication Technology 0417/11 q6 May/June 2013)

7 Optical discs and magnetic tapes are both used to store computer data.

Give **two** uses of optical discs and for each use give a reason why they are preferred to magnetic tapes. [4]

(Cambridge IGCSE Information and Communication Technology 0417/11 q15 May/June 2013)

4 Networks and the effects of using them

Most computers are now connected to networks. For example, in the home your computer may be able to access the internet, in which case it becomes part of a network. You may have a small network at home, which allows you to access the internet on a desktop computer and a laptop at the same time. Networks provide so many more benefits compared to stand-alone computers (i.e. computers not connected to a network).

The key concepts covered in this chapter are:

- Networks (network devices, types of network, and using different types of computer to access the internet)
- Network issues and communication (security issues and network communication)

Networks

A network is two or more computers that are linked together so that they are able to share resources. These resources could be a printer, scanner, software or even a connection to the internet. You can also share data using a network. For example, a pupil database in a school could be accessed from any of the computers connected to the network.

Network devices

To build a network, some or all of the following network devices are needed.

Routers

Each computer linked to the internet is given a number which is called its IP (Internet Protocol) address. An IP address looks like this: 123.456.1.98 and is unique for each device while linked to the internet.

When data is transferred from one network to another the data is put into packets. The packets contain details of the destination address of the network it is intended for. Computers on the same network all have the same first part of the Internet Protocol address and this is used to locate a particular network.

Routers are hardware devices that read the address information to determine the final destination of the packet. From details stored in a table in the router, the router can direct the packet onto the next network on its journey. The data packet is then received by routers on other networks and sent on its way until finally ending up at the final destination network.

Routers can be used to join several wired or wireless networks together. Routers are usually a combination of hardware which act as gateways so that computer networks can be connected to the internet using a single connection.

This wireless router allows all the computers in the home to share an internet connection.

Network interface cards (NIC)

Before a computer can be connected to a network, it will need to have a network interface card. Most modern computers have these when you buy the computer. These cards connect directly to the motherboard of the computer and have external sockets so that the computer can be connected to a network via cables.

Basically a NIC does the following:

- It prepares the data for sending.
- It sends the data.
- It controls the flow of data from the computer to the transmission media (metal wire, optical fibre, etc.).

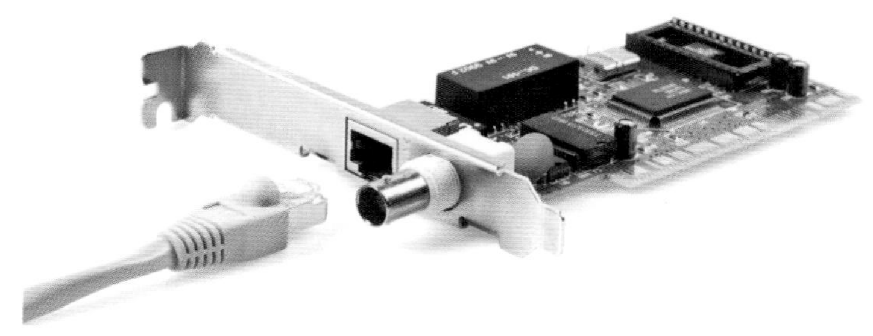

A network card allows a connection to be made between the network cables and the motherboard of the computer

Hubs

A hub is a simple device that does not manage any of the data traffic through it. It is simply used to enable computers on the network to share files and hardware such as scanners and printers.

Data is transferred through networks in packets. A hub contains multiple ports (connection points). When a packet arrives at one port, it is transferred to the other ports so that all network devices of the LAN can see all packets. Every device on the network will receive the packet of data, which it will inspect to see if it is intended for that device or not.

Switches

Like hubs, switches are used to join computers and other devices together in a network but they work in a more intelligent way compared to hubs. Switches are able to inspect packets of data so that they are forwarded appropriately to the correct computer. Because a switch sends a packet of data only to the computer it is intended for, it reduces the amount of data on the network, thus speeding up the network.

Bridges

Bridges are used to connect LANs together. When one of the LANs sends a message, all the devices on the LAN receive the message. This increases the amount of data flowing on the LAN. Often a large LAN is divided into a series of smaller LANs. If a message is sent from one computer in a LAN to another computer in a different LAN then the message needs to pass between the LANs using a bridge. The advantage in subdividing a larger network is that it reduces the total network traffic as only traffic with a different LAN as its destination will cross over the bridge. A bridge therefore usually has only two ports in order to connect one LAN to another LAN.

Modems

Modems have been largely replaced by routers for providing access to the internet as the routers are much faster. A modem converts the data from the computer (digital data) into sound (analogue data) that can be sent along ordinary telephone lines. Once the data arrives at the other end of the line, the sound will need to be converted back to digital data using another modem before it can be understood by the receiving computer.

The use of Wi-Fi and Bluetooth in networks

There are two main technologies that allow wireless communication between computers and other devices.

Wi-Fi

Wi-Fi enables computers and other devices to communicate wirelessly with each other. Areas where the internet can be accessed wirelessly using Wi-Fi are called access points or hotspots and they can be found in many public places, such as coffee bars, hotels, airports, etc.

The range of a Wi-Fi depends on the type of wireless router being used and also if there are obstacles such as walls in the way of the signal. For a home network, the range of Wi-Fi are typically 50m indoors and 100m outdoors.

Advantages of wireless communication:

- Provided you have a wireless signal, you can work in hotels, outside, in coffee shops, etc.
- You are not confined to working in the same place. For example, you can work on trains, buses and even some aircraft provided there is a signal.
- Fewer/no trailing wires to trip over.
- It is easier to keep a working area clean if there are not as many wires in the way.
- There are no network wires so there are no costs for their installation.

Disadvantages of wireless communication:

- The danger of hackers reading messages.
- There are areas where you cannot get a wireless network.
- There is some evidence that there may be a danger to health.
- Limited signal range.

Bluetooth

Bluetooth is a wireless technology used to exchange data over short distances and it makes use of radio waves. The range of Bluetooth depends on the power of the signal and can typically be from 5m to 100m.

Here are some uses for Bluetooth:

- Wireless communication between input and output devices. If you have a wireless keyboard and mouse then they could use Bluetooth. Printers can be controlled wirelessly from a computer using Bluetooth.
- Communication between a wireless hands-free headset and a mobile phone. These are the sort people can use while they are driving.
- Creating a small wireless network where the computers/devices are near to each other and where the small bandwidth is not a problem. For example, creating a small home wireless network.
- Transferring appointments, contacts, etc., between a computer and a mobile device such as mobile phone or tablet.
- Communication using a games controller. Many games consoles use Bluetooth.

Bluetooth is used with hands-free headsets.

Setting up and configuring a small network

To set up and configure a small network is easy. Here are the steps you generally need to take.

1 Choose an Internet Service Provider (ISP), pay them a subscription charge and get a router, often provided by the ISP (which might include wireless access point).

2 Connect the router to the telephone line or cable point. (In some countries, this is connection to a satellite link.)

3 Connect any cabled devices to the router using wires between the ports on the router and the Network Interface Card on the device.

4 Connect any wireless devices - e.g laptops, tablets, printers using wireless network. You will need to open settings in your operating system on each device to create a new connection. You usually need the passcode given to you by the ISP to do this.

5 You will need a browser on each device to access the internet.

Here is some more detail on some of the essential parts of a network.

Web browser software

A web browser is a program that allows web pages stored on the internet to be viewed. Web browsers read the instructions on how to display the items on a web page which are written in a form called HTML (Hypertext Markup Language). You will be learning about HTML in Chapter 21. A web browser allows the user to find information on websites and web pages quickly and it does this through:

- Entering a web address (URL)
- A web/internet portal
- Key word searches
- Links
- Menus.

Web browser software includes email software that allows you to send and receive email.

Access to an internet service provider

Connecting directly to the internet is very expensive and only suitable for large companies and organisations. Most people connect to the internet via an organisation called an internet service provider. This is an organisation that supplies the connection to the internet as well as providing services including:

- Storage on their server, where you can store your website
- Email facilities
- Instant messages where you can send short messages to your friends when they are online
- Access to online shopping
- Access to news, sport, weather, financial pages, etc.

When you establish a network connection with an ISP and pay their subscription charge they generally supply you with a wireless router. This will enable you to link devices wirelessly, thus forming a network that will allow users to:

- Share internet access
- Share access to hardware devices such as printers and scanners
- Share access to files held on different computers.

At the back of the router you will see some sockets into which Ethernet cables can be connected. You can connect part of your network to the router using cables rather than wirelessly.

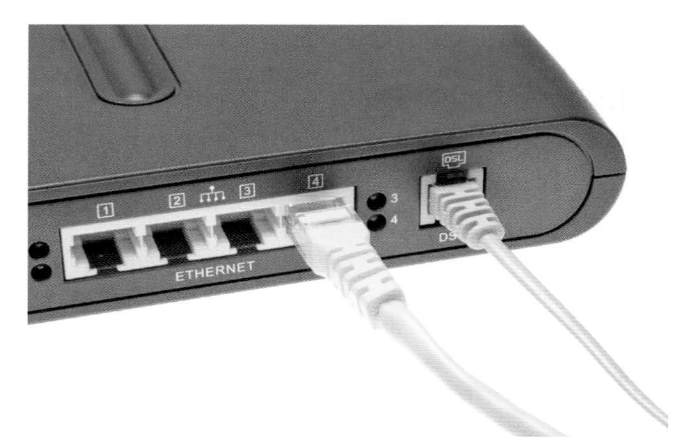

The yellow cable shown here is an Ethernet cable and this allows devices to be connected straight to the router rather than linking them wirelessly

The internet

The internet is a huge group of worldwide networks joined together. Each of these networks consists of lots of smaller networks. When you connect to the internet your computer becomes part of the largest network in the world.

The advantages of the internet:

- Huge amounts of information can be accessed almost anywhere.
- Improved communication systems - this includes the use of text messages, emails, instant messaging, etc.
- Changes in the way we shop - many people prefer to shop online.
- VoIP (Voice over Internet Protocol) - enables cheap international phone calls to be made using the internet rather than having to pay for calls using a mobile or landline telephone. VoIP allows voice data to be transferred over the internet and allows phone calls to be made without the need to subscribe and pay for a phone service.
- Can help people with disabilities to be more independent - because they can order goods and services online.

The disadvantages of the internet:

- Misinformation – there are many bogus or fake sites so you need to check the information you obtain carefully or use only reliable sites.
- Cyber crime – we have to be extremely careful about revealing personal information such as bank and credit card details.
- Addiction to gambling, as there are many casino, bingo, horse racing, etc., betting sites.
- Increased problems due to hacking and viruses.
- Paedophiles look for children using the internet.

E-commerce

Most organisations use websites sometimes for promotional purposes and sometimes to allow people to order goods or services using the site. This is called e-commerce.

QUESTIONS A

1 Give the name of the device that is used so that several computers in a home can all share a single internet connection. *(1 mark)*

2 Explain the main difference between the network devices **hub** and **switch**. *(2 marks)*

3 Bluetooth is a method that allows devices to communicate with each other and pass data.
  - **a** Give the names of **two pairs** of devices that can communicate using Bluetooth. *(2 marks)*
  - **b** Explain **one** advantage in devices communicating using Bluetooth. *(1 mark)*
  - **c** Explain **one** disadvantage in devices communicating using Bluetooth. *(1 mark)*

4 An office is thinking of introducing a wireless network with a wireless connection to the internet.

  Give **two** advantages in using a wireless network rather than a wired one. *(2 marks)*

5 A small office uses a local area network.
  The telecommunications company supplies them with a cable into the building. They would like to set up a wireless network.
  - **a** Give **one** reason they would like their network to be wireless. *(1 mark)*
  - **b** A wireless router is bought. Give the purpose of a wireless router. *(1 mark)*

6 A company has a wireless network installed. Give **one** reason why they might be concerned about the security of their data. *(1 mark)*

Many of these websites allow customers to browse online catalogues and add goods to their virtual shopping basket just like in a real store. When they have selected the goods, they go to the checkout where they decide on the payment method. They also have to enter some details such as their name and address and other contact details. The payment is authorised and the ordering process is completed. All that is left is for the customer to wait for delivery of their goods.

The difference between internet and WWW

There is a difference between the internet and the world wide web (WWW). The internet is the huge network of networks. It connects millions of computers globally and allows them to communicate with each other. The world wide web is the way of accessing the information on all these networked computers and makes use of web pages and web browsers to store and access the information.

Intranets

An intranet is a private network that is used inside an organisation and makes use of web pages, browsers and other technology just like the internet. Schools and colleges use intranets and they can hold all sorts of information from teaching resources, information about courses and adverts, to student personal records and attendance details. Parts of an intranet can be made available to anyone in the organisation, while parts that contain personal details can be made available only to certain people. Restriction to certain parts of the intranet is achieved by using user-IDs and passwords.

The differences between an intranet and the internet

- Internet stands for INTERnational NETwork, whereas intranet stands for INTernal Restricted Access NETwork.
- An intranet contains only information concerning a particular organisation that has set it up, whereas the internet contains information about everything.
- Intranets are usually only used by the employees of a particular organisation, whereas the internet can be used by anyone.
- With an intranet, you can block sites which are outside the internal network using a proxy server.
- Intranets are usually behind a firewall, which prevents them from being accessed by hackers.
- An intranet can be accessed from anywhere with correct authentication.

The advantages in using an intranet:

- Intranets are ideal in schools because they can be used to prevent students from accessing unwanted information.
- The internal email system is more secure compared to sending emails using the internet.
- Only information that is relevant to the organisation can be accessed and this prevents employees accessing sites that are inappropriate or which will cause them to waste time.

The two types of network LAN and WAN

There are two types of network: a local area network (LAN) and a wide area network (WAN).

Basically a WAN is much bigger than a LAN and is spread over different sites. A LAN, however, is within one site or building. This table gives you the main features of each type of network.

LAN	WAN
Confined to a small area	Covers a wide geographical area (e.g. between cities, countries and even continents)
Usually located in a single building	In lots of different buildings and cities, countries, etc.
Uses cable, wireless, infra-red and microwave links which are usually owned by the organisation	Uses more expensive telecommunication links that are supplied by telecommunication companies (e.g. satellite links)
Less expensive to build as equipment is owned by the organisation which is cheaper than renting lines and equipment	More expensive to build as sophisticated communication systems are used involving rental of communication lines

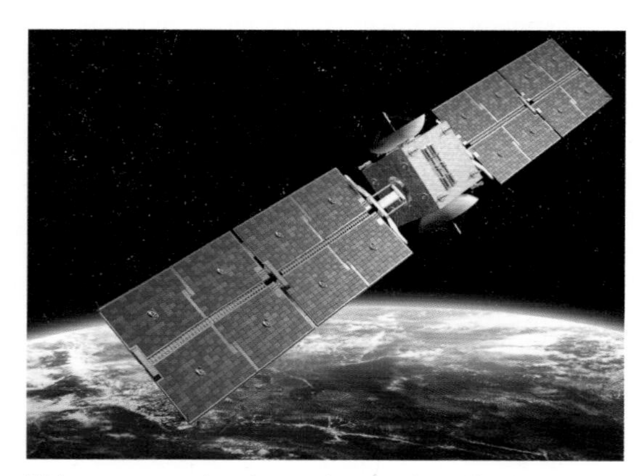

Wide area networks often make use of satellite links.

WLAN

WLAN stands for wireless local area network and means a LAN where the computers are able to communicate with each other wirelessly. WLAN is the type of network you have in your home where several computers are all able to access the internet wirelessly at the same time.

A WLAN allows users to move around an area and yet still stay connected to the internet.

KEY WORDS

Network a group of computers that are able to communicate with each other.
LAN (local area network) a network of computers on one site.
WAN (wide area network) a network where the terminals/computers are remote from each other and telecommunications are used to communicate between them.
WLAN A wireless local area network makes use of infra-red or microwaves as a carrier for the data rather than wires.

Advantages and disadvantages of a WLAN compared to a cabled LAN

A cabled LAN uses either wired cables or optical fibre cables to send and receive data over the network. A WLAN provides links that are wireless between computers and other devices on the network such as printers and scanners.

Advantages of a WLAN:

- It is cheaper as there is less cabling needed.
- It is very easy to link devices such as printers and scanners to the network.
- You can work anywhere in the building in range of the wireless signal.

Disadvantages of a WLAN:

- There may be areas in a building where the radio signal is too weak.
- There are security problems as the radio signals could be intercepted by hackers.
- Wireless signals are not as reliable as wired cables and the data transfer speeds are not as high.
- There can be interference from other electrical and electronic equipment used in the same building and this can disrupt the wireless signal.

The advantages and disadvantages of using different types of computer to access the internet

Many different devices can be used to access the internet and each has its own advantages and disadvantages. These are summarised in the following tables:

Advantages and disadvantages of using a smartphone to access the internet	
Advantages	**Disadvantages**
Can access the internet anywhere there is WiFi or a mobile signal	Not easy for parents to police.
You can stay connected whilst moving around.	Trimmed down websites used with smartphones may not give the information required.
Device is very portable and you are likely to have access to it more than other computers.	Small screen and keyboard are hard to use.

Advantages and disadvantages of using a tablet to access the internet	
Advantages	**Disadvantages**
Lighter and more portable than laptops and desktops.	Smaller screen compared to laptops and desktops can make them harder to use.
Larger screens compared to smartphones makes them easier to use.	On-screen keyboard is more difficult to use compared to the keyboard on a laptop or desktop.
Easier to use when standing up compared to a laptop or desktop.	Using in an awkward position, may lead to health problems.

Advantages and disadvantages of using a laptop to access the internet	
Advantages	**Disadvantages**
Portable compared to a desktop computer.	Not as portable as a smartphone or tablet because of its size and weight.
Almost full-sized keyboard is easier to use compared to those on a tablet or smartphone.	Screen and keyboard are attached and this can lead to backache.
Fairly large screen makes reading web pages easier.	Touchpads are harder to use compared to a mouse when there is no flat surface available.
Can be used when a flat surface is not available.	

Advantages and disadvantages of using a desktop computer to access the internet	
Advantages	**Disadvantages**
Full-sized keyboard and mouse are easier to use.	Large and heavy so not portable.
You can view the full versions of a website rather than a trimmed down version for a tablet or smartphone.	Hard to use whilst standing up.
Screen and keyboard are adjustable so there are fewer health problems.	
Large screen means text is easier to read.	

QUESTIONS B

1. Give **two** benefits of using a local area network. *(2 marks)*
2. Give **one** difference between a local area network (LAN) and a wide area network (WAN). *(1 mark)*
3. Describe **two** advantages and **two** disadvantages in a company using a computer network. *(4 marks)*
4. Explain the difference between the terms internet and intranet. *(2 marks)*
5. Describe what is meant by web browser software. *(2 marks)*
6. A school is currently thinking of replacing their wired LAN with a WLAN.
   - **a** Explain the differences between a wired LAN and a WLAN. *(1 mark)*
   - **b** Give **two** advantages and **two** disadvantages of using a WLAN. *(4 marks)*

Network issues and communication

In this section you will be looking at some of the problems associated with the use of networks. You will also be looking at the different types of network that allow communication to take place between people in different locations.

Security issues regarding data transfer

Once computers are networked together the security risks increase and if the network is connected to the internet, the security risks are even greater. In this section we will look at the security issues of networks.

Here are some of the ways security of a network can be breached.

Password interception

Here a person's password, which should be kept secret to them, becomes known to others. This means that other people can now sign in using their password and gain access to their files. When using the internet you can have lots of different passwords such as those for your email, bank accounts, stores, etc. If people know these passwords then they could read your emails, remove money from your bank account and purchase items using your details.

Virus attack

Viruses pose a major threat to networks. A virus is a program that replicates (copies) itself automatically and may cause damage.
Once a computer or media has a virus copied onto it, it is said to be infected. Most viruses are designed to do something apart from copying themselves. For example, they can:

- Display annoying messages on the screen
- Delete programs or data
- Use up resources, making your computer run more slowly
- Spy on your online use – for example, they can collect usernames, passwords, and card numbers used to make online purchases.

One of the main problems is that viruses are being created all the time and that when you get a virus infection it is not always clear what a new virus will do to the network. One of the problems with viruses is the amount of time that needs to be spent sorting out the problems that they create. All computers should be equipped with anti-virus software, which is a program that is able to detect and delete these viruses. This should be updated regularly, as new viruses are continually being developed and would not always be detected by older versions.

How to prevent viruses

The best way to avoid viruses altogether is to use anti-virus software and adopt procedures which will make virus infection less likely.

Here are some steps which can be taken to prevent viruses entering a network:

- Install anti-virus software.
- Do not open emails from unknown sources.
- Keep anti-virus software up-to-date. Configure the virus checking software so that updates are installed automatically.
- Have a clear acceptable use policy for all staff who use computers.
- Train staff to be aware of the problems and what they can do to help.
- Do not allow programs such as games, video or movies to be downloaded onto the organisation's ICT systems.

- Do not open file attachments to emails unless they are from a trusted source.
- If possible, avoid users being able to use their own removable media (pen drives, portable hard disk drives).

Unknown storage media (e.g. media containing a game or music files someone has given you) should always be scanned by anti-virus software before using. To be sure, it is best not to use:

- A popular way of buying software is to download it from the internet. You must make sure that you are downloading the software from a trusted site as counterfeit software may contain viruses.
- If your browser or other software alerts you that a site is not trusted, then do not visit it as these sites may be a source of viruses.

"The identity I stole was a fake! Boy, you just can't trust people these days!"

How to recognise fake/bogus websites

Fake/bogus websites can be used to distribute viruses and some can con you to pay for goods that never arrive or they steal your personal details to use fraudulently. Fake/bogus sites are set up to deceive you and in many cases they look convincing. Here are a few of the things you should look for when deciding if it is a fake/bogus site:

- Fake/bogus sites are often incomplete which means not all the links work so check some of the links.
- Check the URL (i.e. web address). Check to see if the website has the address you would expect.
- Check for mistakes in spelling/grammar. Not as much checking is done on fake/bogus sites so they frequently contain these errors.
- Check for the padlock symbol and the web address starting with https:// security indicators - Before entering personal information, check that the browser software is showing the padlock symbol next to the web address and that the web address starts with https://. If these security indicators are not present then there is no guarantee of security or encryption of your personal detail. Bogus/fake sites are unlikely to use these security indicators.

Hackers accessing the network

Hacking is the process of accessing a computer system without permission. Hackers often access a network using the internet. In most cases the hacker will try to guess the user's password or obtain the password another way; for example, by using a cracking program which will generate millions of different passwords very quickly trying each of them in turn.

Sometimes special software is installed on a user's computer without their permission and it can log the keys that the user presses, which can tell a hacker about usernames, passwords, emails, etc. This software is called spyware or key-logging software.

Use a firewall – a firewall is hardware, software or both that will look at the data being sent or received by the computer to either allow it through or block it. It logs the outgoing data leaving the network and incoming data from the internet. It therefore prevents access to the network from unknown IP addresses and it can prevent users from accessing unauthorised sites. It also warns a user if software is trying to run a program on the computer without the user's permission.

Use anti-spyware – this is special software that you run on your computer which will detect and remove any spyware programs that have been put on your computer without your knowledge with the purpose of collecting passwords and other personal information.

Change passwords regularly – if someone does manage to obtain your passwords, if they are changed on a regular basis at least they will only have the use of your passwords for a short while.

Protecting access by using user-IDs/usernames and passwords

System security is making sure that only authorised people access a computer system. This is called user authentication and it is normally carried out using a system of user identification (i.e. user-IDs) and passwords. Most computer operating systems provide systems of user-IDs and passwords.

Identifying the user to the system: user-IDs

A user-ID is a name or number that is used to identify a certain user to the network. The person who looks after the network will use this to allocate space on the network for the user. It is also used by the network to give the user access to certain files.

The network manager can also keep track of what files the user is using for security reasons. Network managers can also track improper use of the internet using user-IDs.

Preventing unauthorised access to the system: the use of passwords
A password is a string of characters (letters, numbers, and punctuation marks and other symbols) that the user can select. Only the user will know what the password is. When the user enters the password, it will not be shown on the screen. Usually an asterisk is shown when you enter a character. Only on entry of the correct password will the user be allowed access to the network. Some networks are set up so that if someone attempts to enter a password and gets it wrong three times, the user account is locked. If this happens, the user should contact the network administrator to have the password reset, which may involve answering security questions.

Choosing a password – the differences between strong and weak passwords
There are programs available that are able to crack passwords and if a password is short and contains just letters, then the password is weak and it is very easy to crack. Many people use weak passwords because they find them easy to remember.

Here are some examples of weak passwords:

mypassword

Ahmed

Steve1234

A strong password doesn't need to be long provided it contains symbols, capitals, numbers, and letters. Although strong passwords are not impossible to crack, most hackers would quickly move on to a different computer system where a weak password had been used.

A strong password should be a combination of letters (capital and lowercase), numbers, and symbols that have no relation to each other, just totally random such as 4Hp!9$R;

Here are some examples of strong passwords:

S%7c&X

£D23&!uL

Authentication techniques

As weak passwords can be easily cracked, and strong passwords are hard to remember, there needs to be an easier way to allow users access to networks.

Many networks use magnetic stripe cards or other ID cards containing chips which are used to gain access to the network. These are still used in conjunction with passwords and usernames since the card or ID card could have been stolen.

Biometric scanning systems can be used rather than usernames and passwords to allow access. Retinal scanning uses the unique pattern on the back of the eye to recognise a person and fingerprint scanning uses the unique pattern on a person's finger to identify them. Face recognition can be used to match someone trying to log onto the network with those people who are allowed access.

Iris scanning is another biometric method used for authentication and it uses the coloured area of the eye surrounding the pupil for identification. It has the advantage over retinal scanning in that it can be carried out using regular video cameras and can be performed as far as two feet away and also on people who are wearing glasses.

Biometric methods have the main advantage in that there is nothing to forget such as a card or a code.

ID cards

ID cards are small credit card sized cards used to verify a person's identity. The cards have some information printed on the card as well as information coded into either a magnetic stripe or a chip embedded in the card which enables the data on the card to be read automatically. Typically the ID card would contain a photograph of the person, a unique identification number, their date of birth, address, religion, profession etc.

Some ID cards contain biometric data such as fingerprints and retinal patterns and this data is encoded onto a chip in the card. ID cards are used to gain access to buildings, rooms and computers/networks.

Authentication methods used with passports are covered on page 18.

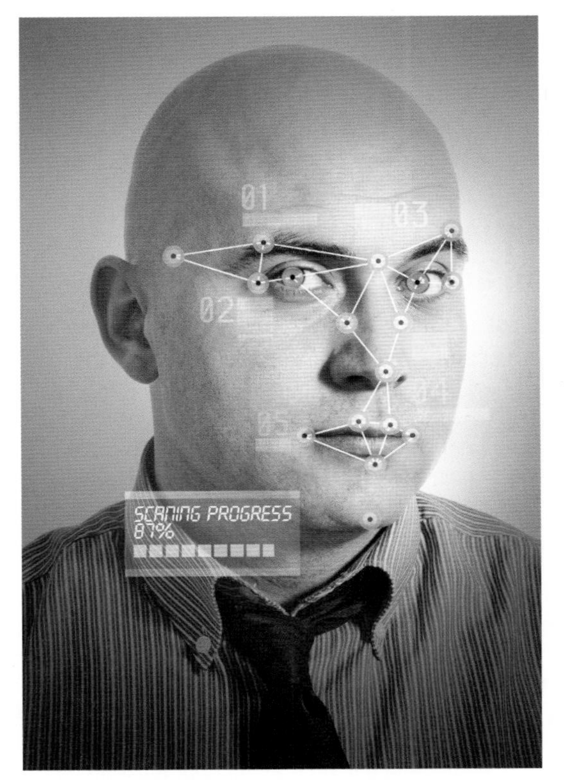

Face recognition can be used to authenticate users of a networks

Encryption

If information needs to be sent over the internet or another network and it needs to be kept secure, then encryption should be used. Encryption scrambles the data while it is being sent and only the true recipient is able to unscramble it. Should the data be intercepted by a hacker, then the data will be totally meaningless.

The process of scrambling data, sending it over the internet and then deciphering it when it reaches the true recipient, is called encryption. This requires encryption and decryption keys.

Encryption should be used for:

- Sending credit card details such as credit card numbers, expiry dates, etc., over the internet.
- Online banking.
- Sending payment details (bank details such as sort-code numbers, account numbers, etc.).
- Confidential emails.
- Sending data between computers when confidentiality is essential.

KEY WORDS

Encryption the process of scrambling files before they are sent over a network to protect them from hackers. Also the process of scrambling files stored on a computer so that if the computer is stolen, they cannot be understood.
Firewall a piece of software, hardware or both that is able to protect a network from hackers. The firewall looks at each item of data to see if it should be allowed access to the network or blocked.
Password a series of characters chosen by the user, used to check the identity of the user when they require access to an ICT system.

The principles of a typical data protection act

The widespread use of ICT has made the processing and transfer of data much easier and to protect the individual against the misuse of data, many countries have introduced a law called a data protection act.

There are a number of problems in organisations holding personal data:

- The personal data might be wrong, which means wrong decisions could be made.
- The organisation may not take care of the personal data it holds so others may find out about it.

What exactly is personal data?

Personal data in a data protection act normally means:

- Data about an identifiable person (i.e. the data must be about someone who can be identified by name, address, etc.).
- Data about someone who is living.
- Data that is more personal than name and address. For example, medical records, criminal record, credit history, religious or political beliefs, etc., are all classed as personal data.

KEY WORD

Personal data data about a living identifiable person which is specific to that person and is more personal than just your name and address.

The effects of wrong personal data

Examples of the effect of wrong information:

- Your medical details could be wrong, meaning you get the wrong treatment – which could be life threatening.
- Wrong decisions might be made; for example, you could be refused a loan.
- Wrong exam results could affect you getting a job.

Rights of the data subject and the holder

The person about whom the personal details are held is called the data subject.

The person in the organisation who is responsible for the personal data held is called the data holder.

A data protection act is a law that protects individuals by placing obligations on the organisations who collect and process the data (i.e. the data holders) in the following ways:

Registration (also called notification)—It requires anyone who uses personal data to register with the person who is in charge of the act/law. They must say what data they intend to hold and what they intend to do with it.

Individuals can see their own personal data—Anyone can apply to see the personal data held about them. Organisations have to show it and if there is any wrong information, then it must be corrected. There are some exemptions where you would be unable to see the information held about you. For example, you could be a criminal and apply to see the information the police have about you.

Data must be kept secure and up-to-date—Data subjects (the people the data is about) could take to court an organisation that does not keep their personal data secure.

The right for a person to claim compensation—If data is processed unlawfully by an organisation then the person could take them to court and claim compensation.

Data protection principles

All data protection acts are different but a typical act would contain the following principles.

1. Personal data shall be processed fairly and lawfully.
2. Personal data shall be obtained only for one or more specified and lawful purposes, and shall not be further processed in any manner incompatible with that purpose or those purposes.

3 Personal data shall be adequate, relevant and not excessive in relation to the purpose or purposes for which they are processed.
4 Personal data shall be accurate and, where necessary, kept up to date.
5 Personal data processed for any purpose or purposes shall not be kept for longer than is necessary for that purpose or those purposes.
6 Personal data shall be processed in accordance with the rights of data subjects under this act.
7 Appropriate technical and organisational measures shall be taken against unauthorised or unlawful processing of personal data and against accidental loss or destruction of or damage to personal data.
8 Personal data shall not be transferred to a country or territory unless that country or territory ensures an adequate level of protection for the rights and freedoms of data subjects in relation to the processing of personal data.

QUESTIONS C

1 Say whether each of the following statements is true or false:

- **a** Passwords should be changed regularly to make it difficult for hackers to gain access to a computer system.
- **b** You should always write your password down in case you forget it.
- **c** Passwords should ideally contain numbers, symbols and a combination of upper and lower case letters.
- **d** Passwords are best if they do not contain a word or phrase.
- **e** A password and user-ID are different words for the same thing.
- **f** Key-logging software is a type of spyware used by hackers to collect your passwords. *(6 marks)*

2 Explain, by giving examples, the differences between a strong password and a weak password. *(4 marks)*

3 Some people are not happy about organisations storing and processing personal information about them.

- **a** Explain, by giving **two** examples, what is meant by personal information. *(3 marks)*
- **b** Give **one** reason why a person might object to an organisation storing personal information about them. *(2 marks)*

4 All schools use computer systems to store details about past and present pupils.

Schools are required to notify their use of the personal data they hold under the terms of a data protection act.

- **a** Give **three** items of personal information the school is likely to store about their pupils. *(3 marks)*
- **b** Pupils in the school are given certain rights under the data protection act. Explain **two** of these rights. *(2 marks)*
- **c** One pupil is worried that the personal information the school holds about her might be incorrect. Explain, with an example, how incorrect information could affect a pupil. *(2 marks)*

5 Give **three** data protection principles that are likely to be included as part of a data protection act. *(3 marks)*

6 Many authentication methods for network access make use of biometrics.

- **a** Explain what is meant by the term biometrics. *(1 mark)*
- **b** Describe **three** biometric methods used for network access. *(3 marks)*

Network communication

In this section we will be looking in detail at the different ICT methods that can be used to enable communication between different people in different locations.

Physical fax

A physical fax consists of a single unit containing a scanner, modem, and printer. The document being faxed is scanned and then the modem converts the data from the scan into a series of sounds that are sent along a telephone line. To send the fax you need enter the number of the fax machine at the receiving end of the line. The modem at the receiving end converts the sounds back into data and the printer prints the fax out. When using a physical fax, only one fax can be sent or received at the same time.

Electronic/internet faxing

Electronic/internet faxing uses internet technology rather than a phone network to send a fax and it is not necessary for each fax to be printed. All you need to send or receive the fax is a computer or mobile device that has access to the internet.

Differences between electronic faxing and physical faxing

- Electronic faxing requires the use of a computer network whereas physical faxing does not.
- No additional phone line needed for electronic faxing.
- All physical faxes are printed; with electronic faxes, you choose to print or not, thus saving ink and paper.
- You do not have the cost of purchasing a physical fax machine.
- Faxes can be sent and received from mobile devices.
- More than one fax can be sent and received at the same time.
- You do not have to pay for the cost of a phone call to send a fax.

Email

An email is an electronic message sent from one communication device (computer, telephone, mobile phone, or tablet) to another. Documents and images in electronic form can be attached to an email. All web browser software has email facilities.

Advantages and disadvantages of using email compared to faxing

Advantages in using email:

- Emails can be viewed without others seeing them. Faxes are usually printed out in an area shared by other employees so they can be seen by others.
- Emails can be viewed on any portable device (computer, smartphone, etc.) that has internet access. Physical faxes can only be obtained using a fax machine.
- The information in fax form is not normally a computer file and cannot be edited by the receiver.
- The receiver's fax machine may be out of paper and this will prevent the fax from being received. With emails this problem does not arise.
- Cheaper to use email. Sending a physical fax means you have to pay for the cost of the phone call as well as the cost of the paper and ink.

Disadvantages in using email:

- Emails can be used to distribute viruses. Faxing does not spread viruses.
- Faxed signed documents are usually legally acceptable whereas emails are not.

Video-conferencing

Video-conferencing allows two or more individuals situated in different places to talk to each other in real time and see each other at the same time. They are also able to exchange electronic files with each other and pass electronic documents around rather than paper documents. It allows people to conduct 'virtual' meetings. If meetings need to be conducted between people in different sites or even countries, then a WAN is used such as the internet.

Desktop video-conferencing systems in addition to a PC include a video compression card, a sound card, a microphone, a web cam, and specialist software.

Advantages of and disadvantages using video-conferencing

Advantages:

- Less stress as employees do not have to experience delays at airports, accidents, road works, etc.
- Improved family life, as less time spent away from home staying in hotels.
- Saves travelling time.
- Saves money as business does not have to spend money on travelling expenses, hotel rooms, meals, etc.
- Improved productivity of employees, as they are not wasting time travelling.
- Meetings can be called at very short notice without too much planning.
- More environmentally friendly as there are fewer people travelling to meetings. This cuts down on carbon dioxide emissions.
- Fewer car journeys means fewer traffic jams and hence less stress and pollution.

Disadvantages:

- The initial cost of the equipment, as specialist video-conferencing equipment is expensive.
- Poor image and sound quality.
- People can feel very self-conscious when using video-conferencing and not communicate effectively.
- Although documents and diagrams in digital form can be passed around, an actual product or component cannot be passed around.
- Lack of face-to-face contact may mean a discussion may not be as effective.
- If the delegates are in distant locations, the picture can be out of synchronisation with the speech, which can be distracting.

Video-conferencing allows virtual meetings where delegates are not at the same location

Web-conferencing

Web-conferencing (sometimes called a webinar) uses a stable internet connection and remote computers with microphones and webcams to allow real-time conferencing to take place. Usually there will be one host and many participants so it can be used for training events, meetings, lectures and presentations. To take part in a web-conference either web-conferencing software needs to be downloaded on all the computers used or everyone involved has to visit a web-conferencing website. During web-conferencing, the host can decide who is allowed to speak and the participants can indicate to the host (usually using a small flag on the host's screen) that they wish to speak.

The facilities offered by web-conferencing include:

- The ability to send and receive any files before or during the meeting.
- The ability to see and hear the host and participants just like at a real physical meeting.
- The sharing of desktops - so if one person wishes to show a video, document, slide presentation etc. then all the participants can view it.
- The ability of participants to edit documents such as contracts, brochures etc. in real time.
- The ability to send instant messages.
- The ability for any participant to draw on an electronic whiteboard so that ideas from the various participants can be built up.

Audio-conferencing

Audio-conferencing (sometimes called a conference call) uses a telephone call that allows a caller (the host) to talk to several people (the participants) at the same time. Usually the host decides on a time for the audio-conference and the host dials the number of the participants and adds them to the call. In other systems the participants dial the number of a special device and enter a PIN which links the telephone numbers. The system allows participants to talk one at a time with the host only if the host allows it.

Policing the internet

Many people think that the internet should be policed because it contains so many undesirable sites, but others consider users should choose what information they access. The problem with policing the internet is that the internet is a global network and you would need agreement from all countries using it, which is highly unlikely.

Here are some of the arguments for and against policing the internet.

Yes, the internet should be policed:

- It is too easy for children to access pornography.
- Films, TV & radio programmes are regulated so the internet should be regulated as well.
- There is a lot of misinformation. Wrong information about a person could potentially ruin their life.
- People can hide through the anonymity of the internet and say things that they would not say to someone's face.
- Other activities such as driving a car, drinking alcohol, etc., are policed so the internet should be policed as well.
- There are many undesirable sites which tell you how to make explosives, produce illegal drugs, etc.

No, the internet should not be policed:

- To police the internet would require resources no law enforcement agency has.
- It is up to people to decide what they want to read or look at.
- What is allowed in one country is not allowed in another so you could not agree on what to police.
- Free speech is a fundamental right and should not be removed.

QUESTIONS D

1 A company has branches all over the world and uses video-conferencing to communicate with its employees.

   a Describe what is meant by video-conferencing. *(2 marks)*

   b Give **two** benefits to the company and its employees of using video-conferencing. *(2 marks)*

2 Many people say that the internet should be policed.

   a Describe **two** reasons for policing the internet. *(2 marks)*

   b Describe **two** reasons for **not** policing the internet. *(2 marks)*

3 There are two ways of faxing: using a physical fax and using an electronic fax.

   Describe the differences between physical faxing and electronic faxing. *(4 marks)*

4 Describe **two** advantages and **two** disadvantages of using email rather than a fax to send a document to another person. *(4 marks)*

5 Describe a difference between audio-conferencing and video-conferencing. *(2 marks)*

REVISION QUESTIONS

1 LAN and WAN are both types of computer networks.

- **a** i What does LAN stand for? [1]
  - ii What does WAN stand for? [1]
- **b** Give **two** differences between a LAN and a WAN. [2]
- **c** Give **three** advantages to computer users of a LAN, rather than working on stand-alone machines. [3]
- **d** Give **one** method which can be used to prevent data from being misused when it is being transferred between computers. [1]

2 When goods are ordered over the internet, payment has to be made.

- **a** Give **one** method of payment used over the internet. [1]
- **b** Describe why some people may not want to put their payment details into a website. [2]
- **c** Describe **one** way an online store can make sure that payment details are safe. [2]

3 Explain briefly how each one of the following helps improve the security of a network.

- **a** User-ID [2]
- **b** Password [2]
- **c** Encryption [2]

4 A company is thinking of installing a new network. They have the choice of a wired network, where cables are used to transmit the data, or a wireless network, where no cables are needed. Describe the relative advantages and disadvantages in using a wireless network. [4]

5 Video-conferencing is used by many organisations to conduct meetings at a distance.

- **a** Explain **two** advantages of video-conferencing. [2]
- **b** Explain **two** disadvantages of video-conferencing. [2]

Test yourself

The following notes summarise this chapter, but they have missing words. Using the words below, copy out and complete sentences **A** to **S**. Each word may be used more than once.

internet intranet router Wi-Fi attachments synchronise local network interface card encryption Bluetooth broadband WLAN wide wireless download bridge spyware anti-virus passwords

A With ______ communication, the data travels through the air rather than through cables.

B An internal network which makes use of web pages and web browser software is called an ______.

C A card that connects directly to the motherboard of the computer and has external sockets so that the computer can be connected to a network via cables is called a ______.

D A hardware device that is used to connect local area networks together is called a ______.

E ______ area networks are those networks that are restricted to a single building or site.

F ______ area networks are situated across a wide geographical area.

G A hardware device that allows several computers to share a single internet connection is called a ______.

H A ______ is a wide area network that uses wireless communication rather than cables.

I A hotspot is an area where ______ is available.

J A hardware device that takes a packet of data and reads the address information to determine the final destination of the packet is called a ______.

K ______ is a method used to transfer data over short distances from fixed and mobile devices.

L Bluetooth can be used to ______ the music on your home computer and your portable music player so the tracks stored on each are the same.

M The letter W in WLAN stands for ______.

N Key logging software records keystrokes made and is an example of ______.

O Credit card details are scrambled before sending over the internet. This process is called ______ and prevents hackers being able to understand the information if intercepted.

P In order to prevent viruses entering an ICT system, ______ software should be used to search for and destroy viruses.

Q Users should be told not to open ______ to emails unless they know who they are from.

R Users should also be told to ______ music and games only from trusted sites.

S To protect the data stored in networks, user-IDs and ______ should be used to prevent unauthorised access.

EXAM-STYLE QUESTIONS

1 Complete each sentence below using **one** item from the list. *[4]*

A hub **An intranet** **A proxy server**
A WAN **A WLAN**

a .... is a device used to connect computers together to form a LAN.
b .... is a network with restricted access.
c .... can allow networked computers to connect to the internet.
d .... is a wireless local area network.

(Cambridge IGCSE Information and Communication Technology 0417/11 q5 May/June 2010)

2 Aftab and his family have three computers in their home. He wants to connect the computers into a network. Explain why he would need:

A router A browser Email An ISP

[4]

Cambridge IGCSE Information and Communication Technology 0417/11 q12 May/June 2010

3 A small office has four stand-alone computers. The office manager wants to connect the computers together to form a LAN.

a Name a network device which would have to be present in each computer before they could be networked. *[1]*
b Give **two** reasons why a WLAN would be preferable to a cabled LAN. *[2]*
c Give **two** reasons why the manager should **not** use Bluetooth technology to create the network. *[2]*
d The company's workers are concerned that their payroll data may not be secure as a result of the computers being networked. Explain why the workers are concerned. *[6]*
e Give **three** actions that the office manager could take to ensure data security. *[3]*

Cambridge IGCSE Information and Communication Technology 0417/13 q14 Oct/Nov 2010

4 Describe **five** differences between a WAN and a LAN. *[5]*

5 **a** Explain what is meant by the term Bluetooth. *[1]*
b Name **three** devices that could be linked to a computer using Bluetooth. *[3]*

6 Companies with offices all over the world use video-conferencing rather than face-to-face meetings.

a Tick **three** advantages of video-conferencing.

It is possible to hold meeting at short notice.	
It is cheaper as companies do not have to pay for travel expenses.	
Fewer workers need to be employed.	
It is possible for employees to work from home.	
You can hand around documents at a face-to-face meeting.	
You get to meet more people using video-conferencing.	

[3]

b Tick **three** disadvantages of video-conferencing.

Companies have to hire a large theatre to hold the meetings.	
Video-conferencing may not be as effective as face-to-face meetings.	
Most people prefer personal contact rather than contact at a distance.	
You cannot hand around documents or show presentations.	
Time differences in different countries can cause problems.	
You can only show the presenter and not the delegates at the meeting.	

[3]

7 Tick **True** or **False** next to each of these statements.

	True	False
A data protection act helps prevent personal data from being misused by organisations.		
A memorable word should be chosen as a password as it is easy to remember.		
Passwords should be changed regularly to deter hackers.		
Data protection principles are part of most data protection acts.		
Passwords should always be written down in case you forget them.		

[5]

8 Compare and contrast the use of a desktop computer and a smartphone to access the internet. *[5]*

9 Most school computer networks have a router and switches.

a Explain what is meant by a switch. *[2]*
b Describe the purpose of a router. *[2]*

(Cambridge IGCSE Information and Communication Technology 0417/11 q13 May/June 2014)

10 Describe the benefits and drawbacks to companies of using video-conferencing. *[5]*

(Cambridge IGCSE Information and Communication Technology 0417/11 q10 Oct/Nov 2012)

5 The effects of using IT

In this chapter you will learn about some of the many effects that the use of IT has on our lives. Most of these impacts are for the good but there are some bad ones, too. You will be looking at the way IT has changed employment and the way that we work. You will look at the impact that using microprocessor-controlled devices in the home has on our lives. You will also be looking at some of the health problems which can develop with the prolonged use of IT equipment.

The key concepts covered in this chapter are:

- Effects of IT on employment
- Effects of IT on working patterns within organisations
- Microprocessor-controlled devices in the home
- Potential health problems related to the prolonged use of IT equipment

Effects of IT on employment

Many businesses now operate 24 hours per day and 7 days a week, so people need to be more flexible in their hours of work. They will no longer spend their whole life doing the same job, so constant retraining will be needed so that they can take advantage of new IT developments. Many organisations operate internationally and some jobs may be transferred abroad if the wage costs are lower in another country. Location is no longer an issue with IT as data can be accessed from anywhere in the world.

Here are the main effects of IT on employment:

- Fewer people needed to complete the same amount of work.
- Increased number of people working from home using IT equipment.
- More automation in factories due to the introduction of robots for assembling products, welding, paint spraying, and packing goods.
- Continual need for training as IT systems change.
- More availability of part-time work as many organisations need to be staffed 24/7.
- More variation in the tasks undertaken and staff need to be flexible and well trained to cope with this.
- Fewer 'real' meetings as 'virtual' meetings using video-conferencing are used to reduce travel time and travel costs.
- Increase in the number of technical staff needed such as network managers.

Areas of work where there is reduced employment

There are many types of jobs where the numbers of people employed have been reduced over the years and some jobs which no longer exist. Here is a summary of these:

- Many manual repetitive jobs such as paint spraying, welding, packing goods, assembly work in factories have been replaced by robots.
- Numbers of people who work in shops have reduced because of an increased number of people shopping online. Online stores are heavily automated (e.g. many use robots to select goods and pack them) and employ fewer staff.
- Call centre automated response.
- Designing and producing CDs/DVDs. Music, games, and software are more likely to be downloaded rather than bought on physical media such as a CD. This eliminates the need for packaging.
- Fewer payroll workers needed. Working out pay and deductions is performed by computers using data from clocking on/clocking off machines which record the days and hours worked. As many organisations pay wages directly into employee bank accounts using electronic transfer, there is no longer a need to prepare wage packets.
- Typists are no longer needed. In the past, large numbers of people were employed as typists who typed documents. Now it is so easy to use a word-processor that most employees type their own documents.

These robots have replaced people in welding panels together in a car factory.

Areas of work where IT has increased employment

New jobs that have been created through IT include:

- Network managers/administrators – these are the people who keep the networks running for all the users and see to the taking of backup copies.
- Website designers – these are the people who design and create websites for others, as well as keep them up-to-date by adding new and deleting old material.
- Development staff – these include systems analysts (who design new IT systems) and programmers who write the step-by-step instructions (i.e. the programs) that instruct the computers what to do.
- As more goods (e.g. groceries, books etc.) are bought online there has been an increase in the number of delivery drivers needed to deliver these.
- New computer programs need to be written as more tasks are performed by computers and existing programs need to be modified so there will be increased demand for programmers.

The effects of IT on working patterns within organisations

Here are the main effects of IT on working patterns:

- **Greater availability of part-time work** – as many organisations operate 24 hours per day, this increases the availability of part-time work. This is good for certain people such as mothers or people caring for elderly relatives who would find it hard to work traditional office hours.
- **Flexible hours** – many organisations offer employees the opportunity to work flexible hours, which can help them fit in with child care or caring for a relative. This allows people with family commitments, such as picking children up from school, to work.
- **Job sharing** – rather than having one employee performing the job, you can have two or more employees sharing the job by each working part-time. Although you may think employers would not like this, they do prefer the flexibility it offers to them as they now have several people to ask to do extra work.
- **Compressed hours** – some employers allow staff to work longer hours over fewer days. So, for example, you could work 8 am to 8 pm over 3 days rather than work shorter hours over a traditional working week.

The effects of microprocessor devices in the home

Microprocessors are computers, usually on a single chip, that are put into electronic devices to check, regulate and control something. If you think of any device in the home that needs to be controlled in some way, then it is likely to contain a microprocessor for this control. Here are some of the many devices used in the home that contain microprocessors:

- Computer systems contain one or more microprocessors for the main processing and this is called the CPU. There are also other microprocessors found in devices such as modems, routers, disk drives, etc.
- Washing machines – used to control valves to let the water in, motors to turn the drum, pumps to pump the water out, heaters to heat water up, and so on.
- Children's toys – use microprocessors to control lights, motors, speakers, etc.
- Heating systems – control the time the heating comes on, keeps the temperature constant by turning the heating on/off, etc. Modern heating systems can control the individual temperatures of each room from a central place.
- Alarm systems – can detect the presence of a burglar and some systems will even contact the police automatically.
- Intelligent ovens and microwaves – some of these incorporate bar code readers which read the bar codes on packaging which the oven uses to find out the ideal time and temperature for cooking.
- Intelligent fridges – a large internet food supply company has developed a fridge that can scan the 'use by' dates on food and then move any food product to the front of the fridge automatically using a series of computer controlled panels. This reduces food wastage.

QUESTIONS A

1 IT has replaced or changed many jobs.

a Give the names of **two** types of job that have been replaced by IT. *(2 marks)*

b Some jobs have changed their nature due to the introduction of IT. Name **two** jobs where this has happened. *(2 marks)*

c The increase in high-speed broadband links has led to cheaper international telephone calls using the internet. Name **one** job that may be transferred to a different country as a result of this. *(1 mark)*

2 The widespread use of IT has had a huge impact on society. One benefit that it has brought is the creation of new and interesting jobs.

a Give **three** examples of jobs that have been created through the introduction of IT. *(3 marks)*

b Many people have had to be retrained to cope with the introduction of new IT systems. Explain why regular retraining is needed in the workplace. *(2 marks)*

3 Describe how the use of IT has had an effect on working patterns in the following areas:

The ability to share jobs.
The ability to work compressed hours.
Greater availability of part-time work.
The opportunity to work flexible hours. *(4 marks)*

KEY WORD

Microprocessor the brain of the computer consisting of millions of tiny circuits on a silicon chip. It carries out program instructions.

The effects of microprocessor-controlled devices on leisure time

Many microprocessor-controlled devices save us time and allow us to have more leisure time. For example, dishwashers can wash the dishes for us while we go out or watch a game of football.

Here are some ways in which microprocessor-controlled devices affect leisure time:

- Mobile phones, laptops, netbooks, tablets, etc., mean employees can do more work on the move and at home, which can mean that they have less time for family life.
- Electrical appliances controlled by microprocessors can complete tasks such as washing and drying clothes, washing dishes, and cooking meals, which can leave a person free to do other things.
- People can become lazy and rely on machines.
- Lack of fitness as people are not spending time doing manual work.
- Can use the extra time saved to go to the gym or exercise more.
- Many people spend their leisure time playing games, surfing the net, downloading and listening to music, etc., on computers and games consoles, which all make use of microprocessors.
- Watching programmes on TV using satellite or cable TV makes use of microprocessors.
- You don't have to be in when washing is being done so you spend time at work or doing other things.

The effects of microprocessors on social interactions

The use of microprocessor-controlled IT equipment such as computers and mobile phones has led to a huge increase in the number of ways we can keep in touch with friends and family. Social interaction is important for a person's well-being and there are a number of ways that microprocessor-controlled devices help and here are some of them:

- Mobile phones, SMS (texting), email, blogs, social networking sites, chat rooms, and instant messaging make it much easier to keep in touch with friends and family.
- It is easy to make new friends in chat rooms and using social networking sites.
- Cheap internet phone calls, made using a service called VoIP, enable people to keep in contact with friends and family who may live in other countries.
- Disabled or elderly people who are confined to their home can make friends with others.
- It is much easier to find out about social activities, e.g. classes, concerts, activities, etc., using the internet.
- Time spent not having to do jobs manually enables the time to be used to do other things.

There are a number of disadvantages and these include:

- People rely on the internet for performing their daily tasks (shopping, banking, entertainment, etc.) and this means they do not meet people, which can lead to social isolation.
- Playing computer games is unhealthy compared to playing proper physical games and can be isolating.
- People can lose some basic housekeeping skills so when the machine breaks down they do not know what to do.

The effect of microprocessors on the need to leave the home

There are many ways in which microprocessor-controlled devices help people perform tasks from home. This may be a convenience for many but it can mean the difference between having or not having your independence if you are disabled and housebound.

Here are some of the ways the use of microprocessor-controlled devices reduces the need to leave the home:

- Online shopping – means you order goods using the internet and have them delivered to your home.
- Downloads – items such as software, music, games, ring tones, videos can be downloaded onto your computer, phone or tablet using the internet. There is no need to go to the shops for these.
- Online banking – you can view statements, transfer money between accounts, pay bills, apply for loans, etc., all from the comfort of your home.
- Research – many people need to find out information in their daily lives such as train times, opening times of shops, reviews of products, etc. All this can be done from home using the internet.
- Entertainment – there are so many ways to entertain yourself using social networking sites, playing games, listening to music, watching videos, etc. All of these are possible using microprocessor-controlled devices.
- Working from home (called teleworking or telecommuting) by making use of computers and telecommunication devices and systems.
- No need to be present to supervise devices like ovens or washing machines.
- Can set TV recording or set oven from a remote place using embedded technology and mobile phones.

QUESTIONS B

1 **a** Give the names of **three** microprocessor-controlled devices you would find in the home (other than a computer). *(3 marks)*

b Describe the positive effects on lifestyle of each of the devices you have named in part **a**. *(3 marks)*

2 Tick **True** or **False** next to each of these statements.

	True	False
Robots in the home can do all the housework.		
Vacuuming robots can vacuum floors without anyone being present.		
Microprocessor-controlled devices in the home free up more leisure time for the occupants.		
The use of robots in the home will reduce the amount of exercise people get doing housework.		
People can become lazy if they rely on machines to perform all their tasks for them.		

(5 marks)

3 Describe the advantages and drawbacks of using microprocessor-controlled devices in the home. *(5 marks)*

The potential health problems related to the prolonged use of IT equipment

As there are potential hazards when using computers and other IT equipment, you need to be aware of what the hazards are. You also need to be aware of the symptoms of the medical conditions they can cause.

The main health problems are:

- Back and neck ache – is a painful condition that prevents you from sleeping properly and doing many activities such as playing sport.
- Repetitive strain injury (RSI) – this is caused by typing or using a mouse over a long period of time. RSI is a painful illness that causes swelling of the wrist or finger joints. It can get so bad that many sufferers are unable to use their hands.
- Eye strain – looking at the screen all day can give you eye strain. Many of the people who use computer screens for long periods have to wear glasses or contact lenses. The symptom of eye strain is blurred vision.
- Headaches caused by glare on the screen, a dirty screen or screen flicker.

Methods of preventing or reducing the risks of health problems

Back and neck ache

The following can cause back and neck ache:

- Not sitting up straight in your chair (incorrect posture).
- Using a laptop on your knee for long periods.
- Working in cramped conditions.

To help prevent back and neck problems:

- Use an adjustable chair (Note: in workplaces in some countries this is a legal requirement but you should ensure that the chair you use at home is adjustable).
- Always check the adjustment of the chair to make sure it is suitable for your height.
- Use a foot support, called a footrest, if necessary.
- Sit up straight on the chair with your feet flat on the floor or use a foot rest.
- Make sure the screen is lined up and tilted at an appropriate angle so you are looking directly across and slightly down – never to the side or up.

"Suspending your keyboard from the ceiling forces you to sit up straight, thus reducing fatigue."

Repetitive strain injury (RSI)

The following can cause RSI:

- Typing at a computer for a long time without taking a break.
- Using a mouse for long periods.
- Not adopting correct posture for use of mouse and keyboard.
- Not having properly arranged equipment (keyboard, mouse, screen, etc.).

To help prevent RSI:

- Take regular breaks.
- Make sure there is enough space to work comfortably.
- Use a document holder.
- Use an ergonomic keyboard/mouse.
- Use a wrist rest.
- Keep your wrists straight when keying in.

- Position the mouse so that it can be used keeping the wrist straight.
- Learn how to type properly – two finger typing has been found to be much worse for RSI.
- Use alternative input methods, e.g. speech recognition.

KEY WORD

RSI (repetitive strain injury) a painful condition caused by repeatedly using certain muscles in the same way for long periods without a break. The condition usually affects wrists and/or fingers.

Eye strain/headaches

The following can cause eye strain/headaches:

- Looking at the screen for long periods.
- Working without the best lighting conditions.
- Glare on the screen/screen flicker.
- Dirt on the screen.

To help avoid eye strain/headaches:

- Take regular breaks (e.g. every hour)
- Keep the screen clean, so it is easy to see characters on the screen.
- Use appropriate lighting (fluorescent tubes with diffusers).
- Use blinds to avoid glare.
- Give your eyes a rest by focusing on distant objects.
- Have regular eye-tests (Note: if you use a screen in your work, then your employer may be required by law to pay for regular eye-tests, and glasses if they are needed).
- Ensure that the screen does not flicker.
- Change to an LCD screen if you have an old CRT monitor.

Revision Tip

Always be guided by the mark scheme to decide how much to write. For example, if you are asked to describe an advantage or disadvantage for one mark, then just a brief statement would be enough. If two marks were allocated, then you would be required to supplement this with further detail or an appropriate example.

Always think out your answer before you start writing it. You need to ensure you make your answer clear, so you need a little time to think about it.

Advantages (sometimes called benefits) and disadvantages are very popular questions. When covering a topic for revision, it is a good idea to list advantages and disadvantages where appropriate.

You need to be clear about the health problems (i.e. what they are called), the symptoms (i.e. how they affect your body) and what can be done to help to prevent them.

QUESTIONS C

1 The use of IT systems has been associated with a number of health problems.

a State **three** health problems that have been associated with the prolonged use of IT systems. *(3 marks)*

b In order to prevent health problems caused by the use of computers, some actions can be taken. Describe **six** such actions that can be taken to avoid the health problems you have identified in part **a** happening. *(6 marks)*

2 An employee who spends much of their time at a keyboard typing in orders at high speed is worried about RSI.

a What do the initials RSI stand for? *(1 mark)*

b Give one of the symptoms of RSI. *(1 mark)*

c Write down **two** precautions that the employee can take to minimise the chance of contracting RSI. *(2 marks)*

3 Copy the table and tick (✓) the correct column to show whether each of the following statements about health risks in using IT is true or false. *(5 marks)*

TRUE FALSE

	True	False
The continual use of keyboards over a long period can give rise to aches and pains in the hands, arms and wrists	☐	☐
RSI stands for repeated stress injury	☐	☐
Wrist rests and ergonomic keyboards can help prevent RSI	☐	☐
Back ache can be caused by slouching in your chair when using a computer	☐	☐
Glare on the screen can cause RSI	☐	☐

Activity 5.1

Repetitive strain injury has become a major worry for those people who use computers continually throughout their working day.

You are required to use the internet to find out more about this condition. You need to find out:

- What are the symptoms?
- Can you make it better?
- What is the likelihood of getting it?
- What can you do to prevent it?

Test yourself

The following notes summarise this chapter, but they have missing words. Using the words below, copy out and complete sentences **A** to **M**. Each word may be used more than once.

payroll **eye-tests** **part-time** **car** **website**
back ache **typing** **delivery** **working** **headaches**
programming **vacuuming** **RSI** **health**
blinds **leisure**

A Robots have replaced many repetitive assembly jobs in manufacturing industries such as ______ assembly, where robots weld panels and paint panels.

B Office jobs have been replaced by computers particularly in the areas of ______ documents and preparing ______.

C The use of IT has created some new and interesting jobs in areas such as ______ and ______ design.

D The increase in shopping online has meant an increase in employment for ______ drivers.

E IT has also changed ______ patterns. There is more opportunity for ______ working which has benefitted people with commitments, such as looking after children or elderly relatives.

F Microprocessor-controlled devices in the home have freed up more ______ time.

G You no longer need to be in the home when ______ floors as you can buy a robot to perform this task.

H The use of IT can lead to a number of ______ problems such as back ache, eye strain, and headaches.

I The symptoms of eye strain include blurred vision and ______.

J Working in cramped conditions and not adopting the correct posture when using a computer can lead to ______.

K ______ is caused by typing at high speed or using a mouse over a long period.

L Adjustable ______ should be used on windows to prevent glare on the screen, and the screen should also be kept free from glare from lights.

M It is important to have regular ______ and use glasses or contact lenses when working with computers, if needed.

REVISION QUESTIONS

1 Describe **one** way in which IT is used in manufacturing in order to reduce the number of people needed to produce goods. [3]

2 There have been a lot of changes in the pattern of employment due to the increased use of IT.
Tick **three** boxes that give sensible reasons why the pattern of employment has changed with increased use of IT. [3]

	Tick three boxes only
Homeworking/teleworking is more popular	☐
Employees are more likely to work more flexibly	☐
It has brought about a huge rise in employment especially among factory workers	☐
Training and retraining are needed regularly	☐
Workers are generally less skilled than they were 20 years ago	☐

Important note

Here are some of the words you will see being used in questions and what they mean:

State/List – single words or phrases are usually enough.
Describe – needs a more in-depth answer to show understanding.

It is important to be able to distinguish between health and safety issues.

Health (person) – RSI, headaches, eye strain and backache.
Safety (incidents) – fire, electrocution, tripping, etc.

EXAM-STYLE QUESTIONS

1 a What do the initials RSI stand for? *[1]*

b RSI is a health problem that may be caused by prolonged computer use.
Write a sentence to show how RSI is caused. *[2]*

c Write down **one** precaution that a computer user can take to minimize the chance of contracting RSI. *[1]*

2 Here is a list of health problems. Write down the names of the ones which can be caused by prolonged computer use.

Backache

Toothache

Stress

Sprained ankle

Repetitive strain injury (RSI)

Eye strain *[4]*

3 People who work with computers for long periods may experience some health problems. These health problems include eye strain and RSI.

a Give the names of **two** other health problems other than eye strain and RSI that a user may experience. *[2]*

b Explain **two** things a user should do in order to prevent future health problems when sitting on a chair at a desk using a computer. *[2]*

4 There are a number of health hazards associated with the use of computers.

a Give the names of **three** health hazards outlining the health problems they create. *[6]*

b For each of the health hazards described in part (a) describe what a user can do to help reduce the risk of their occurrence. *[3]*

5 Describe **two** ways in which the working patterns within organisations have changed owing to the increased use of IT. *[4]*

6 Many workers now use computers a lot in their work and this can cause them to suffer from RSI.

Name **two** other possible health problems and for each one describe a different precaution which could be taken to help prevent it. *[4]*

(Cambridge IGCSE Information and Communication Technology 0417/11 q18 May/June 2014)

7 There are many microprocessor-controlled devices in the modern home.

Describe **five** effects of these devices on people's lifestyles. *[5]*

(Cambridge IGCSE Information and Communication Technology 0417/11 q18 Oct/Nov 2013)

6 ICT applications

ICT is used in almost every area of our lives, so in this chapter you will be looking at the range of ICT applications in everyday life. You will also be looking at the impact of developments in ICT which shape the way we go about our lives.

The key concepts covered in this chapter are:

- Communication applications
- Data handling applications
- Measurement applications
- Microprocessors in control applications
- Modelling applications
- Applications in manufacturing industries
- School management systems
- Booking systems
- Banking applications
- Computers in medicine
- Computers in libraries
- Expert systems
- Computers in the retail industry
- Recognition systems
- Monitoring and tracking systems
- Satellite systems

Communication applications

There are many different ways of communicating using communication services such as phones, email, etc. In this section you will be looking at communication applications and how they allow people to communicate effectively using ICT.

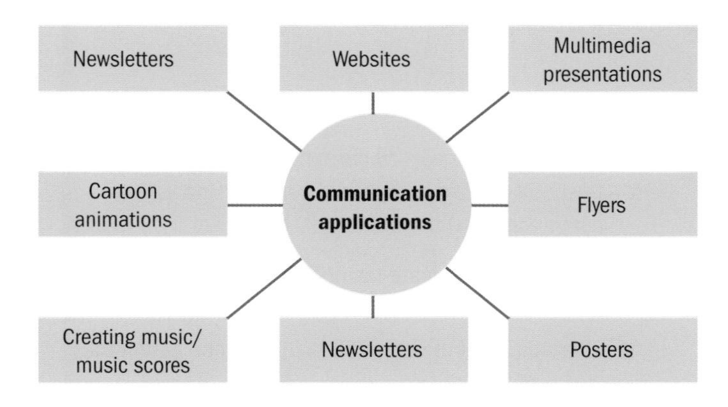

Cartoon animations

Animations can bring the most boring subject to life and they are used extensively in websites and products to help children learn. They are also used in websites to sell goods and services.

Using Flash (and other brands of animation software), when you want an object to move across the screen, you need only to define the key frames where something changes. You leave it to Flash to put in the frames in between. This is called tweening and it enables simple animations to be produced quickly.

Creating music/music scores

Musicians, composers and music producers make extensive use of ICT in the following ways:

- Sequencers can be used. Sequencers are hardware or software used to create and manage electronic music and they include such things as drum machines and music workstations that allow musicians to create electronic music.
- Sound wave editors are used which are software that allows the editing of sound waves. Using the software sound waves can be edited, cut, copied, and pasted, and also have effects like echo, amplification, and noise reduction applied.
- MIDI (Musical Instrument Digital Interface) can be used. MIDI enables a computer and a musical instrument to communicate with each other. For example, when a keyboard is played, the music is transferred using MIDI to the computer where it can be edited if needed and stored. The process can be reversed by the music stored on the computer being fed back and using MIDI it can play the music back on the keyboard or other instrument.
- Notators (music composition software) can be used. A notator is a piece of software that allows you to compose your own music and you do this by entering notes into the computer via:
  - The keyboard
  - A MIDI system
  - Scanning a piece of music on paper using a scanner.

Using ICT equipment to edit sound in a studio.

Mobile phones

New services for mobile phones are being thought up all the time and who knows what they might do in the future. Smartphones offer all sorts of new services and have started to blur the difference between a computer or tablet and a mobile phone.

Here are some of the services available through mobile phones:

- Send and receive text messages using SMS
- Take digital photographs
- Take short video clips
- Access the internet
- Watch live TV
- Send and receive email
- Download and listen to music
- Download and play games
- Send picture messages
- Play videos
- GPS (use your mobile phone as a satellite navigation system)
- Make small value payments.

Advantages and disadvantages of mobile phones

Advantages:

- You can be contacted in case of an emergency.
- Plans can be changed at the last minute.
- Parents like children to have a mobile phone so that they can be contacted in emergencies.

Disadvantages:

- Many people use their phones when walking along and this has caused accidents.
- Calls can disturb other people in the cinema, cafes, etc.
- Many people still use hand-held phones when driving, which is dangerous (and illegal in many countries).
- Long-term use may cause health problems.
- Your calls and your location are recorded and this invades your privacy.
- Fraudsters using smishing to collect personal data about you such as banking and credit card details.

SMS (texting)

SMS (short messaging service) or, as most people know it, texting, allows short low-cost messages to be sent between phones. It is also possible to send text messages to phones using a computer or mobile device such as a tablet, laptop or desktop computer.

VoIP (Voice over Internet Protocol)/internet telephone

VoIP is internet telephony where cheap phone calls can be made using the internet. It is a technology that allows voice signals to be transferred over the internet but it can sometimes be unreliable and the quality of the sound can be poor.

Publicity and corporate image publications

All the documents produced by an organisation should look similar and this will help reinforce the corporate image. It lets everyone know that they are dealing with a professional organisation. There are a number of printed documents that will help with publicity and reinforce the corporate image and these are described in the following section.

Business cards

Business cards are small cards given to people you meet containing the name of the organisation, a logo, your name and role, and contact details such as address, email, and phone numbers.

Letterheads

Letterheads are usually pre-printed on paper containing the name of the organisation, the logo, and contact details. They are put into the printer so that when writing a letter, you can just enter the text of the letter.

Flyers

Flyers are single sheets of paper used to advertise an event or service. They can be handed out to people in the street, posted door-to-door or put on car windscreens in car parks.

Brochures

Brochures are documents several pages long that are used to advertise a product or a service.

QUESTIONS A

1 **a** Explain what is meant by the abbreviation VoIP. *(1 mark)*
b Describe **one** advantage in a person using VoIP to communicate with others. *(2 marks)*

2 People can communicate using a variety of different devices and services.

Give the names of **three** methods of communication that make use of ICT. *(3 marks)*

3 Describe an application for each of the following publicity and corporate image publications.
a Business cards *(1 mark)*
b Letterheads *(1 mark)*
c Flyers *(1 mark)*
d Brochures *(1 mark)*

Data handling applications

Many ICT systems perform simple data handling tasks. In these systems the data is input into the computer into a simple database structure and it can then be processed in some way and the results output in the form of reports.

Surveys

Surveys are used to collect information by asking lots of different people the same set of questions. Because of the huge amounts of data collected, the questionnaires might use OMR (optical mark recognition) for the input of data. The completed questionnaires can be read at high speed by the optical mark reader, which saves time and money. Once the data has been entered it can be stored and processed to produce statistics or graphs and charts. Survey results are often processed using spreadsheet software and displayed as a series of graphs.

Clubs and society records

Clubs and societies have income and expenditures which need to be recorded, and computers with appropriate software can be used for this. They also need to keep membership details such as the contact details of their members. They will have to keep details of membership renewals and use mail merge to send membership renewal reminders out to members.

Address lists

Address lists or contact details are essential for communication by letter, email, SMS, phone etc. Address lists can be prepared using one piece of software and then saved in a file format that can be imported into different software packages so that you do not have to ever input the address details again. Address lists are frequently used for mail merges where a personalised document is sent to a large number of people.

School reports

Every so often your school will need to issue a school report which is used to inform your parents/guardians about your current progress. All these reports need to be kept for each student over their school days. School management software usually has a module which allows the report data to be entered by each teacher for each student. School report systems can allow teachers to insert comments for each student which can then be output and given to students/parents as part of the school report.

Libraries

School and college libraries usually use bar codes to input unique borrower and book numbers (i.e. unique numbers given to each book) into the library system. There are three files used; one to contain book details, one for member details and one which records the borrowings which links books to members. All of these files can be accessed using database software in lots of different ways. For example you can find lists of books on a certain topic, list members who are overdue with their books and so on.

Library systems in general are covered on page 19.

Measurement applications

Computers can be used in conjunction with sensors to measure physical quantities such as temperature, pressure, light intensity, infra-red radiation, etc.

Monitoring

Monitoring involves taking readings/measurements regularly or continuously over a period of time using sensors. There are three things that can then happen:

- They can be stored for later - the readings can be stored and then output in some way, such as graphically.
- Set off alerts - they could be used to issue a warning sound such as the beeper if the heart rate of a patient being monitored changes.
- Auto control - the readings can be used to control a device in some way. For example, the readings sent from a temperature sensor could be used to turn an air-conditioning unit on or off.

Monitoring can be used in science lessons for recording temperature, light, force, etc. In geography lessons monitoring can be used to record the weather. It can also be used in society for monitoring traffic flow, monitoring pollution, keeping track of climate change, for electronic timing, etc.

The main features of monitoring are:

- The readings are taken automatically - there is no need for a human to be present. This means that it is much cheaper than employing a person to do this.
- You can set the period over which the monitoring takes place - this is the total time over which the readings will be collected. You can also set the system to take readings continuously such as in a flood warning system.
- You can set how often the readings are taken. For example, in an experiment to investigate the cooling of boiling water, you might decide to set the frequency with which readings are taken to be every minute.
- The sensors can be put in remote locations - you can put them anywhere in the world and the data can be sent back wirelessly and even using satellites.
- The sent data can be stored and processed by a computer.
- The data can be analysed (you can do calculations such as work out the mean, mode median, range, etc.), and draw graphs and charts. The data can be imported into software such as a spreadsheet package.

Sensors

Sensors are used to detect and measure physical quantities. Here are some examples of sensors:

- Temperature sensors - can be used in school experiments such as investigating the cooling of a hot drink in different thicknesses of cardboard cup. Sensors can be used to control a heating/cooling system in a home or classroom/greenhouse.
- Light sensors - detect the brightness of light. Can be used to see how light levels affect the growth of a plant. They can be used to control lights that come on automatically when it goes dark in a greenhouse.
- Sound sensors - measure the loudness of a sound. They can be used in noise disputes (e.g. monitoring sound at airports).
- Pressure sensors - barometric pressure sensors measure air pressure; other pressure sensors measure depth of liquid or something pressing on them.
- Humidity sensors - these measure the moisture in the air in greenhouses or art galleries, etc.
- Passive infra-red sensors (PIRs) - these are the sensors used in schools and homes to detect movement. They can be used in burglar alarms and also to turn lights on/off automatically in rooms when a person walks in/out.

Advantages and disadvantages of monitoring using ICT

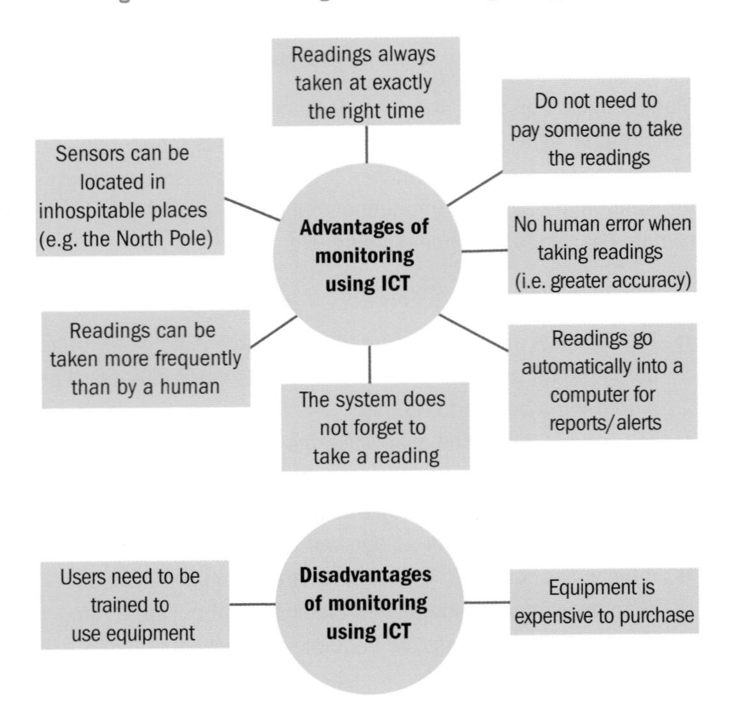

Users need to be trained to use equipment | **Disadvantages of monitoring using ICT** | Equipment is expensive to purchase

Environmental monitoring

Environmental monitoring is used to collect data for weather forecasts, to test water quality in rivers and streams, to collect data about air pollution, etc. Sensors take the readings automatically and these readings can be saved on removable media or the data can be sent back to the computer automatically using a wireless signal.

Pollution monitoring

Water quality in rivers and air quality can be continually monitored for pollutants by remote sensors which send signals back to computers. As the signals from sensors are analogue and computers can only process digital signals, an analogue to digital converter is used to convert the signals from the sensors into ones that can be processed by the computer.

Water pollution

Sensors usually measure indicators such as pH, oxygen level, temperature and the cloudiness of the water. Measurements are collected and processed to produce graphs showing the variation of values over time and in some systems alerts are sounded when values from the sensors fall outside acceptable values called pre-set values.

Air pollution

As with water pollution, sensors measure indicators such as levels of ozone, nitrogen dioxide, sulphur dioxide and fine particles (e.g. dust, smoke etc.).

Analogue to digital conversion

Analogue quantities are continuously variable, which means that they do not jump in steps from one value to another. Temperature is an analogue quantity because temperatures do not jump from one degree to the next as there are many values in between. Digital quantities jump from one value to the next. Computers are nearly always digital devices and can operate only with digital values. If analogue values such as temperature readings need to be input into a computer, they first need to be converted to digital values using an analogue to digital converter.

Microprocessors in control applications

Data from sensors can be used to control devices. For example, the data from temperature sensors are used to turn heaters on or off to maintain a constant temperature. Most household appliances such as washing machines, air conditioning units, kettles, etc., use some form of control, as you will see later in this chapter.

Using a sequence of instructions to control devices

In control systems it is necessary to give a series of commands for the system to obey.

There is an educational package for teaching about control commands called LOGO and it is used to move a turtle or cursor around the screen. The turtle or cursor can be instructed to leave lines to trace its path on the screen. For example, the following set of commands can be used to move an arrow on the screen of a computer. When the arrow moves, it leaves a line.

FORWARD distance
LEFT angle
RIGHT angle

Hence, using the command FORWARD 5 would move the arrow forward 5 units, and LEFT 90 would turn the arrow left through an angle of 90°.

The list of instructions that would draw this shape is as follows:

FORWARD 50
RIGHT 90
FORWARD 30
RIGHT 90
FORWARD 30
LEFT 90
FORWARD 15

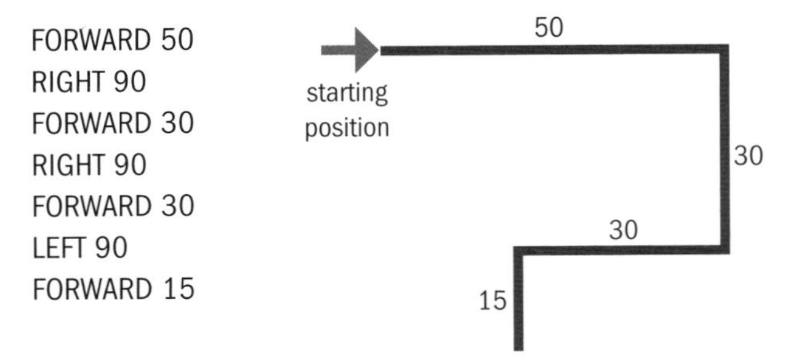

The set of instructions to draw this shape on the screen are:

FORWARD 30
RIGHT 60
FORWARD 27
RIGHT 60
FORWARD 20

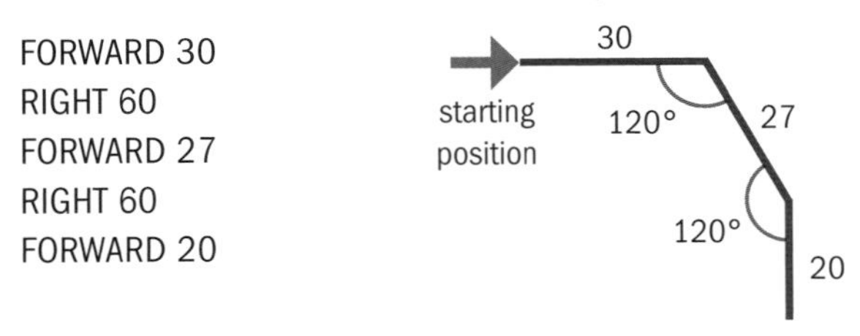

Repeating sets of instructions: instructions can be repeated using the following commands:

REPEAT n
ENDREPEAT

The instructions to be repeated a set number (i.e. n) times are placed in between these two commands.

For example, a square could be drawn using REPEAT n using the following commands:

REPEAT 4
FORWARD 20
RIGHT 90
ENDREPEAT

There are also some other commands:

PENDOWN (this instruction puts the pen down so that a line is ready to be drawn)
PENUP (this instruction raises the pen so the pen can be moved along without drawing a line)

Simple control applications

Simple control systems use data from sensors.

Security light system – uses a PIR (passive infra-red) sensor to sense movement. When a person or animal walks up the path to your house, the PIR senses the heat from the person or animal moving in the sensor's field of view and sends a signal to the processor to turn on a light. After a period of time, if no heat is detected, the light is turned off.

A burglar alarm – works in a similar way to the security light using PIRs as the input into the system. This time the output device is a bell or siren that sounds when the alarm is on and movement is detected.

Automatic cookers – temperature sensors relay temperature data back to the microprocessor to control the operation of the heating elements. The microprocessor also controls when the oven is to come on and how long it stays on for.

Simple control in automatic washing machines

The user selects the program they want to use. The program will be the set of instructions for a particular wash and will include pre-set values such as temperature, spin speed, time spent washing etc.

The control program will then receive data from sensors (e.g. pressure to show how much water has entered the drum, temperature to check that the water is heated to the correct temperature) etc.

The sensors and control program in the computer perform will:

- Control the flow of water into the machine.
- Heat the water up to the correct temperature.
- Control the addition of the washing liquid.
- Control the length of time the drum rotates for.
- Control the pumps which pump the dirty water out of the machine.
- Control the spin speed.
- Control how long the drier cycle is on for by using moisture sensors to sense when the washing is dry.
- Control the release of the door hatch when it is safe to do so.

Computer controlled central heating system

Computer controlled heating systems keep a room at constant temperature. Here is a very simple control system for keeping a room at constant temperature.

- The timer instructs the heating system when to switch on and off. The timer switches the whole system on.
- A temperature sensor measures the temperature of the air and sends a signal to the computer. If the temperature of the air is less than 23°C (i.e. the value pre-set by the user), a signal is sent to the heater to turn it on.
- The air temperature is continually measured by the temperature sensor. Signals are sent back to the computer, which compares the temperature with its pre-set value (i.e. 23°C). As soon as the temperature reaches this value, a signal is sent to the heater to turn it off.
- At any time the temperature drops below 23°C, the heater is switched on again so that the temperature remains constant at 23°C.
- The system continues to keep the temperature of the room constant until the timer switches the whole system off.

Computer-controlled greenhouses

In order to grow plants successfully they need perfect growing conditions. Computer control can be used to monitor the conditions and keep these conditions constant.

The sensors used to collect the input data include:

- Light
- Moisture
- Temperature.

The output devices that are controlled by the computer include:

- Lamps (to make the plants grow faster)
- Heaters
- Motor to turn the sprinkler on/off to water the plants
- Motor to open or close the windows (to cool the greenhouse down if it gets too hot).

The sensors (i.e. light, humidity and temperature) all read analogue values which are converted to digital signals using an analogue-to-digital converter (ADC) before being processed. The processing compares each value with set values to decide whether each of the output devices (i.e. lamps, sprinkler, heater and windows) should be turned on or off.

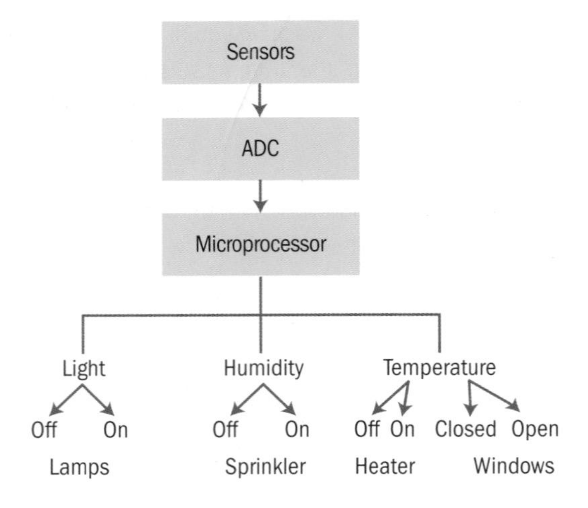

Sensors measure data.

The microprocessor compares the sensor reading to a preset value.

The microprocessor sends signals to actuators.

If temperature reading > preset value, the microprocessor switches the heater off and opens the windows.

If temperature reading < preset value the microprocessor switches the heater on and closes the windows.

If light reading < preset value the microprocessor switches the lights on.

If light reading > preset value the microprocessor switches the lights off.

If humidity reading < preset value the microprocessor switches the sprinkler on.

If humidity reading > preset value the microprocessor switches the sprinkler off.

Advantages and disadvantages of control systems

Advantages:

- Can operate continuously, 24 hours a day, 7 days a week.
- Less expensive to run as you don't have to pay wages.
- Can easily change the way the device works by re-programming it.
- More accurate than humans.
- Can react more quickly to changes in conditions.

Disadvantages:

- Initial cost of equipment is high.
- Equipment can go wrong leading to poor conditions.
- Fewer people needed so leads to unemployment.

QUESTIONS B

1 A turtle that draws on paper uses the following instructions.

FORWARD *n*	Move *n* cm forward
BACKWARD *n*	Move *n* cm backwards
LEFT *t*	Turn left *t* degrees
RIGHT *t*	Turn right *t* degrees
PENUP	Lift the pen off the paper
PENDOWN	Lower the pen onto the paper

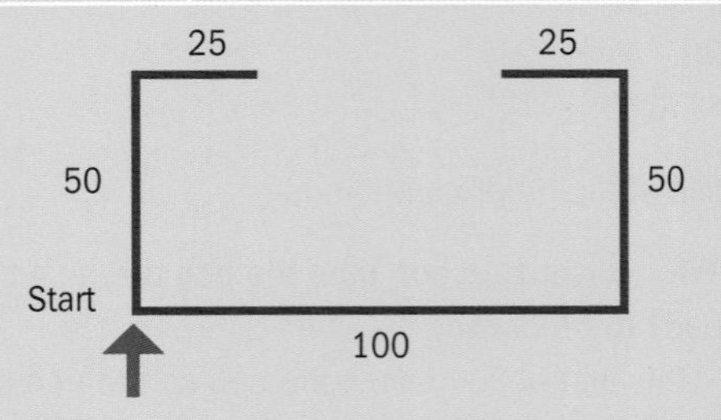

Write a set of instructions that makes the turtle draw the shape shown above.
Assume that the pen is down at the start. *(4 marks)*

2 Computer control is used to control growing conditions in a greenhouse. For example, a temperature sensor will turn on a heater if the temperature inside the greenhouse gets too cold.

a Give the names of **two** other sensors that could be used in the greenhouse and for each one describe why it is needed. *(4 marks)*

b Give **two** advantages in using ICT to monitor and control the growing conditions in the greenhouse. *(2 marks)*

c Describe **one** disadvantage in using monitoring and control to control the growing conditions in the greenhouse. *(1 mark)*

3 Give the names of **three** different output devices that can be controlled by a computer. *(3 marks)*

4 A home weather station consists of a base station that contains the processor and the display. Sensors are also included that are placed outside the house. Readings from the sensors are relayed back to the base station, which processes the data and produces weather information that is displayed on the screen.

a Give the names of **two** different types of sensor that could be used with this system. *(2 marks)*

b Describe **one** method by which the data can get from the remote sensors to the base unit that is situated inside the house. *(2 marks)*

c Once the data has been sent to the base unit it is processed and the information is output. Describe **one** way that the weather information is output from the system. *(2 marks)*

Activity 6.1

Use the internet to research examples of children's toys that use computer control. Produce some notes explaining how the toys use control for their operation.

Modelling applications

Spreadsheet software can be used to construct models and can be used to produce financial models such as the money coming into and going out of a school tuck shop or models on personal finance to help you manage your money. Modelling means producing a series of mathematical equations that are used to mimic a real situation. When values are put into the model or we exercise the model in some way, we are said to be performing a simulation. There are many types of specialist modelling software from games to flight simulators. Models can be created to describe the flow of traffic at junctions and the output from the model can then be used to issue controls to the traffic signals to ensure that the traffic flows as smoothly as possible thus keeping queues to a minimum.

Modelling can also be applied to graphics. For example, you can produce a computer model of a building in 3D using a plan of the building in 2D. Architects can make changes on the plan and then see the effect of the changes on the 3D view.

The advantages and disadvantages of using simulation models

Some advantages:

- Cost – it can be cheaper to use a model/simulation. For example, car engineers can use a computer to model the effect on the occupants during a crash and this is cheaper than using real cars with crash test dummies.
- Safer – flight simulators can model the effect of flying an airplane in extreme situations. Extreme situations might include landing without the undercarriage coming down, or landing with only one engine working, and so on. It would be far too dangerous to expect a pilot to try these in real life.
- It can save time – global warming models can be set up to predict what the likely effects of global warming will be in the future.
- It is possible to experience lots more situations – pilots using simulations can experience all sorts of extreme weather conditions such as sand storms, hurricanes, smoke from volcanic eruptions, and so on. These would be almost impossible to experience any other way.

Some disadvantages:

- The differences between simulation and reality – there will always be some difference between a model/simulation and reality. No model or simulation can ever be perfect because real life situations can be very complex.
- The accuracy of the rules and variables – the person designing the model may have made mistakes with the rules or the variable data.
- Some situations are hard to model – some situations are difficult to model because some aspects of the model are often open to interpretation. For example, experts on the subject may disagree about the rules that apply.

Applications in manufacturing industries

Robots have been widely used in manufacturing for years, especially for painting and welding in car factories. Robots are also used for picking and packing goods in large warehouses.

Advantages and disadvantages of using computer-controlled systems compared to using humans

Compared to humans doing the job, computer-based control systems or robots have the following advantages and disadvantages.

Advantages:

- Can operate continuously, 24 hours per day and 7 days per week.
- Less expensive to run as you don't have to pay wages.
- Can work in dangerous places or places where a human would find it difficult to work (e.g. stacking shelves in a frozen food warehouse).
- Can easily change the way the device works by re-programming it.
- More accurate than humans.
- Can react quickly to changes in conditions.

Disadvantages:

- Initial cost of equipment is high.
- Equipment can go wrong.
- Fewer people needed so leads to unemployment.

Robots assembling car panels in a car factory

School management systems

There are many ways in which ICT is used to help with the administration and management of schools.

School registration

School registration systems are something you will be familiar with. You will know that it is important for the school to keep a record of who is on the premises and who is not.

Computer-based methods of registration

Any ICT system used for student registration in schools or colleges should:

- Capture student attendance accurately
- Capture the student attendance automatically
- Be very fast at recording attendance details
- As far as possible avoid the misuse of the system
- Enable not only morning and afternoon attendance to be recorded but also to record attendance at each lesson and enable attendance patterns for individuals and defined groups to be worked out
- Be relatively inexpensive
- Work with other ICT systems used in the school, such as the system for recording student details.

Optical mark recognition (OMR)

Optical mark recognition works by the teacher marking a student's attendance by shading in boxes using a pencil. The forms are passed to the administration office where they are collected and batched together and processed automatically using an optical mark reader.

Advantages of optical mark recognition:

- Data entry is more accurate than typing.
- Frees up staff from entering attendance marks manually using a keyboard.
- The OMR reader is cheaper than other methods.
- Reader can be used for other purposes such as reading multiple-choice answer sheets.

Disadvantages of optical mark recognition:

- Registration is not done in real time – if a student came in halfway through the morning, this system would not record this.
- Registers need to be passed manually to the administration staff.
- If the forms are folded or damaged, they are rejected by the reader.

ARC

ABSENCE REPORT CARD

CARD NO.

[0]	[0]	[0]	[0]	[0]	[0]
[1]	[1]	[1]	[1]	[1]	[1]
[2]	[2]	[2]	[2]	[2]	[2]
[3]	[3]	[3]	[3]	[3]	[3]
[4]	[4]	[4]	[4]	[4]	[4]
[5]	[5]	[5]	[5]	[5]	[5]
[6]	[6]	[6]	[6]	[6]	[6]
[7]	[7]	[7]	[7]	[7]	[7]
[8]	[8]	[8]	[8]	[8]	[8]
[9]	[9]	[9]	[9]	[9]	[9]

INSTRUCTIONS
Use only pencil when completing this form.
Mark like this ⧺

Student name	Date	Present	Code
		["]	[L] [M] [N] [P] [T] [H] [G] [J] [F]
		["]	[L] [M] [N] [P] [T] [H] [G] [J] [F]
		["]	[L] [M] [N] [P] [T] [H] [G] [J] [F]
		["]	[L] [M] [N] [P] [T] [H] [G] [J] [F]
		["]	[L] [M] [N] [P] [T] [H] [G] [J] [F]
		["]	[L] [M] [N] [P] [T] [H] [G] [J] [F]
		["]	[L] [M] [N] [P] [T] [H] [G] [J] [F]
		["]	[L] [M] [N] [P] [T] [H] [G] [J] [F]
		["]	[L] [M] [N] [P] [T] [H] [G] [J] [F]
		["]	[L] [M] [N] [P] [T] [H] [G] [J] [F]
		["]	[L] [M] [N] [P] [T] [H] [G] [J] [F]
		["]	[L] [M] [N] [P] [T] [H] [G] [J] [F]
		["]	[L] [M] [N] [P] [T] [H] [G] [J] [F]
		["]	[L] [M] [N] [P] [T] [H] [G] [J] [F]
		["]	[L] [M] [N] [P] [T] [H] [G] [J] [F]
		["]	[L] [M] [N] [P] [T] [H] [G] [J] [F]
		["]	[L] [M] [N] [P] [T] [H] [G] [J] [F]
		["]	[L] [M] [N] [P] [T] [H] [G] [J] [F]
		["]	[L] [M] [N] [P] [T] [H] [G] [J] [F]
		["]	[L] [M] [N] [P] [T] [H] [G] [J] [F]

An OMR form which is used to enter student registration information into the Schools Information Management System.

Smart cards

Smart cards look like credit cards and they contain a chip that can be used to hold certain information. Smart cards hold more information than cards containing only a magnetic stripe and can be used in schools in the following ways:

- For registration of students
- For monitoring attendance at each lesson
- For payment for meals in the school canteen
- For access to the school site, buildings, and rooms to improve security
- For access to certain facilities such as the computer network, photocopier, etc.
- To record borrowing and returning of school library books, digital cameras, musical instruments, etc.

Swipe cards

Students are given a swipe card that they use for registration purposes by swiping the card using a card reader. Swipe cards are plastic cards with a magnetic stripe containing a limited amount of data. The swipe card is used to identify the student to the registration system and some other systems such as the library system and the school meals system. The same card can be used for access to school buildings.

Advantages of swipe cards:

- The cost of the cards and the readers is low compared to other methods.
- Readers can be made that are almost vandal proof.

Disadvantages of swipe cards:

- Cards are often lost or forgotten, meaning that students have to be registered using a keyboard.
- Student cards can be swiped in by someone else.

Biometric methods

Biometric methods provide a fast and easy way of recording student attendance. Biometric methods make use of a feature of the human body that is unique to a particular person in order to identify them.

Advantages of biometric methods:

- There is nothing for a student to forget, like a card.
- You have to be there to register so no-one else can do it for you and it cannot be altered by students.
- Performed in real time so the system knows exactly who has registered and when.

Disadvantages of biometric methods:

- Biometric systems are expensive.
- There are privacy issues. Some people object to fingerprinting systems.
- People can vandalise the readers.

School management systems

There are many different pieces of information stored by a school – the student records and the records of attendance are just two. By integrating all the systems, it is possible to extract the data needed from the system in the form of reports. A report might be a list of all those students with 100% attendance so they can be given a certificate.

School management systems are ICT systems that supply school managers and staff with information that can help them make decisions. For example, the attendance system might produce information about those students for whom attendance is poor. The system might produce a report showing when they are absent to see if there is a pattern. The senior teachers can then take action.

Recording learner performance

Included as part of the school management system is a module which records learner performance. In the past, teachers recorded marks for homework, classwork and tests in their paper-based mark books. Opinions about how each student's performance, development of skills, behaviour etc. was often held in the teacher's head or handwritten on paper. Using ICT to record learner performance means:

- Teachers enter marks for homework, classwork, tests and details of skills learnt directly into the computer.
- Results can be analysed using software which can compare the performance of the student against other students in the same class/year or students nationally.
- Good and bad behaviour can be recorded.
- Data is held centrally so all teachers can access it which means a full picture of a student's progress can be obtained across all subjects.
- Students and parents can see a student's progress or otherwise and take appropriate action.

School timetabling

Producing school timetables manually is quite difficult as there are four things to consider, namely:

- Students
- Teachers
- Rooms
- Time slots/periods.

School management systems have a timetable component which will work out a timetable once all the data concerning the above are entered.

Timetabling using a computer system has the following advantages:

- Accurate timetables can be produced in much less time than you could produce one manually.
- It is much easier to make changes to a timetable. For example, if two teachers want to swap rooms, then this is easily altered on the timetable.
- If a classroom cannot be used for any reason (e.g. air conditioning or heating not working), then an alternative classroom can be quickly found using the system.
- If a teacher is off sick, then those teachers who are free at the times for the lessons can be identified by the system so they can cover the class.

Organisation of examinations

School management systems can also help school managers organise examinations by:

- Ensuring that each pupil is entered for correct examinations
- Printing out individual student's exam timetables
- Identifying rooms in which examinations can take place
- Producing lists of students for each exam
- Identifying teachers who are free to invigilate the exams
- Collecting exam results and entering them into the student records.

Here are some other ways school management systems can be used:

- To work out how many students will be in the new intake and to allocate them into tutor groups.
- To decide whether a new teacher should be employed.
- To work out the best way of allocating teachers and classrooms.
- To decide on how best to spend the training budget to keep teachers up-to-date.

The main advantages in using school management systems are:

- They reduce the workload for teachers in the classroom and in the school office.
- They can provide up-to-date information for parents.
- They can support decision-making for school managers.
- They can tackle truancy effectively.
- They can be used to plan timetables.

The main disadvantages in using school management systems are:

- The software is expensive to buy.
- Student data is personal, so there must be no unauthorised access.
- Software is complex, so all staff need training.

Booking systems

The main feature of all online booking systems is that they use online processing. This means that while the transaction is taking place and the user is entering their details, the seats are saved for them. If they choose not to go ahead with their purchase, the seats held are released for sale to others. This immediate updating of files prevents double bookings.

Online booking systems can be used to book:

- Holidays
- Flights
- Train tickets
- Cinema seats
- Theatre seats
- Tickets for sporting events.

The steps for online booking are as follows:

- Find the booking website using a search engine or type in the web address.
- Search the online booking database using dates, times, etc.
- Make your selection (in many cases you can choose where you sit from a plan).
- The seats are now held for you so no-one else can book them.
- Enter your details (names, addresses, etc.).
- Select a payment method (credit card, debit card, etc.) and enter the card details.
- A confirmation appears on the screen telling you that the seats/holiday, etc., have been successfully booked.
- A confirmation email is sent to you, which in many cases can be printed out as it acts as your ticket/reservation.

Advantages of online booking:

- You can book from the comfort of your home at any time.
- There is more time to look for holidays, flights, etc., than when at a travel agents.
- You can make savings for flights/holidays when you book direct as there is no travel agent commission to pay.
- You can read reports from people who have been on the same holiday, seen the same concert, etc.
- There is no need to pick up tickets as you often print these yourself.

Disadvantages of online booking:

- You have to enter credit/debit card details and these may not be kept safe.
- People could hack into the site and know you were away and burgle your house.
- There is no personal service like with a booking agent.
- You could easily enter the wrong information and book the wrong flights or performance on the wrong day.

QUESTIONS C

1 An airline ticket is booked direct with the airline using the internet. The airline booking system uses online processing.
  - **a** Explain what is meant by online processing. *(2 marks)*
  - **b** Describe **two** advantages in booking an airline ticket online. *(2 marks)*
  - **c** Describe **two** disadvantages in booking an airline ticket online. *(2 marks)*

2
  - **a** A school uses a school management system. Describe **two** features of a school management system. *(2 marks)*
  - **b** Explain **one** advantage in the school using a school management system *(1 mark)*
  - **c** Explain **one** disadvantage in the school using a school management system. *(1 mark)*

3 ICT-based student registration systems in schools offer the advantages of speed and accuracy to the form teacher. Discuss **two** other advantages to the school in using ICT to register students. *(4 marks)*

4 Robots are widely used in manufacturing industries.
  - **a** Describe **two** tasks robots would perform in a car factory. *(2 marks)*
  - **b** Describe **three** advantages of robots rather than humans manufacturing cars. *(3 marks)*
  - **c** Describe **three** disadvantages in robots rather than humans manufacturing cars. *(3 marks)*

Banking applications

Many of the facilities offered by banks such as the use of credit/debit cards and online banking would not be possible without the use of ICT. Banks use ICT in many ways which are outlined below.

ATMs (automatic teller machines)

ATMs, commonly called cash points, are the 'hole in the wall' cash dispensers that many people use when the bank is not open or when they do not want to queue inside the branch. In order for you to use the service, the machine needs to check that you are the card holder. You are asked to enter a PIN (personal identification number) that only you should know. If the card is stolen then the thief should have no way of finding this, unless of course you have foolishly written it down!

Here are some of the things you can do using an ATM:

- You can get cash out.
- You can find out the balance in your account.
- You can change your PIN (personal identification number).
- You can make deposits (i.e. put cash, cheques or both into your account).
- You can obtain a mini statement listing your recent transactions (i.e. money in and out of your account).
- Other facilities such as topping up your mobile phone balance might be available.

Type of processing used with ATMs

Real-time transaction processing is used with ATMs. This means that as soon as a customer gets the money out of their account, their balance is updated.

Processes involved in using an ATM

The card is inserted – the ATM connects to the bank.

PIN is entered – if incorrect, it stops and allows a second try or checks to see if the card is stolen.

- If correct the ATM checks if the card is valid/not out of date.

If PIN is OK the system moves onto the next step, otherwise it stops.

Customer selects option.

- ATM checks balance.
- ATM checks if customer is trying to withdraw more than their daily allowance.

If OK the system asks if a receipt is required.

If Yes – receipt is printed

- Card is returned
- Money is given out.

Note that the card is given out first and the money is given out last.

Benefits to banks in using ATMs:

There are some benefits to the banks in the use of ATMs and these include:

- Staff are freed from performing routine transactions so that more profitable sales-oriented work can be done.
- Fewer staff are needed, since the computer does much of the routine work.
- A 24-hour per day service is provided to satisfy their customers' demands.
- The system makes it impossible for a customer to withdraw funds from their account unless they have the money in their account or an agreed overdraft.
- Unusual spending patterns or location of withdrawal can trigger an alert to system not to issue cash without contacting the bank for authorisation using a series of security questions.

Benefits to customers in using ATMs:

- Some customers prefer the anonymous nature of the machine since it cannot think you have stolen the cheque book or think that you are spending too much.
- It is possible to use the service 24 hours per day; ideal for those people who work irregular hours.
- It is possible to park your car or bike near the dispenser of an evening, so getting cash is a lot quicker.
- Fewer queues, since the transactions performed by the ATM are a lot faster.
- If your card is stolen, the thief cannot get money from the ATM unless they know your PIN.

Cheque clearing

Suppose Manisha wants to buy a tablet computer costing $250 from a store called Computer World and pay using a cheque. Manisha banks with Bank AB and the store banks with Bank XY. Here is what happens to the cheque:

Manisha writes out a cheque for $250 payable to Computer World.

- Computer World pay the cheque into their Bank XY branch.
- Bank XY type in the amount of the cheque in magnetic ink characters so that the amount along with other information pre-printed on the cheque in magnetic ink characters, can be read at the clearing house using magnetic ink character recognition.
- All cheques, including this one, are sent to a clearing centre where the cheques are processed. The cheques are passed through sorters/readers where the magnetic ink characters are read automatically and an encrypted file is produced containing customer account number, bank sort code and cheque serial number.
- This file containing details of all the transactions are then sent to the main bank of the country (e.g. the Bank of England or the Bank of Mauritius). They will then transfer the payment from Manisha's bank to Computer World's bank (i.e. XY Bank) along with all the other payments between the banks.
- The cheque is then sent to the XY bank identified by the sort code number, where the amount is deducted from her account.

As you can see the process is complicated and this is why there is a delay of a few days between paying the cheque in and the money arriving in an account. It is also the reason why banks are keen to get more people to use electronic methods such as internet banking or the use of debit/credit cards to pay for goods.

Online banking

Many of the tasks you would have had to go to a bank branch to do, you can now perform at home using internet or online banking. Using online banking you can:

- View bank statements
- Transfer money between accounts
- Make payments for bills
- Apply for loans.

Online banking uses the internet to enable a customer at home to connect to the bank's ICT systems and interact with them. In order to do this the customer has to enter log-in details (username and password) and answer some other security questions.

Any details passed between the bank and their customers are encrypted to ensure hackers cannot access banking details.

Advantages to the bank of online banking:

- The bank can reduce the number of branches, which will reduce costs.
- Fewer bank workers are needed and this means lower total wage bill.
- The bank staff can be less qualified so this reduces the wage costs.

Disadvantages to the bank of online banking:

- Customers may feel lack of personal contact and this may cause some customers to move their account to a different bank.
- It is easier to sell products such as loans face-to-face and this can reduce bank profits.
- Have to spend out large amounts of money on new systems to perform the online banking.
- Have to pay out large amounts of money on redundancies when staff lose their jobs.
- Have to employ and train staff with skills in computer/network development and maintenance.

Advantage to the customer of online banking:

- Do not need to spend time travelling to the bank to perform some banking transactions.
- Can perform banking transactions any hour of the day or night.
- Can pay bills without the need for cheques and money for postage/parking.
- Easier to check on your account balance so less likely to be charged for overdrafts, etc.

Disadvantages to the customer of online banking:

- The worry of hackers accessing your account and stealing money.
- You cannot withdraw cash so will still need to visit a branch or ATM.
- Lack of the personal touch you get when banking at a traditional branch.

EFT (Electronic funds transfer)

Electronic funds transfer (EFT) is a method of transferring money from one bank account to another electronically. Using EFT, data is exchanged rather than physical money between the two accounts and the transfer occurs quickly.

The accounts used for the transfer can be in the same or different countries so there may be some currency conversion involved in the transfer. All the data sent for the transfer is encrypted so if the data is intercepted by hackers they would not be able to understand the data.

When you use a debit card at a store, the money is transferred from your account to the store's account, so this is an example of EFT. Another example of EFT is when an employee is paid by their employer, the employer instructs their bank to make a transfer of money to the employee's account.

Advantages of EFT:

- You can transfer money between accounts on any device with internet access such as desktops, laptops, tablets, and smartphones.
- The transfer of money is fast. Sometimes it can be instantaneous and other times it can take a couple of hours.
- The transfer is secure as it is subject to a high degree of data encryption. Hackers will not be able to understand the data even in the unlikely event of them intercepting it.
- Funds can be sent to anyone who has a bank account anywhere in the world.

Disadvantages of EFT:

- There is a danger that hackers could intercept the transfer.
- The charges for the service can be high especially when the accounts are in different countries.
- A mistake in the entry of one of the very long account numbers could result in the money being transferred to the wrong account.

Phone banking

Some people who are wary of internet banking can still do their banking tasks at a distance using phone banking. In order to use phone banking it is necessary to register with your bank and they will give you a number and a pass-code.

Some of the tasks you can perform using phone banking are automated so they use voice recognition or input from the key pad of your phone to tell the system what you want to do.

Using the part of phone banking where you do not have to wait to speak to an advisor, you can perform some or all of the following:

- Hear the balance of your account.
- Check your latest transactions (i.e. what money has gone into and come out of your account).

- Pay a credit card bill.
- Transfer money between different accounts you hold at the same bank.
- Listen to details of standing orders and direct debits.
- Search for specific payments you have made.

You can usually select that you want to speak to a person. When you speak to an advisor you can perform some or all of the following:

- Pay bills
- Stop cheques
- Order travellers cheques
- Change your details (e.g. address)
- Arrange an overdraft or loan
- Open a new account.

QUESTIONS D

1 Plastic cards, such as the one shown, are often issued to bank customers.

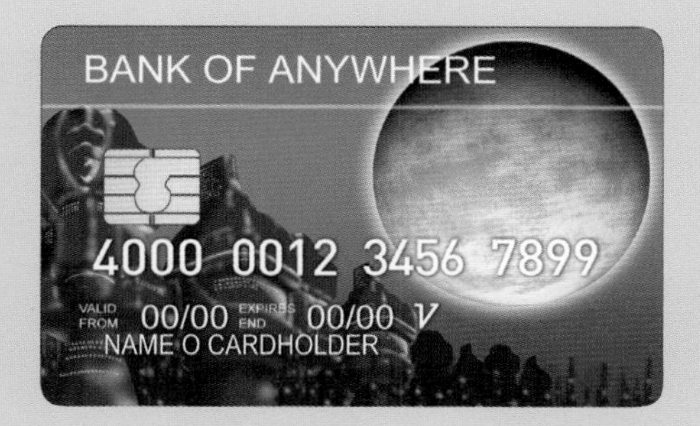

a Give the names of **two** different types of plastic card that are used by bank customers. *(2 marks)*

b These plastic cards often contain data in a magnetic stripe on the card but many of the new cards use chip and PIN.
Explain why these new chip and PIN cards were introduced and what advantages they offer over the older cards. *(4 marks)*

2 One problem in using credit cards is that they can be used fraudulently.

a Explain **two** ways that a credit card could be used fraudulently. *(2 marks)*

b State **one** way such frauds can be prevented. *(1 mark)*

3 Ahmed runs a small business and does most of his banking online. Ahmed uses EFT to transfer money to pay a supplier in a different country.

a Give the meaning of the abbreviation EFT. *(1 mark)*

b Describe **two** advantages and **two** disadvantages of using EFT to transfer money. *(4 marks)*

4 Describe **three** banking services provided by phone banking. *(3 marks)*

Computers in medicine

ICT is used extensively in medicine in the following ways:

- Medical databases – patient records are stored on a database and can be accessed from different places. There is simultaneous access to these records and they are easier to read than handwritten records.
- Patient identification using bar codes – bar codes are on patient wristbands so these can be scanned and the details obtained at the bedside.
- Hospital intranets – only hospital staff are allowed access to this network, which holds patient records. The intranet is also used internally for sending email between staff.
- Patient monitoring – sensors are used to measure vital signs such as temperature, blood pressure, pulse, central venous pressure, blood sugar, and brain activity. These sensors produce analogue signals which need converting to digital signals using an analogue-to-digital converter before they can be processed by the computer. There are acceptable ranges for each reading measured by the sensors and the computer checks that each reading lies within the range. If a reading lies outside the range, the medical staff are alerted by an alarm which sounds. The system also prints graphs so that trends can be spotted. Patient monitoring takes measurements automatically and frees up medical staff. There are no mistakes in the readings and the readings are taken in real time.
- Expert systems – these are used to help less experienced doctors to make a more accurate diagnosis. These are covered later in this chapter.
- Computerised reporting of laboratory tests.
- Digital X-rays can be viewed faster than conventional X-ray film.
- Pharmacy records – generates labels for prescribed medicines, recording patient prescriptions, alerting pharmacist/doctors to allergies or interactions between drugs.

Producing medical aids using 3D printing

3D printers – these are printers than can print 3D objects and can be used in medicine to produce prosthetics (e.g. artificial body parts such as false teeth, teeth implants, artificial limbs, and hearing aids). 3D printers can also be used for the creation of artificial blood vessels.

Tissue engineering is the process of producing layers of cells using 3D printing technology. These layers of cells can be used to replace removed or damaged tissue and in the future it may be possible to produce complete organs using this technology. This technology has been used to replace tissue damaged by arthritis or cancer.

Medical tools and equipment can be designed on a computer using computer-aided design (CAD) software and then printed in 3D using a 3D printer which creates the product by laying down layers of material one at a time. The medical equipment can be tested and the product can be altered using the software and re-printed and tested. The process is repeated until the piece of equipment is perfect.

QUESTIONS E

1 ICT is used in hospitals for patient monitoring.

- **a** Sensors are used to monitor a patient. Give the names of **three** sensors that could be used. *(3 marks)*
- **b** Other than patient monitoring give **two** uses of ICT in a hospital. *(2 marks)*

2 An intensive care unit in a hospital uses sensors connected to computers to monitor a patient's condition.

- **a** Give the names of **three** physical quantities that could be monitored. *(3 marks)*
- **b** Most physical quantities cannot be used directly by a computer without conversion.
  - **i** Give the name of the device that does the conversion. *(1 mark)*
  - **ii** Explain why this device is necessary. *(2 marks)*

3 Computer-controlled patient monitoring systems have many advantages compared to using nursing staff to monitor a patient's condition. Tick which of the following are advantages of computer-controlled patient monitoring systems.

	✓
Nurses cannot take readings accurately.	
Nurses cannot take readings at the correct time.	
Sensors attached to computers can take readings more regularly.	
Computer monitoring can take place all the time and not at certain intervals.	
Printouts from the computer make it easier to spot trends in the patient's condition.	
Computers can measure more than one physical quantity at the same time.	
It is cheaper to use a computer system rather than employ lots of nurses.	
Continual monitoring is safer than monitoring only now and again.	

(6 marks)

4 A computer-controlled heating system is used to keep a room at constant temperature.

This constant temperature is the pre-set value.

A temperature sensor measures the temperature, and the analogue signal is converted to a digital signal that the microprocessor is able to understand.

The microprocessor determines whether the heater should be turned on or off.

- **a** Give the name of the device used to convert analogue signals to digital signals. *(1 mark)*
- **b** Explain how the microprocessor uses the pre-set value to keep the room at a constant temperature. *(5 marks)*

Computers in libraries

Most libraries are computerised and the systems usually make use of bar codes and a relational database.

Each member is given a unique member number. Rather than type this number, the number is coded in the form of a bar code. This is faster and more accurate than typing it in.

Books are also bar coded with a unique number and when they are borrowed the member's ticket bar code is scanned along with the bar codes of all the books that have been borrowed. On their return the books are scanned thus telling the computer that the borrower has returned the books. The system can identify overdue books and it is possible to use mail merge to send letters, or email members asking them to return the books.

The system can be used to locate a specific book and to find out whether it is on the shelf or has been borrowed. It is possible to automatically generate lists of books (e.g. by author). Books can be reserved using the system so when the book arrives back, it is kept aside rather than being put back on the shelf.

Most libraries are still using bar codes for recording loans and returns but there is a better more up-to-date method which makes use of RFID chips.

RFID being used to help manage a library

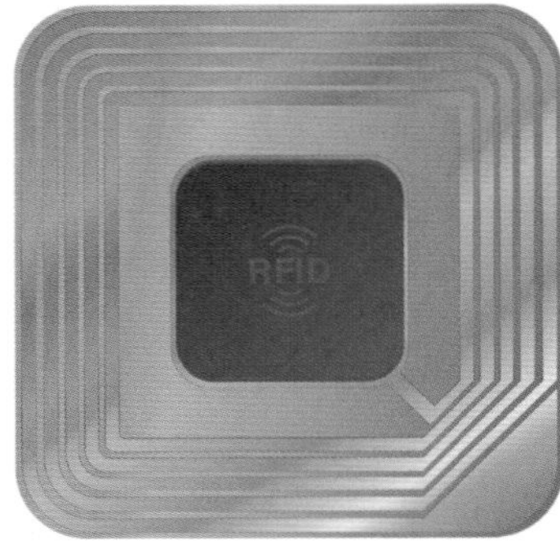

RFID can be used to help manage a library

RFID uses radio waves to identify people or objects. In a library the RFID tag on a book contains a serial number which identifies the book. Each book in the library will have a different serial number.

RFID tags are stuck onto the back of a book's cover and the data contained in the tag can help manage the loan and return of books. The self check in/out enables people to use the tag on the book to borrow and return books and this frees up library staff to do other tasks. Alternatively there is the option to check books in and out at the desk.

When the library is closed, there is a book drop which means you simply drop the book into the box where the RFID tag will be read automatically and you are issued with a receipt to prove you have returned the book.

QUESTIONS F

1 Tick **True** or **False** next to each of these statements.

	True	False
RFID uses radio signals to communicate with the reader		
The reader and the RFID must be in contact for the reader to read the data off the chip		
RFID is used in library management systems		
RFID does not make use of an aerial		
RFID is replacing bar codes in library management systems		
Using RFID, books can be returned when the library is closed		
RFID makes use of magnetic ink characters		

(7 marks)

2 a Explain how RFID is used in libraries to keep track of books. *(4 marks)*

b Older library systems made use of bar codes.

Describe **two** advantages in using RFID rather than bar codes in library management systems. *(2 marks)*

Expert systems

An expert system is an ICT system that uses artificial intelligence to make decisions based on data supplied in the form of answers to questions. This means that the system is able to respond in the way that a human expert in the field would to come to a conclusion. A good expert system is one that can match the performance of a human expert in the field.

Expert systems consist of the following components:

- Knowledge base – a huge organised set of knowledge about a particular subject. It contains facts and also judgemental knowledge, which gives it the ability to make a good guess, like a human expert.
- Rules base – this is made up of a series of IF, AND, OR and THEN statements that form a set of rules to closely follow human-like reasoning.
- Inference engine – this part uses the rules base along with the knowledge base to make a decision/conclusion based on the facts being entered by the user.
- User interface – this uses an interactive screen (which can be a touch screen) to present questions and information to the operator and also receives answers from the operator.

Stages of designing/making an expert system

The stages of designing/making an expert system are as follows:

Experts are interviewed.
Data is collected from experts.
Knowledge base is designed and created.
Rules base is designed and created.
Input and output format/screens are designed and made.
Expert system is checked by using known results.

The processes involved in a typical expert system such as a medical expert system for diagnosis are as follows:

1 An interactive screen appears and asks the user questions.
2 Answers are typed in or options are selected on a touch screen.
3 The inference engine matches the data input with the knowledge base, using the rules base until matches are found.
4 The system suggests the probable diagnosis and suggests treatments.

Applications of expert systems

Expert systems can be used for all sorts of applications and here are some of them.

Medical diagnosis

A health service website enables users to determine whether they need to call a doctor or visit a hospital. An expert system guides them through a series of questions, which they answer by clicking on a choice of answers. Such expert systems can determine, for example, whether someone with pains is having a heart attack or simply has indigestion.

Prospecting for minerals and oil

Using geological information, an expert system can use the information to determine the most likely places to choose for further exploration. This reduces the cost of mineral or oil exploration.

For giving tax advice to individuals and companies

Tax is complex and a lot of expertise is needed in order to give the correct advice. This is where expert systems come in. They are able to store a huge amount of data and they can ask the user a series of questions and come up with expert advice on how to pay less tax.

Car engine fault diagnosis

Modern car engines are very complex and when they go wrong it is hard for engineers to know what the problem is. Using an expert system created by the car manufacturer the engineers can be guided through a series of tests until the exact fault is identified.

Chess games

Chess game software is an expert system because it mimics an expert human chess player.

Advantages of expert systems:

- Fewer mistakes – human experts may forget but expert systems don't.
- Less time to train – it is easy to copy an expert system but it takes many years to train a human expert.
- Cheaper – it is cheaper to use an expert system rather than a human expert because human experts demand high wages.
- More expertise than a single expert – many experts can be used to create the data and the rules, so the expert system is a result of not one but many experts.
- Always asks a question a human expert may forget to ask.

Disadvantages of expert systems:

- Lack common sense – humans have common sense, so they are able to decide whether an answer is sensible or ridiculous. Human experts can make judgements based on their life experiences, and not just on a limited set of rules as is the case with computer expert systems.
- Lack senses – the expert system can react only to information entered by the user. Human experts have many senses that they can use to make judgements. For example, a person describing a type of pain might use body language as well, which would not be detected by an expert system.
- The system relies on the rules being correct – mistakes could be made that make the system inaccurate.

QUESTIONS G

1 Expert systems are used in medicine to help diagnose illnesses.
   a An expert system consists of four components. Name **two** of these components. *(2 marks)*
   b Give **one** benefit to the patient in using an expert system. *(1 mark)*
   c Give **one** benefit to the doctor in using an expert system. *(1 mark)*
   d Give **one** possible disadvantage in using this type of expert system. *(1 mark)*

2 Doctors and hospital consultants often make use of expert systems in their work.
   a Explain what is meant by an expert system. *(2 marks)*
   b Describe **one** way in which a doctor or hospital consultant can make use of an expert system. *(2 marks)*

3 Expert systems are becoming popular uses for ICT.
   a Name **two** jobs that are likely to use an expert system. *(2 marks)*
   b For each of the jobs you have named in part **a** explain how the expert system would be used. *(2 marks)*

4 One example of where an expert system is used is to help doctors diagnose illnesses. Give **one** other use of an expert system. *(2 marks)*

5 Expert systems have many uses. For example, one expert system called MYCIN is used by doctors to pinpoint the correct organism in blood that is responsible for a blood infection.
   a By referring to the above example give **one** advantage of using an expert system rather than a doctor. *(1 mark)*
   b By referring to the above example, give **one** disadvantage in using the expert system rather than a doctor. *(1 mark)*

Computers in the retail industry

Point of sale (POS) and electronic funds transfer at point of sale (EFTPOS) systems

Point of sale terminals are the computerised tills where you take your goods for payment in a shop. Many of these terminals also allow payment from your bank account to the store's bank account directly using a debit card. These terminals are called EFTPOS systems. EFTPOS systems can also allow the customer to get 'cashback' which moves money from their bank account to the store and this money is given to the customer as cash.

Because the terminals are networked together, when an item is sold and its bar code is scanned, the system looks up the price and description details to print out an itemised receipt. At the same time the system will deduct the item from stock so that the stock control system is updated.

POS/EFTPOS terminals consist of the following hardware:

- Bar code reader/laser scanner – this is used to input a number that is coded in the bar code as a series of thick and thin lines.
- Keyboard – the keyboard is used to enter codes on items if the bar code is damaged.
- Touch screens – these are often used in restaurants where there are no goods to scan.
- Swipe card readers – these are used to swipe the magnetic stripes on loyalty cards.
- Chip and PIN readers – these are used by customers to insert their credit/debit cards containing a chip. The system then asks them to enter their PIN (personal identification number) which is a number only they know. This is compared with the number encrypted and stored on the card. If the number entered matches that stored, it proves to the system that they are the genuine owner of the card.

Point of sale terminals are connected via networks to other systems such as:

- Loyalty card systems – where customers are given loyalty points according to how much they spend.
- Accounts systems – where the money coming into the shop is accounted for.
- Automatic stock control systems – the system knows what has been sold, so that it can automatically reorder more items once the stock falls below a certain level.

Automatic stock control

When an item is sold at the POS/EFTPOS terminal the number of that particular item in stock is reduced by one. This means that the computer knows how many items are in stock. Once the number of items has fallen below a certain level, the computer system will automatically order more stock from the supplier. This means that stores should theoretically not run out of fast-selling items. Good stock control systems are very important in supermarkets as customers will go elsewhere if the shop keeps running out of basic necessities such as bread, milk, etc.

Touch screens are used as input devices in restaurants and bars.

Internet shopping

Internet shopping means purchasing goods and services using the internet.

Most businesses have websites to show the products and services available. Lots of these websites allow customers to browse online catalogues and add goods to their virtual shopping basket/trolley just like in a real store. When they have selected the goods, they go to the checkout where they have to decide on the payment method. They also have to enter some details such as their name and address and other contact details. The payment is authorised and the ordering process is completed. All that is left is for the customer to wait for delivery of their goods.

Advantages of internet shopping to customers
Goods/services are cheaper because of lower costs of internet business
Goods are delivered to your home – ideal if people cannot get out because they are elderly or disabled
Wider range of goods to choose from
Cost savings are passed to customers with cheaper goods
No travelling costs to go shopping
Worldwide marketplace – you can order goods from anywhere in the world

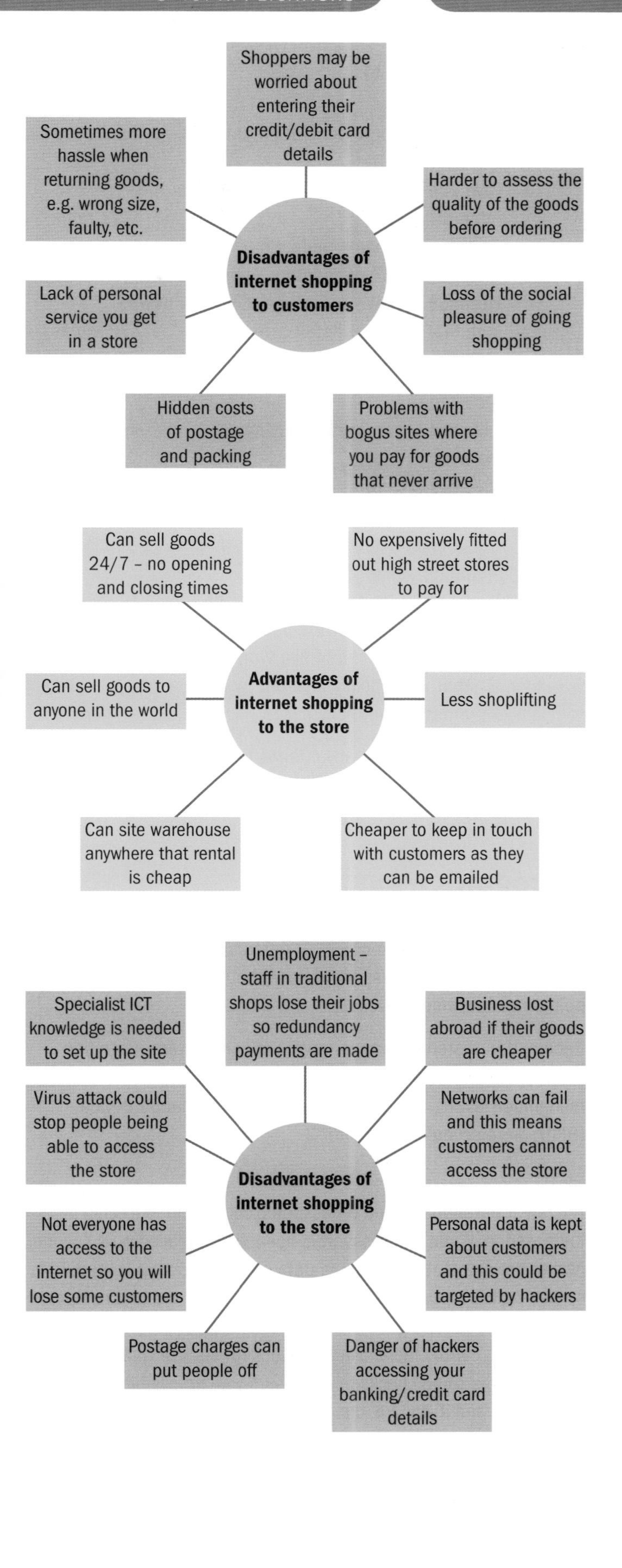

Disadvantages of internet shopping to customers
Shoppers may be worried about entering their credit/debit card details
Sometimes more hassle when returning goods, e.g. wrong size, faulty, etc.
Harder to assess the quality of the goods before ordering
Lack of personal service you get in a store
Loss of the social pleasure of going shopping
Hidden costs of postage and packing
Problems with bogus sites where you pay for goods that never arrive
Advantages of internet shopping to the store
Can sell goods 24/7 – no opening and closing times
No expensively fitted out high street stores to pay for
Can sell goods to anyone in the world
Less shoplifting
Can site warehouse anywhere that rental is cheap
Cheaper to keep in touch with customers as they can be emailed
Disadvantages of internet shopping to the store
Unemployment – staff in traditional shops lose their jobs so redundancy payments are made
Specialist ICT knowledge is needed to set up the site
Business lost abroad if their goods are cheaper
Virus attack could stop people being able to access the store
Networks can fail and this means customers cannot access the store
Not everyone has access to the internet so you will lose some customers
Personal data is kept about customers and this could be targeted by hackers
Postage charges can put people off
Danger of hackers accessing your banking/credit card details

QUESTIONS H

1 When goods are ordered over the internet, payment has to be made.

- **a** Give **one** method of payment used over the internet. *(1 mark)*
- **b** Describe why some people may not want to put their payment details into a website. *(2 marks)*
- **c** Describe **one** way the online store can make sure that payment details are safe. *(2 marks)*

2 Internet shopping has changed the way that most people shop.

- **a** Give **two** advantages of buying goods over the internet. *(2 marks)*
- **b** Give **two** disadvantages of buying goods over the internet. *(2 marks)*
- **c** Some organisations will benefit by the use of internet shopping, while others will not.
  - **i** Give the names of **two** types of organisation that would benefit by the use of internet shopping.
  - **ii** Give the names of **two** types of organisation that would likely lose out by the use of internet shopping. *(4 marks)*

3 Discuss in detail the advantages to shoppers in shopping online. *(6 marks)*

Recognition systems

A recognition system is a computer application that can automatically identify an object or person. The recognition system can automatically process the data without a human operator being present, which saves time and money.

Recognition systems using magnetic ink character recognition (MICR), optical mark recognition (OMR), optical character recognition (OCR), and radio frequency identification device (RFID) were covered in Chapter 2.

RFID used in passport control

The latest passports, called e-passports, contain an RFID chip inside the passport. When the passport is opened and placed on a scanner next to the automatic barrier, the data on the RFID chip is read. The system runs a face-recognition check using a camera and compares this with your photograph encoded on the chip in your passport. The system uses certain facial characteristics to produce a match and then performs some checks to make sure you are eligible to enter the country, and if you are, the gate opens automatically.

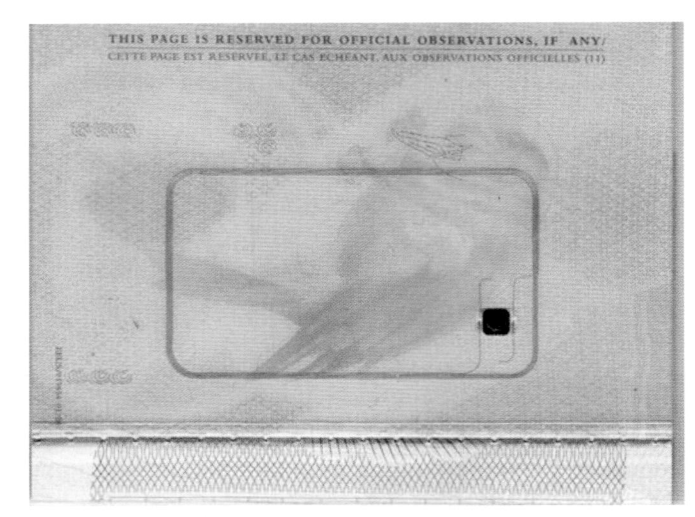

This is an inside page of a passport containing an RFID chip. Notice the aerial (the large rectangle of copper wire), which is used to transmit and receive data stored in an integrated circuit (i.e. chip), which is the small black square in the picture.

RFID used for contactless payments

Contactless payment systems are credit cards and debit cards, key fobs, smartcards, or other devices that use (RFID) radio frequency identification for making secure payments. The embedded chip and antenna enable consumers to wave their card or fob over a reader at the point of sale to make a payment. The payment system is available only for low value payments where it is designed to replace having to carry cash.

There is another method for contactless payments where you use an app on your mobile phone to make a payment. Contactless payments can be made using a special watch.

Monitoring and tracking systems

ICT systems are able to monitor and track people and vehicles. Many people are worried that all these systems can be used to check up on us and erode our privacy. Other people see these systems as a necessary evil as they are an important tool in the fight against crime and terrorism. This section covers the different types of monitoring and tracking systems and their applications.

Number plate recognition systems

Automatic number plate recognition (ANPR) is a method that uses optical character recognition (OCR) on images obtained from video cameras to read car registration plates.

The camera takes the image and then uses OCR software to record the registration plate details which can be used in a number of ways depending on the application. Applications include the following.

Road enforcement:

- Speed cameras recognise the plate of cars speeding and use this to find details of the registered keeper/driver to send them fines.
- Cameras recognise cars parked illegally so fines can be sent.
- Cameras recognise cars whose tax is out of date so fines can be sent.

Car park management:

- Number plates are read by a camera and if allowed into the car park, the barrier is raised.
- Can be used for pre-booked spaces at airport car parks. No ticket is issued. (Tickets can easily be lost.)

Electronic toll collection:

- Car numbers using toll roads, bridges or tunnels are recognised so that payment can be collected from the driver automatically. This saves them having to keep change for the tolls if they are regular users.

Public monitoring/tracking

Your web browsing activity is recorded by your internet service provider (i.e. the organisation that provides your internet connection) and it is possible to identify the computer from which an email or message is sent. It is therefore almost impossible to remain anonymous on the internet.

Most websites use cookies. Cookies are pieces of text that websites put onto your hard drive when you visit them. Internet cookies record details of the websites you have visited and how long you spent on them, what pages were looked at, if you have visited the site before, etc. Many people consider this to be an invasion of their privacy.

It is possible to locate the position of a person using their mobile phone. In fact with some mobile phones it is possible to determine all the places they have been over a period of time.

Worker monitoring/tracking

Some employers keep a constant check on their employees. This monitoring/tracking is usually for commercial purposes but it can be used to check up on their employees which can be used to discipline employees if they have not obeyed company rules.

- For example, supermarket checkouts can monitor and log how many items are put through the checkout each hour by each shop assistant.
- Internet use is monitored by network managers. They can record what sites are visited and how long an employee spent on them. They can therefore spot employees who are spending too much time browsing the internet when they are supposed to be working.
- Emails can be read. This means that employers could read work-related emails as well as personal ones that have been sent during work time.
- Delivery drivers can have their location tracked by their employers using GPS (global positioning systems). Usually they are tracked to determine where they are in case a customer wants to know when a delivery is likely to be made. Taxi firms track their taxis so that the nearest available taxi to a customer can be found.

Key logging

Key logging is the process of someone recording/monitoring the keys you press when you are using your computer using a key logger which can either be hardware or software. As it can record keystrokes, someone can record your passwords and banking details. They can then use the information they collect fraudulently.

Call monitoring/recording

Many employers monitor the calls employees make using phones provided by the company in order to detect whether the system is being abused.

Mobile phone calls are always recorded by the mobile phone companies and kept for a certain period as part of the fight against terrorists and criminals.

Companies record calls in case of a dispute over what was said.

QUESTIONS I

1. Employers often monitor the use of the internet by their staff and they may read their emails and check what they have been looking at.
   - **a** Give **one** reason why staff have their internet use monitored. *(1 mark)*
   - **b** Give **one** reason why staff have their email use monitored. *(1 mark)*
2. The police use automatic number plate recognition systems.
   - **a** Describe briefly how the automatic number plate recognition system works. *(3 marks)*
   - **b** Give **one** advantage that the police have in using this system. *(1 mark)*
   - **c** Give **two** uses by the police of the number plate recognition system. *(2 marks)*
3. Car park management systems often make use of number plate recognition.
   - **a** Give the name of the input device used with this system. *(1 mark)*
   - **b** Explain how the system uses optical character recognition (OCR) to allow a car to enter the car park by raising the barrier. *(3 marks)*
   - **c** Give an advantage of this system compared to having a manned barrier. *(1 mark)*

Satellite systems

Satellite systems are an essential part of communication systems. In this section you will be looking at the uses of some satellite systems.

Global positioning system (GPS)

GPS stands for global positioning system. GPS is a satellite-based system which uses at least four satellites to work out the distance between the receiver and each of the satellites, in order to work out the position (i.e. latitude, longitude, and altitude) of the receiver on the Earth. When the position is combined with a map, the system becomes a satellite navigation system with which most new cars are equipped.

GPS is used by surveyors for marking out plots of land and positions of homes accurately so there can be no future disputes over boundaries.

GPS is used by walkers so they don't get lost.

Geographic information systems (GIS)

A geographic information system, or GIS for short, is an ICT system that is used to capture, manage, analyse, and display geographically referenced information.

What can you use a GIS for? Using a GIS you can:

- Determine how far it is from one place to another
- See a bird's eye view of your house and its surroundings
- Plan the quickest route to school/college
- View the surroundings when you go to a new place or go on holiday
- Look at the surrounding area when you are thinking of buying or renting a property.

Examples of GIS:

- Satellite navigation systems - you can get navigation instructions as you drive, be directed to the nearest petrol station, locate hotels, etc.
- Google Earth.
- Multimap - useful for maps and aerial views.
- The AA - useful for finding route details from one place/ postcode to another.
- The Energy Saving Trust - this site allows you to enter your postcode and it will tell you whether you could use a wind turbine to generate your own electricity.

Satellite navigation systems

Satellite navigation systems make use of GPS and a map to accurately position a car on the road in order to issue traffic directions. These traffic directions are given on screen as well as verbally and can give all sorts of other information such as where there are speed cameras, estimated time of arrival, and alternative routes avoiding, for example, toll roads or motorways.

Advantages of satellite navigation systems

There are many advantages in using satellite navigation systems and here are just some of them:

- Reduces fuel consumption which is therefore greener because you do not get lost.
- You can arrive at your destination without delay, as you can be warned in advance of road-works and so take an alternative route.
- Can save money by choosing the shortest route.
- You can find where the nearest petrol station is.

Disadvantages of satellite navigation systems

Some of the disadvantages of satellite navigation systems include:

- They can send you down very small and winding roads.
- They are sometimes difficult to use.
- They can cause accidents if people start using them while driving.
- Sometimes the information is out-of-date.

Media communications systems

Media communications systems are those systems that use satellite signals rather than terrestrial (i.e. land-based) signals to communicate. They are used by media companies (e.g. TV, newspaper, and radio) to send stories, pictures, and video from remote locations such as up a mountain such as Mount Everest or in a wilderness such as Antarctica.

QUESTIONS J

1 Here are some statements about GPS and you have to decide whether each of them is **True** or **False**.

	True	False
GPS stands for Global Positioning Service.		
GPS uses a single satellite.		
GPS is only accurate to 10 m.		
GPS cannot be used for the navigation of aircraft.		
GPS cannot be used for walkers as it does not work in remote places.		
GPS is used by surveyors for accurately marking out plots of land.		
It is only possible to use GPS in satellite navigation systems.		

(7 marks)

2 **a** Give the meaning of the abbreviation GPS. *(1 mark)*
b Explain what is needed for a GPS system to become a satellite navigation system as used in cars. *(1 mark)*
c Give the name of the input device used with a satellite navigation system. *(1 mark)*
d Describe **three** uses for a GPS system other than for a satellite navigation system. *(3 marks)*

3 Most new cars are equipped with satellite navigation systems.
a Describe the advantages in using a satellite navigation system in a car. *(3 marks)*
b Describe the disadvantages in using a satellite navigation system in a car. *(3 marks)*

4 **a** Explain what GPS is and how it works. *(3 marks)*
b Describe **two** ways in which GPS is used. *(2 marks)*

Test yourself

The following notes summarise this chapter, but they have missing words. Using the words below, copy out and complete sentences **A** to **O**. Each word may be used more than once.

internet **cheaper** **smartphones** **experts**
VoIP **knowledge base** **rules base**
pre-set **expert** **emails** **phone sensors**
question **ATM** **digital** **diagnosis**
analogue-to-digital converter **user interface**

A There has been a huge increase in the way people can communicate with each other owing to the introduction of the ______.

B Use of the internet means that cheap phone calls can be made to anywhere in the world by using a service called ______.

C Mobile communication has been made possible by the use of ______ which can be used to make phone calls, send texts, send ______ and browse the internet.

D An ______ system is an ICT system that uses artificial intelligence to make decisions based on data supplied in the form of answers to questions.

E A ______ is a huge organised set of knowledge about a particular subject used by an expert system.

F The set of rules on which to base decisions on used by expert systems is called the ______.

G The part of an expert system where the user interacts with the system is called the ______.

H Expert systems have the advantage that they are usually built using the knowledge of many ______.

I Expert systems have the advantage that they will not forget to ask a ______ nor will they forget things.

J One application of expert systems is medical ______ where the expert system is used by a less experienced doctor to help make an expert and accurate diagnosis.

K Expert systems have the advantage in that it is much ______ than consulting one or several human experts.

L Banking has changed and it is no longer necessary to visit a branch to perform routine transactions owing to the introduction of ______ and ______ banking.

M If you want to obtain cash from your account out of normal banking hours then you can use an ______.

N Control systems make use of data obtained from ______. As these record quantities which are analogue these need to be converted to ______ data by using an ______.

O In control systems such as an air conditioning system to keep a room at a constant temperature, the reading from a temperature sensor is compared with a ______ value in order to decide what action to take.

REVISION QUESTIONS

1 a Give the name of a household device that uses a control system. [1]
b Explain how the control system controls the device you have named in part **a**. [3]

2 A school is using a traditional paper-based registration system.
The school uses registers to record morning and afternoon attendance details for each student.
a Describe **three** advantages in using ICT systems for registration. [3]
b Describe **three** disadvantages in using ICT systems for registration. [3]

3 Give **two** examples of tasks that are completed by robots. [2]

4 Online banking is very popular with home users of ICT.
a Name and describe **three** services offered by online banking. [3]
b Some people are sceptical about online banking. Describe **two** worries that people might have with online backing. [2]
c Describe **one** way that the banks can address one of the worries you have described in part **b**. [1]

5 Many people use the internet to access booking systems.
By referring to a relevant example, explain how an internet booking system works and the advantages to the home user in being able to book tickets/seats online. [5]

EXAM-STYLE QUESTIONS

1 A floor turtle can use the following instructions:

INSTRUCTION	MEANING
FORWARD *n*	Move *n* mm forward
BACKWARD *n*	Move *n* mm backward
LEFT *t*	Turn left *t* degrees
RIGHT *t*	Turn right *t* degrees
PENUP	Lift the pen
PENDOWN	Lower the pen
REPEAT *n*	Repeat the following instructions *n* times
END REPEAT	Finish the REPEAT loop

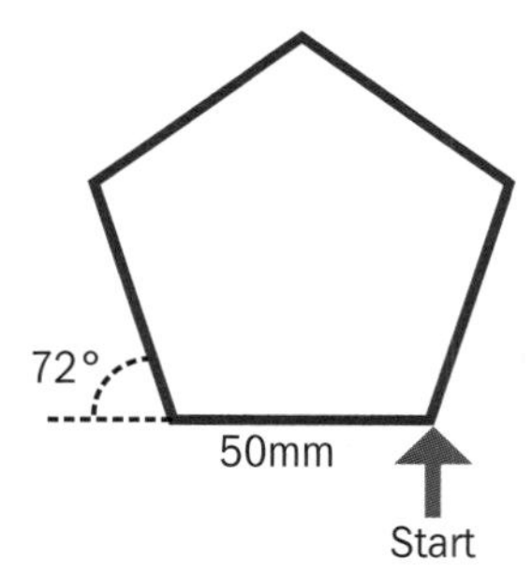

Complete the set of instructions to draw this shape by filling in the blank lines. *[5]*

PENDOWN

.... 90

REPEAT ....

FORWARD ....

.... 72

....

(Cambridge IGCSE Information and Communication Technology 0417/13 q6 Oct/Nov 2010)

2 An oil company is investigating whether they are likely to find oil at a certain site. They will use an expert system to help them. There are a number of inputs and outputs used with such a system. List **four** examples of these. *[4]*

(Cambridge IGCSE Information and Communication Technology 0417/13 q12 Oct/Nov 2010)

3 A car repair centre uses an expert system to help diagnose car engine faults.

a Describe the inputs, outputs and processing of this system. *[6]*

b Give **two** other examples of situations where expert systems might be used. *[2]*

(Cambridge IGCSE Information and Communication Technology 0417/01 q11 Oct/Nov 2009)

4 A computerized greenhouse is used to grow tomatoes.

a Give the names of **three** sensors that would be used to send data to the computer. *[3]*

b *i* Explain the difference between analogue and digital data. *[2]*

ii Explain why it is necessary for analogue to digital conversion when a computer is used to control the conditions in a greenhouse. *[2]*

c Give **three** reasons why a computer rather than people is used to control the conditions in the greenhouse. *[3]*

5 Automatic washing machines use microprocessors for control.

a Sensors send data to the microprocessor for processing. Give the names of **three** sensors that would be used with an automatic washing machine and for each sensor explain its purpose. *[6]*

b Describe how the data from the sensors is used by the microprocessor to control the operation of the automatic washing machine. *[5]*

6 Computers are used in libraries to record the borrowing and return of books.

Two methods of input are used to input details of borrowers and books.

Give the names of two suitable methods and explain the relative advantages and disadvantages of each. *[5]*

7 Patients are continually monitored in the intensive care unit of a hospital.

a Name **three** physical quantities that would be monitored. *[3]*

b Give **three** advantages in using computerized monitoring of patients rather than using nursing staff for monitoring. *[3]*

8 Describe the difference between phone banking and internet banking. *[4]*

9 A central heating system uses a temperature sensor, a microprocessor, and a heater to keep a room warm in the winter.

a Explain why the microprocessor is not able to use the data from the temperature sensor directly and give the name of the extra device that would be needed. *[3]*

b With reference to this central heating system, describe what is meant by a pre-set value and explain how it is used by the system to keep the temperature of the room constant. *[4]*

10 The Geography teacher in a school has set up an automatic weather station connected to a computer.

a Name **three** physical variables which will be measured by the sensors. *[3]*

b *i* Explain why the computer is unable to read the data directly from the sensors. *[2]*

ii Name the device that is needed to enable it to do so. *[1]*

c Give **three** advantages to the Geography department of having a computerised weather system rather than a manual weather system. *[3]*

(Cambridge IGCSE Information and Communication Technology 0417/11 q10 May/June 2014)

11 A company uses robots to manufacture cars.

a Tick **four** advantages to the company of using robots rather than humans to manufacture cars. *[4]*

	✓
Robots are cheap to buy	
Running costs are lower as humans have to be paid wages	
Robots never need maintenance	
Humans cannot work continuously	
Robots can work in hazardous conditions	
There is lower productivity with robots	
Robots produce the same standard of finished product every time	
Humans have greater accuracy than robots	

b Describe **three** tasks that humans will have to do when robots are used to manufacture cars. *[3]*

(Cambridge IGCSE Information and Communication Technology 0417/11 q16 Oct/Nov 2012)

12 A shop owner uses a spreadsheet to calculate his profits. This is part of the spreadsheet.

	A	B	C	D	E	F	G
1	**Producer**	**Food type**	**Number in stock**	**Cost Price**	**Selling Price**	**Profit**	**Total profit**
2	Logekks	Potato flakes	168	$1.80	$1.90	$0.10	$16.80
3	Squarebranch	Chocolate bar	202	$0.75	$0.80	$0.05	$10.10
4	Roofs	Beefburgers	88	$2.05	$2.25	$0.20	$17.60
5	Kapats	Gravy	120	$3.20	$3.45	$0.25	$30.00
6	Startle	Yoghurt cream	122	$1.50	$1.65	$0.15	$18.30
7							
8		**Total in stock**	700		**Overall profit**		$92.80

a Give the cell reference of the cell that contains $3.45. *[1]*

b Give the cell reference of a cell that contains a label. *[1]*

c How many columns are shown in the spreadsheet? *[1]*

d Write down the formula which should go in cell G2. *[1]*

e Formulae similar to that used in cell F2, that is to say = E2 – D2, have been used in cells F3 to F6. These were not typed. Describe how these were entered. *[2]*

f Spreadsheets can be used for modelling. Give **three** reasons why computer models are used rather than the real thing. *[3]*

(Cambridge IGCSE Information and Communication Technology 0417/11 q12 Oct/Nov 2013)

13 Many schools use ICT to record learner performance. Explain **three** ways in which ICT can be used to record learner performance in a school. *[3]*

14 Here are some statements concerning cheque clearing. Tick whether these statements are true or false.

	True	False
Cheque clearing uses magnetic characters which are read automatically using optical mark recognition.		
Cheques are sent to a clearing centre where they are processed.		
It takes over two weeks to clear a cheque.		

[3]

7 The systems life cycle

The systems life cycle is the series of stages that are completed when developing a new system or improving an old one. The stages are carried out in order and this ensures that the system is developed properly. The existing system might be paper-based or, more probably, uses computers but is no longer good enough or up-to-date. In this chapter you will learn about the systems life cycle and the tasks that are completed for each stage.

The key concepts covered in this chapter are:

- Analysis
- Design
- Development and testing
- Implementation
- Documentation
- Evaluation

The purpose of the systems life cycle

It is important that new systems being created are fit for purpose. Systems developed without too much thought tend to have serious faults. To prevent this happening, most systems are developed in a series of set stages completed in a set order. These are called the stages of the systems life cycle.

Starting with Analysis, these stages are shown in the diagram:

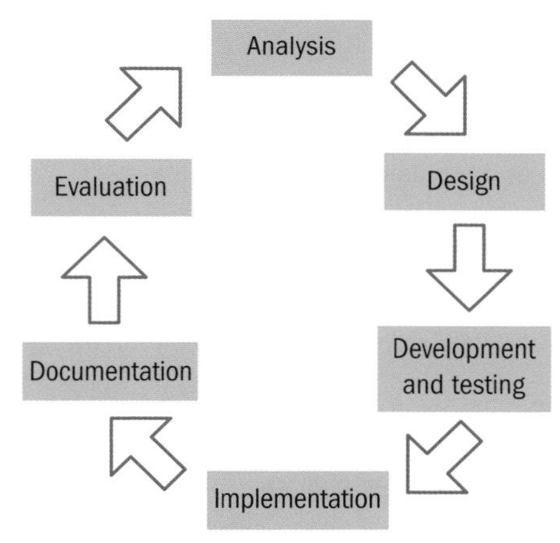

The systems life cycle.

Analysis

Analysis looks in detail at the current system or the requirements for a task that has never been performed before.

Analysis will normally involve the following:

- Identifying the problems that need solving.
- Collecting facts about the old system or the required system using questionnaires, interviews, etc.
- Identifying the inputs, outputs, and processing of the current system, or the new system if there is no existing system.
- Identifying problems with the existing system.
- Identifying the user and information requirements necessary to solve the problem(s).

The different methods of researching a situation

The starting point of analysis involves finding out what people want from the new proposed system or looking at the existing system to find out how it works and might be improved. There are several ways this can be done:

- Using questionnaires
- Using interviews
- Using observation
- Examination of existing documentation.

Using questionnaires

A questionnaire could be given to each user and left with them for completion. Questions should be about how the job is done now and not about the overall running of the business. It could also be about the information the new system needs to give them.

Advantages of using questionnaires

- More honest answers as they can be completed anonymously.
- You do not have to pre-arrange appointments so easier to collect information.
- Much less time-consuming method of getting information from lots of people.
- Questionnaires can be analysed automatically by making use of optical mark recognition (OMR).
- Less time needed to collect the information than by interviews.

Disadvantages of using questionnaires

- People often do not return the questionnaire.
- Questions may be misunderstood, giving wrong answers.
- It is hard to design a questionnaire that will collect all the information needed.
- There is no-one to ask if a question is unclear.

Questionnaires can be used to collect information from lots of people.

Using interviews
Interviews take longer than questionnaires, so this method is good if there are only a few users of the system. People at the different levels in the organisation who will use the new system should be interviewed. At these interviews you can find out how the existing system works and what things are required from the new system.

Advantages of using interviews

- Questions can be explained if they are not understood.
- No need to work to a set script so can ask questions in response to answers to other questions.
- People are likely to take an interview more seriously than a questionnaire.
- Questions can be changed to suit the member of staff being interviewed.

Disadvantages of using interviews

- People may find it intimidating and not give honest answers.
- Very expensive as people need to be taken away from their normal work.
- Person being interviewed cannot remain anonymous.
- The analyst has to conduct them and this makes the technique expensive.
- Very time consuming, especially if lots of people need to be interviewed.

Using observation
Here you sit with someone who is actually doing the job the new system is designed to do. You then see the problems encountered with the old system as well as chat to the user about what the new system must be able to do.

Advantages of using observation

- Can immediately see the processes involved in the system.
- It is possible to gain a more accurate view of the existing system.
- It is fairly inexpensive as you are not taking the person being observed away from their work.
- There is someone to ask about the old and new systems.

Disadvantages of using observation

- Person being observed may work in a different way from normal.
- Person being observed may feel as though they are being spied on.

Examination of existing documentation
This involves looking at any of the paperwork involved with the current system. This would include documents such as order forms, application forms, lists of stock, and so on. You can also look at the records that are kept in filing cabinets.

Advantages of using existing documentation

- Saves time as there may be copies of previous analysis.
- It is easy to understand the data flows through the system.
- Allows the person doing the analysis to determine the size of system needed using the volume of orders, invoices, etc., as a guide.
- Can see existing input and output designs.

Disadvantages of using existing documentation

- It may waste time if the existing documentation is not relevant to the new system.
- It is expensive because of the time it takes the analyst to perform.
- It is very time consuming.

System specification
The system specification is a document which identifies and justifies each of the following:

- Hardware for the new system
- Software for the new system.

Hardware for the new system will include the following:

- Storage requirements of the system – this will influence the choice of storage media, e.g. hard disk, storage on file servers, storage in the cloud, etc.
- Processor – some applications need very fast processors.
- Memory – large amounts of memory results in programs running faster.
- Input devices – for example, a keyboard and mouse are standard but other forms of input could be used such as voice recognition, touch screen, optical mark recognition (OMR), optical character recognition (OCR), etc.
- Output devices – these can include screens, printers, speakers, etc.

Software for the new system will include the following:

- The operating system to be used – this is often determined by the application software. For example, if you want a particular piece of application software it may run only using a particular operating system.
- The application software – this could be obtained off-the-shelf or it could be specially written by programmers.

Design

After the analysis stage, the requirements for the new system should be thoroughly understood and work can start on how best to design the new system.

You would never start building a house without proper designs and plans. ICT systems are no different, as it is much easier to design systems carefully rather than have to change them at a later stage. ICT systems can be very different but it is usually necessary to design documents, files, forms/inputs, reports/outputs and validation methods.

Producing the designs

All ICT systems are different. For example, you might be creating a website or a database. Each must be designed carefully by completing the relevant design tasks, which would include:

- Designing data capture forms (these are forms used for the input of data).
- Designing screen layouts (these are part of the user interface).
- Designing validation routines that help prevent invalid/unreasonable data from being processed.
- Deciding on the best form of verification for the system.
- Designing report layouts (this is the output from the system that is printed).
- Designing screen displays for the output (this is the output from the system that is displayed on the screen).
- Designing the required data/file structures (for example, if a database needs to be produced then the tables will need to be designed).

Designing validation routines

Validation is a check performed by a computer program during data entry. Validation is the process that ensures that data accepted for processing is sensible and reasonable. For example, a living person's date of birth could not be before 1895, because in 2015 this would make them 120 years old (the current oldest person is 115). Validation is performed by the computer program being used and consists of a series of checks called validation checks.

When a developer develops a solution to an ICT problem, they must create checks to reduce the likelihood of incorrect data being processed by the computer. This is done by restricting the user as to what they can enter, or checking that the data obeys certain rules.

Validation checks

Validation checks are used to restrict the user as to the data they can enter. There are many different validation checks, each with their own special use including:

- Boolean checks – data is either: True or False, Y or N.
- Data type checks – these check that data being entered is the same type as the data type specified for the field. This would check to make sure that only numbers are entered into fields specified as numeric, or only letters in a name.
- Presence checks – some database fields have to be filled in, while others can be left empty. A presence check would check to make sure that data had been entered into a field. Unless the user fills in data for these fields, the data will not be processed.
- Length checks – there is a certain number of characters that need to be entered. For example, in one country a driving licence number has a length of 15 characters. Without the correct number of characters, the data would be rejected.
- Consistency checks – checks to see if the data in one field is consistent with the data in another field. For example, if gender is M then there should not be Miss in the title field.
- Range checks – are performed on numbers. They check that a number being entered is within a certain range. For example, all the students in a college are aged over 14, so a date of birth being entered which would give an age less than this would not be allowed by the range check.
- Format checks – are performed on codes to make sure that they conform to the correct combinations of characters. For example, a code for car parts may consist of three numbers followed by a single letter. This can be specified for a field to restrict entered data to this format, e.g. DD/MM/YY for a date field.
- Check digits – are added to important numbers such as bank account numbers, International Standard Book Numbers (ISBNs), etc. Check digits are placed at the end of the block of digits and are used to check that the digits have been entered correctly into the computer. When the large number is entered, the computer performs a calculation using all the digits to work out the check digit. If the calculation reveals that the check digit is the same as that calculated by the other digits, it means that the whole number has been entered correctly.

KEY WORDS

Check digit a number (or alphanumeric character) added to the end of a large number such as an account number for the purpose of detecting the sorts of errors made on data entry. All the digits are used in a calculation which calculates the check digit. This calculated number is then compared with the check digit to see if the two are the same. If so, then the large number has been input correctly.

Range check data validation technique that checks that the data input to a computer is within an allowed range. If the data is outside the range, then it is rejected.

Validation checks checks that a developer of a solution set creates in order to restrict the data that a user can enter, so as to reduce data entry errors. Validation checks check that the data being entered is reasonable.

Verification checking that the data being entered into the ICT system perfectly matches the source of the data. Verification can involve visual checking or double entry of data.

QUESTIONS A

1 a What is meant by a check digit? *(3 marks)*

b Give **two** different examples where check digits are used. *(2 marks)*

2 Here are some dates of birth that are to be entered into an ICT system:

a 12/01/3010

b 01/13/2000

c 30/02/1999

Assume that all the dates are in the British format dd/mm/yyyy. For each one, explain why it cannot be a valid date of birth. *(3 marks)*

3 When an employee joins a company they are given an employee code.

a Here is an example of an employee code: LLLNNNNNN, where L is a letter of the alphabet and N is a number.
Describe **one** type of validation that could be used with this field. *(2 marks)*

b Employees are given an annual salary.
Describe **one** type of validation that could be used with this field. *(2 marks)*

4 A computer manager says, 'data can be valid yet still be incorrect'. By giving **one** suitable example, explain what this statement means. *(3 marks)*

5 An online form for ordering DVDs uses a presence check for some of the fields.

a Describe what a presence check is and why some fields have them while others don't. *(3 marks)*

b Give **one** field that might have a presence check and **one** field that would not need a presence check. *(2 marks)*

6 During the development of a new system it is important to find out how the existing system works (if there is one). The usual way to do this is to carry out a fact search/fact find.

Name and describe **three** different ways of collecting facts about an existing system. *(6 marks)*

Development and testing

Development of the new system means creating the new system following the designs created in the previous stage. The new system will need to be thoroughly tested during and after the development.

Developing the system from the designs and testing it

Creating the data/file structure and testing it

In the case of a database this would involve creating the database tables and entering the field names, the type of each field, the length of the field, etc., and setting up key fields. The links between the tables will need to be created. Test data can then be entered for each field to check that the data can be extracted in the way required.

Creating validation routines and testing them

The validation routines/checks will have been decided in the design stage. These designs will be used with the software to create the checks that the data being entered is reasonable or acceptable. Each validation check will need to be thoroughly tested by entering some test data, which will be discussed later.

Creating input methods and testing them

There are many different ways of inputting data into a system. For example, share prices can be input directly from a website and put into a spreadsheet. Many systems make use of a keyboard to enter data into a form, spreadsheet, etc. As well as creating these methods, they will need to be tested.

Creating output formats and testing them

This involves designing of output such as reports from databases, output on screen, etc. Testing of the output involves checking that the information output is complete and that it produces the correct results.

Testing strategies

Each module of the ICT solution created needs to be thoroughly tested to make sure it works correctly. The modules are tested with data that has been used with the existing system so that the results are known.

Testing strategies often make use of a test plan. A test plan is a detailed list of the tests that are to be conducted on the system, when it has been developed, to check it is working properly. These test plans should be comprehensive. A good way of making sure that they are indeed comprehensive is to make sure that:

- Tests are numbered
- Each test has the data to be used in the test clearly specified
- The reason for the test is stated
- The expected result is stated.

Space should be left for:

- The actual result and/or comment on the result
- A page number reference to where the hard copy evidence can be found.

Testing using normal, abnormal, and extreme data

Testing should always be performed with the following three types of data:

- Normal data – is data that should pass the validation checks and be accepted.
- Abnormal data – is data that is unacceptable and that should be rejected by the validation check. If data is rejected then an error message will need to be displayed explaining why it is being rejected.
- Extreme data – is data on the borderline of what the system will accept. For example, if a range check specifies that a number from one to five is entered (including one and five) then extreme data used would be the numbers one and five.

Here is a test plan to test a spreadsheet for analysing the marks in an examination. A mark is input next to each candidate's name. The mark is a percentage and can be in the range 0% to 100%.

Test no	Test mark entered	Purpose of test	Expected result	Actual result
1	45	Test normal data	Accept	
2	0	Test extreme data	Accept	
3	100	Test extreme data	Accept	
4	123	Test abnormal data	Reject and error message	
5	-3	Test abnormal data	Reject and error message	
6	D	Test abnormal data	Reject and error message	

The 'Actual result' column would be filled in when the test mark was entered. If the expected results and the actual results are all the same then the validation checks are doing their job. If the two do not agree, then the validation checks will need to be modified and re-tested.

Testing using live data

Rather than creating artificial data for the testing of a new system it is sometimes easier to test a system using live data. Live data used for testing is a copy of the existing data files currently being used for the running of the existing system. For example, a copy of the customer file can be used for testing the new system. The advantage in using live data is that test data does not have to be created so the testing can take less time; however, it is a shortcut and is not as thorough as testing using test data.

To test a new system, the live data is input into the system and the results this data produces are recorded. These results can then be compared with results obtained from the previous system to see if there are any discrepancies.

Testing the whole system

Once all the modules have been tested separately and any problems with them solved, the modules can be joined to create the whole system. The whole system is then thoroughly tested and any problems addressed.

Improvements needed as a result of testing

Testing is performed to identify problems with the new system. For example, it may not be possible to produce the output the client has asked for and this may result in the need to change the data/file structures. The test data may reveal that some of the validation routines need to be changed in order to trap certain errors in data. Input or output formats may also need changing.

Implementation

In order to change from one system to another it is necessary to have a way of doing this and this is called a method of system implementation. There are several methods used and these are outlined here.

Methods of implementation

Direct changeover

With direct changeover you simply stop using the old system one day and start using the new system the next day.

Advantages of direct changeover include:

- Fastest method of implementation.
- Benefits are available immediately.
- Only have to pay one set of workers who are working on the new system.

Disadvantages of direct changeover include:

- If the new system fails, you might lose all the data.
- All staff need to be fully trained before the change, which may be hard to time/plan.
- The old system is removed so there is no system to go back to if things go wrong.

Parallel running

This method is used to minimise the risk in introducing a new computer system. Basically, the old computer system is run alongside the new computer system for a period of time until all the people involved with the new system are happy it is working correctly. The old system is then switched off and all the work is done entirely on the new system.

Advantages of parallel running include:

- If the new system fails then no data will be lost.
- It allows time for the staff to be trained gradually.
- You can improve the system before full implementation.

Disadvantage of parallel running:

- It is expensive if paying two sets of workers.

Phased implementation

One module at a time can be implemented in phases until the whole system is implemented.

Advantages of phased implementation include:

- Only need to pay for the work to be done once.
- Training can be gradual as staff need only to train in the module required each time.

- If the new system fails, the company/organisation can still use the old system.
- If the new system fails, you only need to go back to the latest phase and do not have to review the whole system.
- IT staff can deal with problems caused by a module before moving on to new modules.

Disadvantages of phased implementation:

- If there is a problem, then some data may be lost.
- There is a cost in evaluating each phase before implementing the next.
- It is only suitable for systems consisting of separate modules.
- It can take a long time before the whole system is implemented.

Pilot running

This method is ideal for large organisations that have lots of locations or branches where the new system can be used by one branch and then transferred to other branches over time.

Advantages of pilot running:

- The implementation is on a much smaller and more manageable scale.
- There is plenty of time available to train staff.

Disadvantages of pilot running:

- It can take a long time to implement the system across the whole organisation.
- If the system fails in one of the branches, data can be lost.

Which method should be chosen?

- For organisations or departments within organisations which need a quick changeover, direct changeover is best. However, this method is risky as some data could be lost if the system malfunctions but it is also relatively inexpensive because the work is not having to be done twice as with parallel running.
- For organisations or departments within organisations which cannot afford to lose data, parallel running, phased implementation or pilot running can be used. Parallel running offers less risk because the old system can still be used if there are problems, but it takes much longer and is more expensive because everything is done twice.

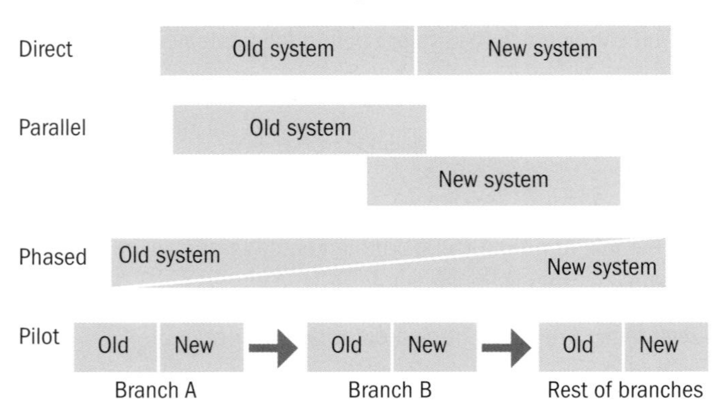

Documentation

Once a new system has been produced it needs to be documented. Two sets of documentation are produced with one set aimed at the user and the other set at the developers who may need to alter the system in the future.

User documentation

User documentation is documentation that the user can turn to for learning a new procedure or for dealing with a problem that has cropped up. User documentation will normally consist of the following:

- The hardware requirements to run the system
- The operating system required
- How to run the program/use the system
- How to log in and log out of the system
- How to perform tasks such as enter data, sort data, search for data, save data, produce printouts, etc.
- Details of sample runs
- Tutorials to help a user become familiar with using the system
- Details of input and output formats (e.g. screen layouts and print layouts)
- Error messages and how to deal with them
- Trouble-shooting guide
- Frequently asked questions/ FAQ/help guide
- Purpose of the system/program
- Limitations of the system.

Technical documentation

Technical documentation is aimed at people who may maintain the new system or develop systems in the future. Technical documentation will normally consist of the following:

- Purpose of the system
- Hardware requirements and software requirements
- A copy of the system design
- Copies of all the diagrams used to represent the system (program flowcharts, system flowcharts, network diagrams, etc.)
- Program listings/program coding
- Lists of variables used
- Details of known bugs
- Sample runs (with test data and results)
- File structures (e.g. structure of database tables, etc.)
- Validation routines used
- User interface designs
- Test plans
- Meaning of error messages
- Limitations of the system.

Evaluation

Evaluation takes place soon after implementation. It is only then that the users and others involved in the development of the system will find out about any problems with the new system.

KEY WORD

Evaluation the act of reviewing what has been achieved, how it was achieved and how well the solution works.

The need for evaluation

Evaluation of a new system will look at each of the following:

- The efficiency of the solution – this looks at how well the system works such as, is it fast? Does it give people the information they require?... and so on.
- The ease of use – is the system easy to use? Has a good user interface been developed? Are training costs minimised because it is so easy to use?
- The appropriateness of the solution – this can involve checking that the original user requirements have been fully met by the new system and assessing how happy the clients are with the development of the new system.

Evaluation strategies

Ways of evaluating a system include:

- Comparing the system developed with the initial requirements.
- Evaluating the users' responses to using the system.
- Comparing the performance of the new system with that of the old system.
- Identifying any limitations that need to be made to the system.
- Identifying any improvements that need to be made to the system.
- Interviewing users to gain an insight into problems with the system.
- Seeing if cost efficiencies have been met (e.g. a person getting through more work than with the old system, etc.).

Once they are developed, systems may still need to be changed in some way. These changes can be changes in hardware or software. Here are some of the changes that may need to be made:

- Changes in the way the business or organisation operates may require that the system be altered. For example, a change in the rate of sales tax or changes in income tax could mean that software will need altering.
- Users of the new system may ask for extra functions to be added. The existing software will need to be changed to do this.
- Users may report system crashes which need to be investigated and are often solved by changes to the software.
- Poor performance or bugs in the software will be reported during evaluation meetings so they can be corrected.
- Slow performance may be reported and corrected by changes to hardware.

Revision Tip

You may be asked to discuss the different implementation strategies for the organisation mentioned in a question. Remember that you need to discuss this in relation to this organisation and not a general discussion that could apply to any business. Also, remember that a 'discuss' question needs to be answered in complete sentences. When you do this, try to ensure that each sentence contains one or more points.

QUESTIONS B

1. Here is a list of the steps that are stages of the systems life cycle. At present these are in the wrong order. Put the steps in the correct order. *(5 marks)*

   Documentation
   Evaluation
   Analysis
   Development and testing
   Implementation
   Design

2. Here are some of the steps that are stages in the systems life cycle.

   Development and testing
   Implementation
   Design
   Analysis
   Evaluation

   Write down the name of a step from the list above where the following tasks would be carried out:

   a Planning the construction of the new system. *(1 mark)*
   b Planning the testing of the new system. *(1 mark)*
   c Getting the user to answer a questionnaire to find out what is required from the new system. *(1 mark)*
   d Asking users what they think of the new system that has been developed. *(1 mark)*
   e Putting data into the computer to check if the output is what was expected. *(1 mark)*

3. When one ICT system is being replaced by another there are a number of different ways of doing this. One way is direct changeover.

   a Describe the system strategy for implementing systems called direct changeover. *(2 marks)*
   b Give **one** advantage and **one** disadvantage of direct changeover. *(2 marks)*

Activity 7.1

Produce a mind map (either hand drawn or produced using mind mapping software) that can be used to summarise what is involved in each of the system implementation strategies shown here. Your mind map should also show the advantages and disadvantages for each method:

- Direct changeover
- Phased implementation
- Parallel running
- Pilot running.

Test yourself

The following notes summarise this chapter, but they have missing words. Using the words below, copy out and complete sentences **A** to **K**. Each word may be used more than once.

development and testing **normal** **input** **evaluation** **documented** **training** **implementation** **whole** **direct changeover** **facts** **parallel running**

A During the analysis stage, the person developing the system will collect ______ about the system using techniques such as questionnaires, observation and examination of existing documents.

B During the design stage the ______, processes, and output will be designed.

C With the design stage complete, the working version of the solution can be produced. This is called the ______ stage.

D Testing involves testing each module with ______, abnormal, and extreme data.

E Testing also involves testing the ______ system when all the modules are put together to form the whole system.

F ______ involves stopping using the old system and starting to use the new system.

G There are four main ways of implementing a system called ______, parallel running, pilot running, and phased implementation.

H ______ involves the old ICT system being run alongside the new ICT system for a period of time until all the people involved with the new system are happy it is working correctly.

I Users need to know how to use the new system. The processing of helping them to understand the new system is called user ______.

J Once a new system has been developed it will need to be ______ and there are two types called user documentation and technical documentation.

K A review of the development of the project is completed at the end and this is called ______.

REVISION QUESTIONS

1 A person's date of birth is entered into a database. State **three** things the validation program could check regarding this date as part of the validation. [3]

2 When a new member joins a fitness club they are given a membership number. The membership number is made up in the following way:

- Date of birth as six numbers.
- The final two figures of the year in which the customer joins the fitness club.
- A letter which is either J or S depending on whether they are a junior or senior member.

a Write down the membership number for a junior member who joined the club in 2010 and was born on 21/05/98. [1]

b When the membership number is entered into the database, it is validated. Describe what is meant by data validation. [2]

c Two examples of data validation are:

- Range check
- Format check.

Describe how these two methods could be used on the membership number field described above. [2]

3 State the names of **three** different methods of implementation from one ICT system to another and describe an advantage for each method. [6]

EXAM-STYLE QUESTIONS

1 Identify **three** methods which could be used to implement a new system. *[3]*

(Cambridge IGCSE Information and Communication Technology 0417/13 q17 Oct/Nov 2010)

2 When a new system is implemented, documentation is provided with it. Identify **four** items which would be found in technical documentation but **not** in user documentation. *[4]*

(Cambridge IGCSE Information and Communication Technology 0417/13 q18 Oct/Nov 2010)

3 Joan owns a small company. She wishes to replace the existing computerised system with a new one. She has employed a systems analyst, Jasvir, to plan this.

a Before Jasvir decides on a system he must collect information about the existing system. Tick whether the following statements about the various methods of information collection are **TRUE** or **FALSE**.

	TRUE	FALSE
Examining documents has to be done in the presence of all the workers	☐	☐
Appointments have to be made with a worker in order to complete a questionnaire	☐	☐
It is possible to change questions in the course of an interview	☐	☐
Observing the current system can provide a detailed view of the workings of the system	☐	☐

[4]

b After Jasvir has completed the analysis of the existing system, he will need to design the new system. Tick **four** items which would need to be designed.

- Inputs to the current system ☐
- User and information requirements ☐
- Data capture forms ☐
- Validation routines ☐
- Problems with the current system ☐
- File structure ☐
- Report layouts ☐
- Limitations of the system ☐

[4]

Cambridge IGCSE Information and Communication Technology 0417/11 q14 May/June 2010

4 A systems analyst has been asked by a librarian to develop a computer system to store information about books and borrowers. After the existing system is analysed the new system will be designed. The first item to be designed will be the input screen.

a Name **four** items of data about **one** borrower, apart from the number of books borrowed, that would be input using this screen. *[4]*

b Describe **four** features of a well-designed input screen. *[4]*

c The librarian will need to type in data about each book from existing records. In order to prevent typing errors the data will be verified. Describe **two** methods of verification which could be used. *[4]*

d After the system is designed it will need to be implemented and then tested.
No borrower can take out more than 6 books.
Describe the **three** types of test data that can be used, using a number of books as an example for each. *[6]*

e The system must now be evaluated. Tick **three** reasons why this is done.

	✓
Improvements can be made.	
The hardware and software can be specified.	
Limitations of the system can be identified.	
To see how many books are required.	
To make sure the user is satisfied with the system.	
So that program coding can be written	

[3]

f After the system is implemented the librarian will be given technical documentation and user documentation. Name **three** different components of each type of documentation (Technical/User). *[6]*

(Cambridge IGCSE Information and Communication Technology 0417/01 q9 Oct/Nov 2009)

5 A new system has been developed in a school: 25 multiple choice questions are being marked for each student using optical mark recognition. The mark is then processed so that the teacher can see where revision needs to be focused.

The system is to be tested using test data. The maximum mark for the test is 25. Test data can be normal, abnormal or extreme. Place a tick if the data items shown in the table are examples of **Normal**, **Abnormal** or **Extreme** data.

	Normal	Abnormal	Extreme
25			
eighteen			
15			
30			
-5			
0			
10			

[7]

6 Tick whether each of the following tasks is carried out in the Analysis or Evaluation phase of the systems life cycle.

	Analysis	Evaluation
Carrying out research on the current system		
Comparing the solution with the original task requirements		
Producing a system specification		
Identifying any limitations of the solution		
Justifying suitable hardware and software for the new system		

[5]

7 A new computerized system is being developed for a school library.

A screen input form is being developed which will be used for entering the details of each book into the system in turn. The librarian will input the details into the Books file for each book using this form.

Design a screen input form that can be used for the input of book details.

Your form should have at least 8 appropriate fields with enough space for their entry.

There should also be some navigation included on the form to make it easy to use. *[10]*

8 A company is going to replace its existing computer system with a new one.

Put the following steps into the correct order that the systems analyst would follow:

Implement the new system.

Develop the new system.

Collect information about the existing system.

Evaluate the new system.

Design a file structure. *[5]*

(Cambridge IGCSE Information and Communication Technology 0417/11 q11 May/June 2014)

9 Mario has asked Louise, a systems analyst, to create a new database system for keeping records of books he sells in his bookshop.

a Louise will collect information about the existing system.

Describe **three** methods she would use to do this. *[3]*

After collecting information, Louise noticed that Mario sells both non-fiction and fiction books in hardback and paperback. She also discovered that no books cost more than $20.
She wrote down some of the questions that customers ask, such as:
Have you got any non-fiction books by Arthur C Clarke?
Have you got the hardback version of 'Harry Potter and the Philosopher's Stone'?
Have you got any books for less than $10?

b Complete the design table below filling in the field names and **most** appropriate validation checks to create a database which would answer these questions.

Field name	Validation Check
	none
	none
Price	

[7]

c Identify **three** items of test data which could be used with the Price field giving reasons for your choice. *[6]*

(Cambridge IGCSE Information and Communication Technology 0417/11 q14 Oct/Nov 2012)

10 Iqbal wants to test the new computerised payroll system he would like to introduce to his company. No company worker is paid less than $100 and no worker is paid more than $500.

Explain what is meant by the following three types of test data using examples of the wages paid to workers.

Normal

Abnormal

Extreme *[6]*

(Cambridge IGCSE Information and Communication Technology 0417/11 q16 May/Jun 2013)

Safety and security

This chapter looks at safety and security when using computer systems. There are a number of ways in which the use of computers can physically harm you and steps should be taken to reduce or eliminate these dangers. When you are using the internet you may encounter people you don't know and these people could harm you. Again, you must be able to recognise the dangers and take measures to protect yourself. As well as ensuring your safety, you must also ensure the safety of your computer system to protect it from hackers, who may try to steal your identity, and from viruses, which can cause you to lose files.

The key concepts covered in this chapter are:
- Physical safety
- e-safety

Physical safety

Physical safety is about ensuring that you do not come to any harm while working with computer equipment.

Physical safety issues and measures for preventing accidents

There are a number of physical safety issues related to using computers and these include the following:

- Overheating
  Causes – computers give out large amounts of heat and rooms containing them can become unbearably hot in the summer.
  Prevention – install air-conditioning or ensure adequate ventilation.
- Fire
  Cause – ICT equipment uses a lot of power sockets and if multi-sockets are used then it is easy to overload the mains circuit. This is dangerous and could cause a fire.
  Prevention – computer rooms should be wired specially with plenty of sockets. Fire extinguishers should be provided and many large organisations use sprinkler systems that activate automatically. Smoke detectors should also be used.
- Tripping over cables
  Cause – trailing wires present a tripping hazard.
  Prevention – the cables need to be managed by sinking into floors or by using floor covering, or attaching to walls.
- Electrocution
  Cause – faulty equipment, users tampering with the inside of computers or users spilling drinks over computer equipment.
  Prevention – any malfunctioning equipment must not be used and should be reported to the technician. Drinks should be kept away from computers.
- Heavy equipment falling
  Cause – equipment not positioned properly.
  Prevention – ensure that the equipment has enough space to be safely positioned. Use furniture designed to do the job.

Evaluating your own use of IT equipment and developing strategies to minimise the potential dangers.

When using computers you need to be checking regularly that you are working safely by:

- Checking to ensure there are no trailing wires that could cause a tripping hazard.
- Ensuring that you and others do not drink near computers and risk electric shocks/electrocution.
- Checking that cooling vents on computers, printers etc. are not covered by paper, clothing etc. which could cause overheating with the possibility of fires.
- Ensure that any faulty equipment is removed immediately from use.
- Making sure that too many plugs are not plugged into a multi-socket which could cause overloading and a fire hazard.
- Taking regular breaks so that you don't risk some of the health problems associated with using computers for lengthy periods without breaks such as eye strain, back ache and RSI (Repetitive Strain Injury).
- Assessing the ergonomic way you are working by checking you comply with all the health and safety regulations and advice about chairs, screens, seating position, lighting etc. The relevant information about this can be found on health and safety websites.

Personal data

Personal data is data about a living identifiable person such as their date of birth, the name of the school they attend, their religion, details of medical conditions they may have, banking details and so on.

How to avoid inappropriate disclosure of personal data

Sometimes you have to disclose personal data (e.g. at a doctors, when filling in details for holiday insurance etc.). You should not disclose personal information to strangers or people who you do not trust as they could use these details to cause you harm.

QUESTIONS A

1 Tick whether each of the following problems concerning health and safety when using computer equipment or working in computer rooms is a health problem or a safety problem.

	Health	Safety
Backache caused by incorrect posture while using a computer.		
Overloaded multi-sockets being used.		
Tripping over a trailing wire.		
RSI caused by using a mouse for long periods.		
Electric shock caused by spilling a drink over a computer.		
Headaches caused by glare on the screen.		
Eye strain caused by looking at the computer screen for long periods.		

(7 marks)

2 The head teacher of a school wants to make working in computer rooms as safe as possible. A number of safety risks have been identified. For each one of the following risks, describe a measure which the head teacher can put in place so as to reduce the danger.
   a Tripping *(1 mark)*
   b Electrocution *(1 mark)*
   c Fire *(1 mark)*
   d Heavy equipment falling *(1 mark)*

3 Working in ICT exposes you to a number of health and safety risks.
   a Explain the difference between a **health** risk and a **safety** risk. *(2 marks)*
   b Describe **two** safety risks that you are likely to encounter in a computer room and explain what can be done to help prevent the risks. *(4 marks)*

The details you should not disclose to strangers include the following:

- Own name
- Address
- School name
- A picture of yourself in school uniform

You must ensure you do not give these details in chat rooms or on social media sites or gaming sites as these details can be used to identify you.

Evaluating your own use of the internet and using strategies to minimise potential dangers

You need to consider how you can use the internet so as not to expose yourself to e-safety issues such as:

- Accessing web pages with offensive, violent or undesirable content.
- Accessing sites promoting inappropriate behaviours such as eating disorders and drug use.
- Accessing sites which are known to be a source of viruses.

The strategies you can use to minimise these e-safety issues include:

- Ensure that parental controls are used to restrict access to certain sites.
- Use only those sites that are recommended by teachers.
- Only use a learner-friendly search engine.

Evaluating your own use of email and using strategies to minimise potential dangers

You need to consider how you use email and how you might be exposing yourself to the following e-safety issues:

- Revealing personal data in emails to people you do not know or trust.
- Looking at the content of emails from people you do not know or trust.
- Opening file attachments from people you do not know or trust.
- Including details such as school name or a picture of you in school uniform.

The strategies you can use to minimise these e-safety issues include:

- Not including personal data in any email.
- Not opening emails from people you do not know.
- Not opening file attachments from people you do not know as they may be offensive pictures or viruses that can infect your computer.
- Not including details of your school or pictures of yourself in school uniform included in or attached to emails.

Evaluating your own use of social media/networking sites, instant messaging, internet chat rooms and using strategies to minimise potential dangers

You need to consider how you use social media/networking sites, instant messaging and internet chat rooms and how you might be exposing yourself to the following e-safety issues:

- Following links which lead you to undesirable content or fake websites designed to steal personal data or money.
- Sharing personal information from which you can be identified.
- Downloading music, films etc illegally.
- Not keeping your parents aware of what you are doing online.

The strategies you can use to minimise these e-safety issues include:

- Know how to block and report unwanted users.
- Never arranging to meet anyone on your own and tell an adult and meet in a public place.
- Never include images that are in any way rude as these could be used to threaten you in the future.
- Only use appropriate language (i.e. no swearing that can offend).
- Respect confidentiality. If someone tells you something private then you should not repeat it to others.

e-safety

e-safety is about using the internet in a safe and responsible way. There is an unpleasant/dangerous side to the internet and this leaves individuals vulnerable to accidental or deliberate harm. e-safety is needed so that:

- you are kept safe from physical harm
- you are not frightened or worried by bullying or stalking
- you are not exposed to offensive or violent images/material
- you are not tricked into committing a crime such as downloading copyrighted material illegally

There are a number of precautions to take when using the internet and these include the following:

- Never arrange to meet an online 'friend' on your own – always take a responsible adult with you.
- Never reveal personal information such as your name, address, school name, photograph, or any other information from which you can be identified.
- Use only those websites that are recommended by teachers or your parents.
- Parents should supervise internet use where possible.
- Do not open an email from an unknown person.
- Do not open email attachments from strangers.
- Send emails only to people you know.
- Do not use your real name when playing games online.
- Use a search engine which has a parental guidance setting which your parents can set so that it filters out any unsuitable content.
- Do not email a picture of yourself in school uniform or give out the name of your school.
- Know how to block and report unwanted users in chat rooms.

Precautions to take when playing games on the internet

Many computer games are played with others over the internet. Most people use a fictitious name when playing games so you are unaware of who they really are. There are a number measures you can take to protect yourself when playing online games and these include the following.

- Don't reveal personal information such as your name, addresses and passwords to anyone when playing online games.
- Playing online games if you have anti-virus/antispyware software included on your computer. This will help protect against identity theft.
- Always buy new games and downloadable games from reputable sources as illegal copies of games are often sources of computer viruses.

How to recognise when someone is attempting to obtain personal data

There are a number of ways you can recognise that someone is attempting to obtain personal data from you, and these include the following:

- They may send an email saying that your inbox is full or that your email account will be suspended unless you validate your account by sending some personal details. If you click on the link in the email you will be taken to a fake site where you could be tricked into entering personal information.
- You may receive an SMS message (i.e. a text) containing a web address (i.e. a URL) or a telephone number to ring and if you reply you will be asked for personal details which can be used to commit fraud. This is referred to as smishing.
- You may be asked to enter payment details into a fake website. The fake website looks similar to a genuine website except not all the links work and a secure link which starts with https is not included in the URL (i.e. web address). Also look for the padlock symbol showing that the site is a trusted site.

When you find any of the above you should report them usually to the internet service provider or the organisation they are using for their internet connection.

To avoid the disclosure of personal information you should not:

- Enter personal details in response to emails or SMS messages.
- Follow links from emails and then enter personal details.
- Enter personal details into sites which do not start with https in the URL.
- Give out personal details to people you do not know or trust.

Security of data

Hacking and its effects

Hacking is the process of accessing a computer system without permission. Hackers often access a computer system using the internet. In most cases the hacker will try to guess the user's password or obtain the password another way.

Some hackers just hack to see if they can enter a secure system while others do it to commit fraud by making alterations to the data, etc. In most countries hacking is a crime but it is quite hard to prove as it is necessary for it to be proved that the hacker deliberately hacked into the system.

Sometimes special software is installed on a user's computer without their permission and it can log the keys that the user presses, which can tell a hacker about usernames, passwords, emails, etc. This software is called spyware or key-logging software.

The effects of hacking

Once a hacker has gained access to a computer system there are a number of things they might do, including:

- Stealing credit card and bank account details for fraud or identity theft.
- Stealing customer information or other important business data.
- Stealing passwords for access to your internet service provider. This means they will be able to use your internet connection for free.
- Stealing your email addresses which can then be used for spamming.
- Using your computer for the sending of spam without your knowledge.

Use of firewalls

Firewalls were introduced on page 37. Firewalls work like a filter between your computer/network and the internet and you can control what comes in and what goes out. A firewall blocks other computers from accessing your computer/network using the internet. Without a firewall someone could hack your computer and access the data held or load a malicious program that can be used to steal passwords, personal data including banking details. In addition to keeping out hackers, here are a few of the other things firewalls do:

- They can be programmed to prevent access to certain websites such as social networking sites.
- The firewall can be used to block access to certain undesirable sites.
- Firewalls can restrict searches so that searches made using certain words are not allowed.

Authentication techniques

Authentication techniques are those techniques that are used to check that a person accessing a network or communications system is the genuine person. They can also be used to ensure that an email sent by a person is genuinely from them and not somebody pretending to be them.

The main authentication techniques are:

- Usernames and passwords
- Digital signatures
- Biometrics.

Identifying the user to the system: usernames

A username is a series of characters that is used to identify a user to the network. A username must be unique, meaning that no two users will have the same username. The person who looks after the network will use this to allocate space on the network for the user. It is also used by the network to give the user access to certain files. The network manager can also keep track of what files the user is using for security reasons.

Preventing unauthorised access to the system: the use of passwords

A password is a string of characters (letters, numbers, and punctuation marks) that the user selects. Only the user will know what the password is. When the user enters the password, it will not be shown on the screen. Only upon entry of the correct password will the user be allowed access to the network.

Characteristics of good passwords

A good password that will keep out hackers should have the following characteristics:

- Do not use a word that can be found in a dictionary.
- Do not use your own or any other name or surname even if you put numbers after it.
- Include numerals as well as letters.
- Include a mixture of upper and lower case letters but try not to make the first letter a capital letter.
- If the password system allows, put other characters in your password like £, &, %, $, @, etc.

When using passwords:

- Do not write your password down.
- Change your password regularly (usually the system will prompt you to do this automatically).
- Do not tell anyone else your password.
- Do not use your user-ID as your password.
- Never respond to an email that asks you for your password.

Digital signatures

Ordinary signatures can be used to check the authenticity of a document. By comparing a signature you can determine whether a document is authentic. When conducting business over the internet you also need to be sure that emails and electronic documents are authentic. Digital signatures are a method used with emails to ensure that they are actually from the person they say they are from and not from someone pretending to be them.

KEY WORD

Digital signature a way of ensuring that an email or document sent electronically is authentic. It can be used to detect a forged document.

Authentication using biometric data

Many systems use biometric data to authenticate someone as an allowable user of a computer system. These biometric methods include:

- Fingerprint recognition
- Retinal scans
- Iris scans
- Face scans.

Biometric methods for the authentication of a user have the following advantages:

- There is nothing to forget such as a key fob, card, etc.
- It is almost impossible to forge a fingerprint, face, etc.
- It is easy for someone else to access a card, fob, etc., but with a biometric method, the person allowed access has to be present.

The security of data online

Online services such as email, banking, shopping, etc., all need to be used with care and there are a number of ways in which we can protect against malicious actions when we use these services. This section looks at the problems and ways in which we can ensure the security of our online data.

Digital certificates

A digital certificate is an attachment to an electronic message used for security purposes. A digital certificate verifies that a user sending a message is who he or she claims to be and it also provides the receiver with the means to encode a reply.

Secure Socket Layer (SSL)

SSL is a standard used for security for transactions made using the internet. SSL allows an encrypted link to be set up between two computers connected using the internet and it protects the communication from being intercepted and allows your computer to identify the website it is communicating with.

The encryption prevents your details, such as credit card and banking details, from being stolen and used maliciously.

Using most browser software, you can tell that you are on a SSL on a secure server as the web address begins with 'https' and there is a padlock shown on the page.

Phishing, pharming and smishing

There are a number of methods fraudsters will use to try to trick you into revealing personal information that may include credit card and banking details, and these methods are outlined here.

Phishing

Phishing is fraudulently trying to get people to reveal usernames, passwords, credit card details, account numbers, etc., by sending them emails that appear to be from a bank, building society, credit card company, etc. The message in the email typically claims that there has been a problem with your account and it asks you to update (reveal) information such as passwords, account details, etc. Under no circumstances should you reveal this information. If you do then these details will be used to commit fraud.

Pharming

Pharming is where malicious programming code is stored on a computer. Any users who try to access a website which has been stored on the computer will be re-directed automatically by the malicious code to a bogus website and not the website they wanted. The website they are directed to looks almost the same as the genuine website. In many cases the bogus website is used to obtain passwords or banking details so that these can be used fraudulently.

Pharming is similar to phishing but instead of relying on users clicking on a link in a fake email message, pharming re-directs victims to a fraudulent website even if they type the correct web address of their bank or credit card company into their web browser.

Smishing

Smishing is a combination of the terms SMS and phishing and it uses text messaging (SMS) rather than email to send fraudulent messages that try to steal your credit card, banking or other personal details. Some of these messages may say you have won a prize or that your account details need updating and if you click on the link you will be taken to a fraudulent website where you will be asked to enter your personal details. If you enter your details then it is likely that your details would be used to commit fraud.

Moderated and un-moderated forums and the security offered by them

An online forum is a discussion on the internet where people can join the discussion and add their comments.

Moderated forums have a person called a moderator who checks the comments before posting them on the forum. The moderator will check that a post on the forum is not spam, is not rude or offensive and does not wander off the topic being discussed.

Un-moderated forums allow people to post whatever they want on the forum. This means you can get rude and offensive messages. Some of these forums may be used by people pretending they are someone they are not and they may use the forum maliciously. They may use the forum to try to obtain personal details about you.

There are dangers lurking in un-moderated forums as they are not policed by a moderator and they should be avoided.

Spam

Spam is email that is sent automatically to multiple recipients. These emails are unasked for and in the main about things you are not interested in. Spam is also called junk email and in most countries, the sending of spam is not illegal despite organisations wasting huge amounts of time getting rid of it.

There are a number of problems with spam:

- It can take time looking at the email names of spam before deleting it.
- Spam can sometimes be used to distribute viruses.

People who collect email addresses for the purpose of sending spam are called spammers. They collect email addresses from chat rooms, mailing lists, social networking sites, etc.

Spam filters are software used to filter out spam from legitimate email. The problem is that some spam gets through and some legitimate email ends up as spam. This means you cannot just delete all the emails in the spam folder in one go.

KEY WORDS

Phishing tricking people into revealing their banking or credit card details using an email.
Spam unsolicited bulk email (i.e. email from people you do not know, sent to everyone in the hope that a small percentage may purchase the goods or services on offer).

Activity 8.1

Identity theft is a real problem for all computer users. It puts off many people from banking online or buying goods and services over the internet where they have to input their card details.

Perform some research using the internet so you can write an article in your own words entitled 'What we can all do to protect ourselves from identity theft'.

You will need to mention the following in your article along with your own ideas of what you feel the readers should know:

- What identity theft is
- The consequences of identity theft
- Phishing
- Pharming
- Smishing
- Un-moderated forums
- What you can do to ensure that you are not the next victim.

Recognizing spam email and avoiding being drawn into it

It is important to be able to recognise spam from legitimate email. Here are some of the ways you can identify spam:

- Look at the email address. If contains strings of alphanumeric characters before the @ symbol, then it is probably spam.
- Look at the email address to see if you are familiar with it or it has the person's name followed by @name of organisation.com. For example, stephendoyle@oup.com
- If there is a link in the body of the message where you have to click on a link to correct something (e.g. keep your continued use of a service) then it is likely to be spam.
- Asking you to do something immediately or quickly indicates spam. This is because the email you will be replying to is changed after a couple of days so the spammers do not get caught.

QUESTIONS B

1. a Explain the meaning of the term spam. *(1 mark)*
   b Describe **two** ways in which spam causes annoyance to computer users. *(2 marks)*
   c Give **one** method that can be used to reduce the amount of spam. *(1 mark)*
   d Give **two** ways of identifying a spam email. *(2 marks)*

2. Tick **True** or **False** next to each statement.

	True	False
Unsolicited email is called spam.		
Pharming uses SMS to collect card details.		
Pharming uses programming code which has been put on the user's computer without their knowledge.		
Spam is fake messages trying to steal your personal details.		

(4 marks)

3. Discuss the different ways of authenticating a computer user. *(5 marks)*

4. A parent is worried about the e-safety of their young child when they are using the internet.
   a What is meant by the term e-safety? *(2 marks)*
   b Describe **four** actions the parent can take to improve the e-safety of their child's computer use. *(4 marks)*

5. Many internet users participate in forums.
   a Explain what is meant by a forum. *(1 mark)*
   b Forums can be moderated or un-moderated.
      i Explain the difference between moderated and un-moderated forums. *(2 marks)*
      ii There are more dangers when using un-moderated forums. Explain why. *(2 marks)*

- Spelling and grammatical errors often indicate spam. Often they originate from different countries so they are not as familiar with spelling and grammar.
- Look for emails that say that they are from your bank, credit card company, internet service provider, etc. If they were genuine, they would never ask for your account or other personal details including your PIN.
- Genuine emails would address you by your name or have the first or last few digits of your credit card number or account number. If the email addresses you generically (e.g. dear member, dear customer etc.), then it is likely to be spam.

Encryption

Because so many people use a network it is important to ensure that the system is secure. If there is access to the internet via the network, then there needs to be protection against hackers. These hackers could simply intercept and read email or they could alter data or collect personal details and credit card numbers to commit fraud.

Encryption is used to protect data from prying eyes by scrambling data as it travels over the internet. Encryption is also used when saving personal data onto laptops or removable storage devices. If any of these gets lost or stolen then the data cannot be understood. Only the authorised person being sent the information will have the decryption key that allows them to unscramble the information.

Encryption should be used for the following:

- Sending credit card details such as card numbers, expiry dates, etc., over the internet.
- Online banking.
- Sending payment details (bank details such as sort code numbers, account numbers, etc.).
- Confidential emails (e.g. with personal or medical details).
- Sending data between computers on a network where confidentiality is essential (e.g. legal or medical).
- Storing sensitive personal information on laptops and portable devices and media.

KEY WORD

Encryption the process of scrambling files before they are sent over a network to protect them from hackers. Also, the process of scrambling files stored on a computer so that if the computer is stolen, the files cannot be read.

Computer viruses

Viruses pose a major threat to ICT systems. Once a computer or media has a virus copied onto it, it is said to be infected. Most viruses are designed to do something apart from copying themselves. For example, they can:

- Display annoying messages on the screen
- Delete programs or data

- Use up resources, making your computer run slower
- Spy on your online use – for example, they can collect usernames and passwords, and card numbers used to make online purchases.

One of the main problems is that viruses are being created all the time and when you get a virus infection it is not always clear what it will do to the ICT system. Apart from the damage many viruses cause, one of the problems with viruses is the amount of time that needs to be spent sorting out the problems that they create. All computers should be equipped with anti-virus software, which is able to detect and delete the majority of these viruses.

Preventing a virus attack

To prevent a virus attack you should:

- Install anti-virus software
- Not open file attachments to emails unless you know who they are from
- Not allow anyone to attach portable drives or memory sticks to your computer unless they are scanned for viruses first
- Not download games and other software from a site on the internet unless it is a trusted site.

Anti-virus software should be kept up-to-date and scans should be scheduled so that they are performed automatically on a regular basis.

KEY WORDS

Virus a program that replicates (copies) itself automatically and can cause harm by copying files, deleting files or corrupting files.

Anti-virus software a program that detects and removes viruses and prevents them entering a system.

The dangers of using a credit card online

Credit cards are designed to make purchasing online easy. The trouble is possession of your credit card details along with some personal information about you can make it easy for criminals to make purchases using your card. Using this information they can also apply for new credit cards or loans without your knowledge.

Here are some of the ways fraudsters can obtain your credit card and personal details:

- The use of key logging software. Key logging software could be put on your computer without your knowledge and this records your keystrokes and allows a fraudster to obtain your passwords, credit card details, and personal information. Once they have enough information they can steal your identity and commit fraud.
- The use of bogus sites. Bogus sites are similar to the real sites except their purpose is to steal your credit card and personal information when you make purchases.
- The use of pharming, phishing, and smishing to trick you to reveal credit card and personal details in response to messages you receive on your computer or mobile phone.
- Hacking into secure sites to obtain the details.

Here are some precautions you can take to help prevent fraudulent use of credit cards online.

- Do not follow a link and then enter your credit card details as the site could be phony.
- Always check that your credit card details will be encrypted by checking that the web address of the page starts with https.
- Do not write your credit card details down and ensure that you do not leave your credit cards lying around for others to see.

The security of data in the cloud

People use all sorts of computers to access their data. For example, they may do most of their work on a desktop but when they are out they may want to access their data using a tablet or smartphone. The problem is that copying the data from one device to another takes time and you have to be careful about which files are the latest version otherwise you can end up losing work.

Storage in the cloud is a way of storing data in one place so that it can be accessed by all your devices as long as you have internet access. Basically you store your files on a file server in a remote place (called the cloud) and you access your files using a user-ID and password. Sometimes the organisation providing the storage in the cloud provides it free and other times you pay a subscription.

There are a number of security issues with data stored in the cloud and these include:

- The company providing the storage could go out of business. If this happens then you have to consider what would happen to your data.
- Some of the data could be personal data, in which case it should be encrypted so that if it is hacked into, hackers would not be able to understand and use the data.
- You have to trust that the organisation providing the storage puts in all the security measures to keep your data safe.
- Who owns the data when it is stored in a cloud? When photographs are stored using some cloud storage organisations, you have to give permission that anyone can access and use the photographs you store.

Physical security

Physical security involves:

- Locking doors to offices and computer rooms.
- Not leaving your computer logged in when you are away from your desk.
- Ensuring that there are no shoulder surfers looking at your screen when you enter a password.

The effectiveness of different methods of increased security

There are lots of different methods by which you can increase the security of your data and it is best when all these methods are used together. No matter how much security is in place there are always weaknesses that can be taken advantage of by a skilled and determined hacker. For example firewalls prevent people outside using the internet to access your computer/network. However it would not prevent someone who is internal to an organisation (and therefore is able to access the network), from accessing some data for which they are not allowed access.

Anti-virus software will check against known viruses and new viruses which have certain characteristics but people who produce viruses are constantly finding other ways to distribute viruses.

Encryption methods have improved but hackers are still able to access some of the most secure systems such as those of banks and the security services.

QUESTIONS C

1 **a** Explain what is meant by a computer virus. *(2 marks)*

b Describe steps a computer user can take to prevent computer viruses from infecting their computer. *(3 marks)*

2 **a** Explain what is meant by hacking. *(2 marks)*

b Describe the steps a computer user can take to help prevent hacking. *(3 marks)*

3 Tick **True** or **False** next to each of the statements.

	True	False
Email attachments can contain viruses.		
Viruses consist of programming code designed to cause harm to your computer or annoyance.		
It is impossible to hack into a computer without using the internet.		
Anti-virus software must be kept up-to-date.		

(4 marks)

4 Computer viruses pose a serious threat to all computer systems.

a Give the name of the piece of software that can be used to check for viruses and also remove them. *(1 mark)*

b Describe **two** actions that may be taken, other than using software, to prevent viruses from entering a computer system. *(2 marks)*

5 Many people are now storing their data in the cloud.

a Explain what the above statement means. *(2 marks)*

b Describe **one** advantage in storing your data in the cloud. *(1 mark)*

c Describe the security issues when you store your data in the cloud. *(4 marks)*

Test yourself

The following notes summarise this chapter, but they have missing words. Using the words below, copy out and complete sentences **A** to **K**. Each word may be used more than once.

files **download** **encrypted** **spam**
passwords **hacking** **phishing** **hackers**
firewall **fake** **pharming** **biometric**
digital signatures **anti-virus**

A To protect data from deliberate damage caused by hackers illegally gaining access to a computer network via the internet, a ______ should be used.

B In order to prevent viruses entering an ICT system, ______ software should be used to search for and destroy viruses.

C Users should be told not to open ______ attached to emails unless they know who they are from.

D Users should also be told not to ______ music and games illegally off the internet from file sharing sites.

E Unauthorised use of an ICT system with a view to seeing or altering the data is called ______.

F ______ means sending emails which falsely say they are from banks or other organisations in order to trick people into revealing their banking or credit card details.

G It is possible to use special software to filter out unwelcome emails, which are popularly called ______.

H ______ is where program code is deposited onto a user's computer or onto a server in the case of a network. When the user types in the web address of a real website such as a bank's website, they are instead re-directed to a ______ website.

I When credit card details are sent to make a payment for goods bought from an online retailer, the details are ______ and this prevents them being understood by ______ if they are intercepted.

J User-IDs and ______ are used to authenticate users of computer systems and are methods which make use of properties of the human body. These methods are called ______ methods and include fingerprint and retinal scanning.

K ______ are used to ensure documents sent over the internet are authentic.

REVISION QUESTIONS

1 Many people now store their data (documents, photographs, music, etc.) in the cloud.

a Explain what is meant by storage in the cloud. [1]

b Give **one** advantage and **one** disadvantage of storing data in the cloud. [2]

2 For each statement, tick the box indicating whether it refers to smishing, phishing or pharming.

	Smishing	Phishing	Pharming
Uses emails			
Involves fake programming code being loaded onto a computer			
Does not involve being directed to a website			
Uses SMS messages			

[4]

3 Tick whether the following statements are **True** or **False**.

	True	False
Tripping accidents can be caused by trailing wires.		
It is safe to use lots of multi-sockets connected to the one plug.		
Overloaded power sockets can cause fires.		
Waste paper bins should be emptied regularly to reduce the fire risk.		

[4]

4 A school uses a website which students and parents can access from home. There is an area on the website where students can access the marks they obtained for homework and tests. Only the student or their parents should be able to access this area.

a Describe **three** ways the school could ensure that there is no unauthorised access to the area containing the marks. [3]

b The school is worried about viruses infecting their computer system. Describe **three** things the school should do to reduce the likelihood of a virus attack. [3]

5 A user is ordering goods from an online supermarket.

They have ordered the goods and are on the checkout web page where they are asked to enter their payment details.

a The web page says it uses SSL.

i Give the meaning of the abbreviation SSL. [1]

ii Explain how SSL is used to make it safe to enter credit card details. [1]

b Apart from the site saying it uses SSL, how else could you tell that the site was using a secure server for entering credit card details? [1]

6 Young people must be careful about using email services.

Describe **three** actions you should take in order ensure your e-safety when using email services. [3]

7 A bank customer is using online banking to pay an electricity bill.

Describe **three** methods of ensuring that the data being transferred is secure. [6]

EXAM-STYLE QUESTIONS

1 There are many examples of abuses of the internet. Explain what is meant by the following terms. *[2]*

a Pharming

b Phishing

2 Spam causes annoyance to computer users. Explain what is meant by spam and give **one** example of why it causes annoyance to computer users. *[2]*

3 There are a number of health and safety issues caused by the use of computers. *[4]*

a Describe **two** health issues that are caused by computer use.

b Describe **two** safety issues that are caused by computer use.

4 Here are some statements. You have to tick whether the statements refer to pharming or phishing.

	Pharming	Phishing
Malicious programming code is stored on a computer.		
Emails are sent pretending to be from your bank.		
You are re-directed to a bogus/fake website.		
Uses emails for identity theft.		

[4]

5 Young people need to be aware of e-safety when using the internet.

a Explain the meaning of the term e-safety. *[2]*

b Discuss why e-safety is important when young people use the internet. *[5]*

6 Chat rooms and social networking sites present many dangers to young people.

Describe **three** precautions a young person can take to ensure their safety when using chat rooms and social networking sites. *[3]*

7 Banks keep their customer account records on their file servers.

a Describe **three** ways in which hackers could misuse this data. *[3]*

b Describe the purpose of **three** different authentication techniques. *[3]*

(Cambridge IGCSE Information and Communication Technology 0417/11 q14 May/June 2014)

8 Explain how a firewall could be used to secure the data in a computer connected to the internet. *[3]*

(Cambridge IGCSE Information and Communication Technology 0417/11 q5 Oct/Nov 2012)

9 Tick **True** or **False** next to each statement.

	True	False
Using a password always prevents unauthorised access to data.		
A strong password is one that is difficult for a hacker to guess.		
Giving your password to a friend is a good idea in case you forget it.		
If you forget your user ID you can still gain access to data using you password.		

[4]

(Cambridge IGCSE Information and Communication Technology 0417/11 q9 May/Jun 2012)

10 The head teacher of a school is concerned that students may be hacking into the exam scores database to change their marks.

Other than passwords, describe ICT methods that could be used to prevent this from happening. *[4]*

(Cambridge IGCSE Information and Communication Technology 0417/11 q17 May/Jun 2013)

11 Working with ICT equipment is fairly safe but you do need to check the way you are working to ensure you are working safely and not exposing yourself to possible health problems.

Describe **three** checks you can make on the way you work with ICT equipment to minimise any potential dangers. *[3]*

12 a Describe what is meant by personal data. *[2]*

b Give **two** items of personal data you should never give to a stranger. *[2]*

c Describe **two** reasons why e-safety is needed when you are online. *[2]*

13 Here are some statements about the effectiveness of increased security.

Tick whether these statements are true or false.

	True	False
Firewalls prevent people from inside and outside the organisation accessing a computer/network.		
Anti-virus software will detect all viruses.		
It is impossible for hackers to understand encrypted data.		

[3]

14 Most computers have internet access.

Explain **two** reasons why the internet is so popular. *[2]*

9 Audience

When developing ICT solutions you must have a clear understanding of who your audience is for the solution. This is important as it will influence your choice of design, reading age of text, complexity of the user interface, font and font size, the use of images, complexity of language used, etc. When developing an ICT solution you will need to also be aware of the legal, moral, ethical and cultural needs of your audience. You need to be aware of all these so that you do not offend any of your intended audience. You also need to consider the legal, moral, ethical and cultural implications that might apply to your proposed solution.

The key concepts covered in this chapter are:

- Audience appreciation
- Legal, moral, ethical and cultural appreciation

Audience appreciation

Your audience are the people at whom your ICT solution is aimed.

The age of your audience is very important, because a website designed for children will not meet the needs of an older audience. The following diagram shows how you might categorise the age of your audience.

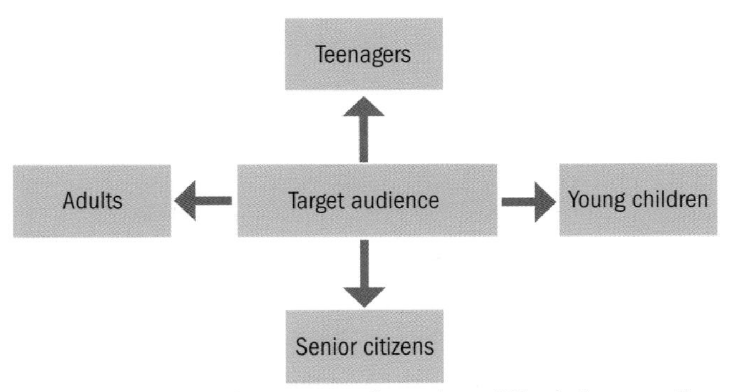

It is common to classify a target audience for an ICT solution according to their age.

Finding out about your target audience

Once you have identified who your audience is, you still need to find out more about them. You must ask yourself questions such as:

- **What are their needs?**
  Your audience will look at your document for a reason. Think about what this reason is and what the audience will need from your document. For example, if you have designed a poster advertising a school/college night out, think about what information your audience will need. Satisfying your audience's needs is an important part in creating any document.
- **How much do they know already?**
  It is essential to assess how much your audience knows about the subject of your document. If they have a good knowledge of the subject, you will not have to start at the beginning.
- **Knowledge of the reader about the subject**
  If you were writing an article on a subject such as the Data Protection Act for computer users, then it would need to be different from the same article aimed at lawyers who are able to untangle the intricacies of the law.
- **What level of literacy do they have?**
  Not everyone is good at English. Any document aimed at the whole of the adult population will need to be written simply. Generally, to make a document as readable as possible, you would need to make sure that only well-known words are used, sentence length is kept short and punctuation marks are used only where necessary.
- **How much specialist vocabulary can they handle?**
  Most subjects have a series of words that tend to be used only within that subject. The subject is said to have a specialist vocabulary. For example, doctors and nurses have special medical terms for illnesses. You need to make sure that if you are using any specialist words to people not familiar with the area, then you will need to carefully explain them.

Analysing the needs of an audience

Suppose you are asked to design and produce a website. You would need to consider the needs of the audience and analyse how your solution can meet these needs. The two diagrams below show the needs for two different audiences for a website:

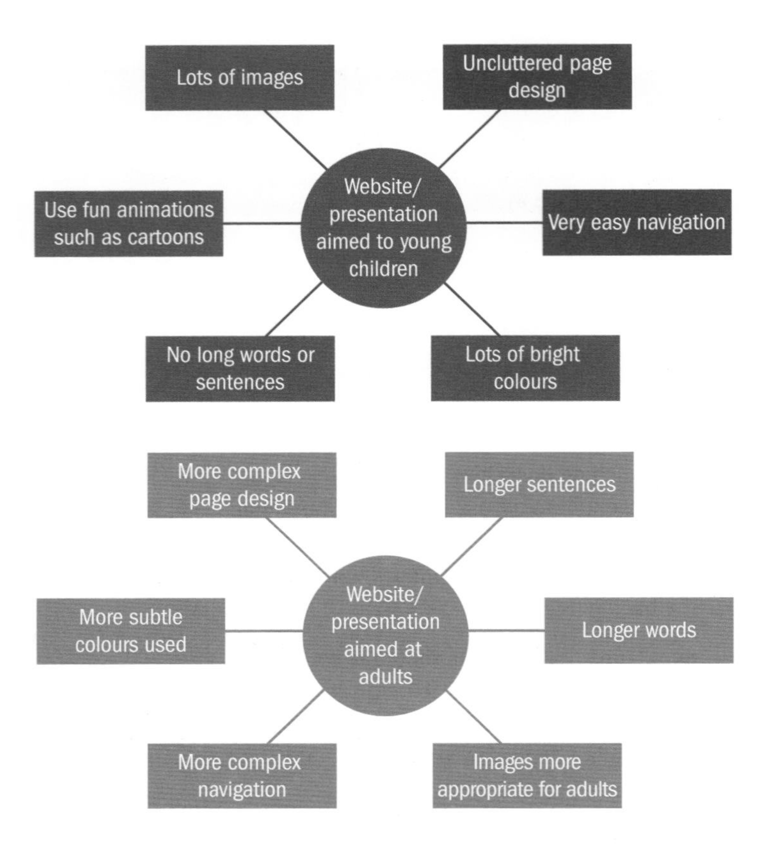

Reasons why ICT solutions must meet the needs of the audience

Many ICT solutions are developed which do not meet the needs of the audience. You have probably visited websites that are frustrating to use or do not enable you to obtain the information you want. Here are some reasons why any ICT solution should always meet the needs of the audience:

- The whole point of an ICT solution is to solve a problem and if it fails to provide a solution it will be of no use.
- People will not use an ICT solution that does not meet their needs.
- It will be frustrating to use and annoy the users.
- If the solution is too complex, then the users it is aimed at will not be able to use it.
- If users have paid for the solution, they are entitled to their money back if it fails to do what it is supposed to do.

Legal, moral, ethical and cultural appreciation

When creating ICT solutions, you must consider the legal, moral, ethical and cultural implications of your solution.

Legal implications

Any ICT solutions must not break any laws. Hence they should:

- not breach copyright
- do what they are supposed to do, otherwise they would breach trading standards legislation
- be safe to use otherwise it would breach health and safety legislation.

Copyright and software piracy

When people produce software, their software is protected by copyright, which makes it an offence to copy the software in ways that are not allowed by the software licence. Software piracy is the illegal copying of software. It is important to note that it is not illegal to copy all software. For example, if you have bought some software, then you may be allowed to copy it onto more than one computer if the software licence allows it. You may also be allowed to take a backup copy of the software for security purposes.

The need for copyright legislation

Many people spend a lot of time and money creating original work such as a piece of music, a picture, a piece of software, a photograph, a newspaper article, etc. Many of these people do it for a living, so it is only fair that their work should not be illegally copied.

Most countries use copyright legislation, which makes it illegal for intellectual property to be copied such as:

- Software
- Text (books, magazine articles, etc.)
- A new innovative human-computer interface
- Hardware (a flexible screen, the design of a power saving chip, etc.)
- Books and manuals
- Images on websites.

Here are some actions that would be illegal:

- Copying software and music without permission or license.
- Copying images or text without permission or license.
- Copying sections of websites without permission.
- Sharing digital music illegally using file sharing websites.
- Running more copies of software than is allowed by the site licence.

KEY WORD

Copyright legislation – laws making it a criminal offence to copy or steal software.

The moral and ethical implications of software piracy

You may think that software piracy does not really hurt anyone as it is only the software companies (that have lots of money) that are affected. The truth is it affects everyone and here is how:

- It is against the law and is a crime punishable by a fine or a jail sentence.
- It is theft as it deprives the owner of payment for their work.
- The money from software piracy can be used to fund other illegal activities such as drug dealing, people trafficking, etc.
- Not everyone who produces software is wealthy and many rely on this income for their everyday living expenses.
- Companies will not invest in new software, if they think they will lose potential income through illegal copying.

Internet service provider (ISP) denying service

An internet service provider is the organisation that provides most people with their internet connection. If you are caught illegally downloading music, video and other files then your ISP could deny you connection to the internet. The aim of this is to prevent persistent offenders from gaining internet access.

How software producers attempt to prevent copyrighted software being illegally copied

Preventing copyrighted software from being illegally copied is very difficult. There are a number of methods used such as:

Encryption of the execution code – the execution code enables the software to run. This code is encrypted and a key is needed to unlock the code to enable the software to be used.

Use of a digital signature on the DVD/CD containing the software – this prevents the software being copied exactly.

Use of an activation key – when you purchase software, after loading it on your computer, you will be asked to activate the software by going onto the software producer's website and then entering a product activation key (a long code).

Use of a dongle – a dongle is a piece of hardware that attaches to a computer using the USB port and allows a secured piece of software to run. The dongle contains an electronic key that unlocks the program on the computer and allows it to run. Making a copy of the dongle is hard.

Use of programming code – the program code can be altered to help block copying.

Use of guards – guards are hardware or software modules that keep a check on the program as it is being run to ensure that it has not been tampered with in any way.

Moral implications

Morals are norms based on definitions of right and wrong. Most people accept that telling lies is morally wrong so a website where people are able to say what they like about people even when it is incorrect would be morally wrong. A computer game that encouraged violence would be morally wrong.

Any ICT solution must not:

- Offend the majority of people
- Encourage violence
- Encourage people to lie about others.

Cultural implications

Culture is the way beliefs and history influence society. An ICT solution which is acceptable in one country may not be acceptable in another country because of cultural differences.

- Make sure that any documentation and training manuals have an easy-to-understand language for any group of individuals.
- Determine a common language that all users of the solution can understand.
- Address someone correctly – addressing someone using the wrong title may not matter in one culture but could offend someone from a different culture. For example, the use of a forename before you have met the person is acceptable in some cultures but not others.

Policing the internet

Many people think the internet should be policed so that all the inappropriate sites (how to make bombs, how to grow drugs, pornography, paedophilia, etc.) could be closed down. Others see policing the internet as an infringement of human rights. The reasons for and against policing the internet are covered in Chapter 4.

An attempt has been made to police the internet in certain countries. In some countries all internet access is controlled which means you are not free to access all parts of the internet.

Test yourself

The following notes summarise this chapter, but they have missing words. Using the words below, copy out and complete sentences **A** to **I**. Each word may be used more than once.

legislation dongle money encrypt piracy
digital viruses needs audience

A When designing an ICT solution to a problem you should always think about the requirements of your intended ______.

B The audience will have ______ which have to be satisfied by your ICT solution.

C The illegal copying of computer software is called software ______.

D Copyright ______ are laws which make it a crime to illegally copy software and other copyrighted material such as text, images, music, etc.

E Copying software illegally is no different from stealing, as it deprives someone of ______.

F You have to be careful of illegally copied software as it often contains ______ that can be copied onto your computer.

G One method that software producers use to prevent illegal copying of software is by the use of a ______.

H Some software producers ______ the execution code of the software so that a key is needed to unlock the code and allow the software to run.

I A ______ signature can also be used to help prevent illegal copying.

REVISION QUESTIONS

1 Explain the term software piracy. [2]

2 When creating an ICT solution you must first analyse the audience's needs.

Describe why this is necessary. [2]

3 Tick **True** or **False** next to each of these statements. [5]

	True	False
The illegal copying of computer software is an example of software piracy.		
Copying software is always illegal.		
Software piracy is acceptable as it helps us obtain software cheaper or sometimes even for free.		
Software licences usually allow you to run a certain number of computers with the same software.		
Copying deprives small software producers of money.		

EXAM-STYLE QUESTIONS

1 Most countries have copyright laws which make it a crime to copy software illegally.

- **a** Describe why copyright laws are needed. *[2]*
- **b** A copyright law will make it illegal to copy software in certain situations.
  - **i** Describe a situation where the copying of computer software would be allowed. *[1]*
  - **ii** Describe **two** terms a copyright law would have in it. *[2]*

2 A website is being designed for children aged 4 to 6 years old to help them read.

Explain, using this example, why the audience's needs should be taken into account in the design of this website. *[4]*

3 It is reckoned that about 40% of software being used by businesses has been illegally copied.

- **a** Explain what is meant by software piracy. *[1]*
- **b** Give **one** reason why copyright laws are necessary. *[1]*
- **c** Software producers use methods to prevent software copyright being broken.

  Describe **two** methods used to help software producers prevent their software from being copied illegally. *[2]*

10 Communication

The main methods of communication using ICT are email and the internet. This chapter looks at the constraints that affect the use of email. Email is widely used as a form of communication and overall it is a highly effective communication tool. Email is inexpensive, as you require only an internet connection that is generally already present. Although a printout of emails is possible, emails often stay as soft copies because archiving and retrieving email communications is easy.

This topic also looks at the fundamentals of the internet as well as the advantages and disadvantages in using the internet as a source of information.

The key concepts covered in this chapter are:

- Communicating with other ICT users using email
- Effective use of the internet

Communicating with other ICT users using email

Communication by email has taken over from hand-written letters as the way people in businesses communicate with each other.

What is an email?

An email is an electronic message sent from one communication device (computer, telephone, mobile phone, or tablet) to another. All web browser software has email facilities. There are many email facilities but those shown here are the main timesaving ones.

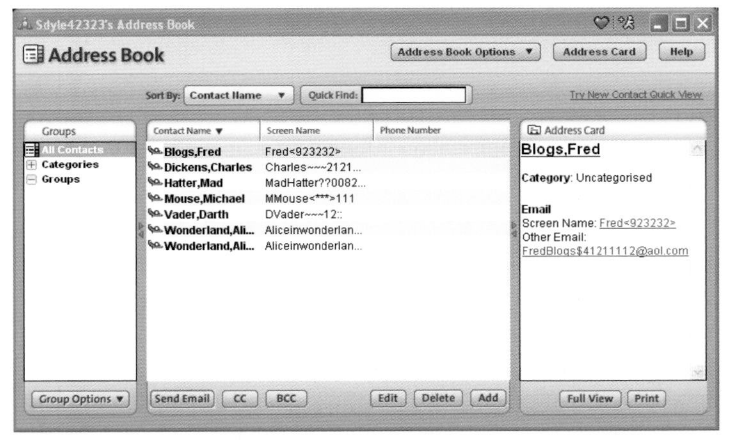

The screenshot shows an address book. Rather than type in the email address of the recipients and maybe make mistakes, you can simply click on their address. Notice the facility to create groups.

Using the advanced features of email, you can create groups.

Groups

Groups are lists of people and their email addresses. They are used when an email needs to be distributed to people in a particular group. For example, if you were working as part of a team and needed to send each member the same email, then you would set up a group. Every time you needed to send the members of the group email, you could then just send the one email to the group, thus saving time.

Using cc (carbon copy)

cc means carbon copy and it is used when you want to send an email to one person but you also want others to see the email you are sending. To do this you enter the email address of the main person you are sending it to and in the box marked cc you enter all the email addresses, separated by commas, of all the people you wish to receive a copy.

The constraints that affect the use of email

When communicating using email, there are a number of constraints:

The laws within a country – free communication using email is taken for granted in most countries but this is not always the case. In some countries, email is monitored and some email is read, meaning that you cannot criticise the government or talk about certain topics freely.

Acceptable language – you should always be polite in an email and avoid the use of bad language. If you send an offensive email, it is easy for the police to track it to your computer since internet service providers are required by law to keep the emails for a certain period for purposes such as law enforcement.

Copyright – copyright laws apply to text in emails. This means that if you pass on a block of text or send a picture to others without the copyright owner's permission, then this would be illegal.

Local guidelines set by an employer – most employers will set out in your contract of employment what you can and cannot do using email. If you do not obey these rules then you would be in breach of your contract and could be dismissed.

The need for security – emails containing certain words will be routinely examined and investigated. Many criminals use emails to communicate with each other so the law enforcement agencies continually monitor emails in most countries.

Password protection – emails often contain personal information. If you can access someone else's email, you can find out a lot of information about them. Email services are protected by passwords and it is up to the user to protect this password and ensure that is not disclosed to anybody.

Netiquette – when using email, you must learn to abide by certain rules and these include:

- Not typing in all capitals – excessive use of capitals is considered to be the written equivalent of shouting.
- Ensuring that any attachments are not too large – the person downloading the file may not be able to download the file if it is too large. Also large files will clog up their storage space and will also load slowly.
- Never divulging personal information – always consider your e-safety. Emails are easy to hack into.
- Emails should be concise and to the point. The reader should not have to waste time reading unnecessary information.
- Use plain text and do not use any fancy graphics as they load slowly and waste storage space.
- Remember to attach file attachments; people often mention them in the email and then forget to attach them.

Spam

Sending the same email message to large numbers of people is a fast, easy, and cheap form of communication. This means that you can get lots of spam email sent to you. Spam is email sent to you that you didn't ask for (unsolicited email) – usually advertisements for services that don't interest most people, such as medications and products of an adult nature. People who send spam emails are called spammers. It is has been estimated that about 75% of all email messages are spam.

Reasons why spam needs to be prevented

The following are reasons why spam should be prevented:

- Spam clogs up your inbox making legitimate email harder to spot.
- Spam takes time to delete.
- Spam can be used to spread computer viruses and malicious software.
- Unnecessary email slows the internet down.
- Young children could view inappropriate message or images.

Methods used to help prevent spam

It is impossible to eliminate spam completely but here are some methods that can reduce the amount you receive.

Use spam filters – this is software that is able to recognise what is and isn't spam. Spam emails are put into a separate folder and can then be deleted.

Avoid publishing your email address on websites, blogs, forums, and suchlike – spammers often use these to collect email addresses.

Check the privacy policy when signing up for things online – check they won't sell your email address to a spammer.

Change your password for internet access regularly and use strong passwords – this will keep out hackers who may steal all the email addresses in your address book to sell to spammers.

Web browsers and search engines

A web browser is an application installed on your computer that provides you a graphical user interface for accessing the internet (the HTTP protocol among others).

A search engine is an application typically installed on a server at a company's data centre that you access via your web browser. A search engine 'crawls' the internet categorising and storing information about sites (metadata) so when you use your web browser to access a company's search engine to perform a search, it provides you with relevant results (for the most part).

You download and install a web browser (Google Chrome, Internet Explorer, Safari, etc.) to your computer and use it to access a search engine on the internet.

Terms used when talking about the internet

There are a number of terms used when talking about the workings of the internet. Here are the ones you must understand and remember.

Protocol – there are different ways, called protocols, for transferring data using the internet. A protocol is a set of rules governing the format of the data and the signals to start, control, and end the transfer of data.

HTTP (HyperText Transfer Protocol) – a protocol that defines the process of identifying, requesting, and transferring multimedia web pages over the internet.

HTTPS (HyperText Transfer Protocol secure variant) – a protocol that defines the process of identifying, requesting, and transferring multimedia web pages over the internet, except unlike HTTP it uses encrypted connections to hide passwords, bank information, and other sensitive material from the open network. This is to prevent hackers from accessing and hence using banking details fraudulently.

Uniform Resource Locator (URL) – a web address. The system used to identify the location of a web page or document stored on the internet.

Hyperlink – a feature of a website that allows a user to jump to another web page, to jump to part of the same page web page or to send an email message. Hyperlinks can be a word, phrase or image.

Internet service provider (ISP) – a company that provides users with an internet connection.

FTP (File Transfer Protocol) – a common protocol for the movement of files across the internet. Users can access a server that contains the file they want and then download it to their computer using FTP. Alternatively, they can upload a file to the server from their computer using FTP. An example of this would be when you upload your web pages to a web server where your website is hosted.

Storage in the cloud/cloud storage – where your data is not stored locally on your computer or attached storage devices, but is instead stored on servers in remote locations. When you save a file it is transferred using the internet to one of the servers which are owned and managed by a host company who you pay to keep your data accessible and safe. Sometimes they provide a small amount of storage free and you pay extra for additional storage. Storage in the cloud means you can access your files from any device such as desktops, laptops, tablets and smartphones.

The differences between an intranet, the internet, and the world wide web (www)

An intranet is a private network that uses the same technology as that used by the internet for the sending of messages/data around a network. The main use of an intranet is to share organisational information and resources. The main feature of an intranet is that only employees of the organisation are able to use it. It is important to note that an intranet need not be confined to a single site and it is still possible for people on an intranet to access the internet.

The internet is a huge network of networks. Each computer on the network, provided it has permission, can access the data stored on other computers on the network and also transfer messages between computers.

world wide web (www) – the world wide web is a means of accessing information contained on the internet. It is an information sharing model that is built on top of the internet. The world wide web uses HTTP, which is one of the languages used over the internet, to transmit information. The world wide web makes use of web browser software to access documents called web pages.

The internet provides services other than access to web pages. Using the internet you have:

- Email facilities
- Instant messaging
- FTP (File Transfer Protocol) which is a way of exchanging files between different computers connected to the internet.

All of the above services require protocols different to that required by the world wide web (which uses HTTP).

The internet is therefore the actual network whereas the world wide web is the accessing of web pages using the internet. It is important to realise that the world wide web is only one of the facilities based on the internet.

Effective use of the internet

The internet is the largest store of information in the world and being able to access relevant information quickly is an important skill which you will need to demonstrate in the examination.

There are a number of different ways of locating information using the internet and these are described here.

Locating specified information from a given website URL

URL stands for Uniform Resource Locator, which is a complicated way of saying a website address. One of the ways of accessing a website is to type the URL (website address) into a web browser or search engine. Each page of a website has its own URL and the URL quoted in books/magazines, etc., will usually take you to the homepage of the website. You can then search within the website for the information you need by making use of links and search facilities.

Finding specific information using a search engine

Searching for information can be performed using a search engine such as Google or Yahoo. The search results are displayed on the basis of relevance or who has paid the most money to get in the top results. Clicking on a search result takes you to a website and may take you straight to the information you want or you may go to the home page of the website. It may then be necessary to perform a search of the content of the website. This is done by entering key words into a search box.

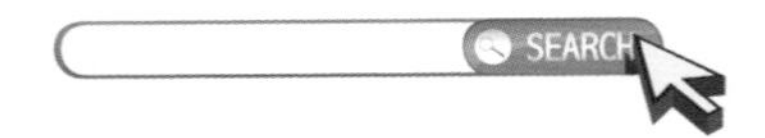

A key word search on a website.

The structure of a web address

Here is the web address of the publishers of this book, Oxford University Press:

http://www.oup.com

There are a number of different parts to this web address.

http:// stands for HyperText Transfer Protocol and this helps your web browser locate the website. You do not have to type it with some web browsers as it is added automatically.

www. stands for world wide web which means the page you are looking for is on the world wide web.

oup. is the name of the server or website where the website or web page is located. In many cases it will be the name of the organisation or an abbreviated version.

.com is the domain which tells you where the web page is registered and also what type of website it is. For example .com tells you that it is a commercial site. Other common domains include .org (sites for organisations), .edu (sites for schools and colleges), .ac (sites for universities), and .gov (sites for government departments).

Sometimes there are suffixes such as .co.uk which tells you that the site resides in the UK. There are suffixes for most other countries as well.

Sometimes there will be a slash (/) following the domain name. For example, http://www.cie.org.uk/programmes-and-qualifications. The / followed by another word or words lets you know that you are going to a different area or page on that website. In the example, you will be taken to the programmes-and-qualifications page on CIE's (Cambridge International Examinations) website. You can have slashes after this showing pages that are linked in a hierarchy.

Blogs/web logs

Blogs/web logs are online diaries of events or journals. Blogs can be about anything. Groups, singers, and celebrities have blogs which let people know about their life and what they are doing. Blogs are also used by politicians, and for collecting public opinion about certain topics.

Features of a blog:

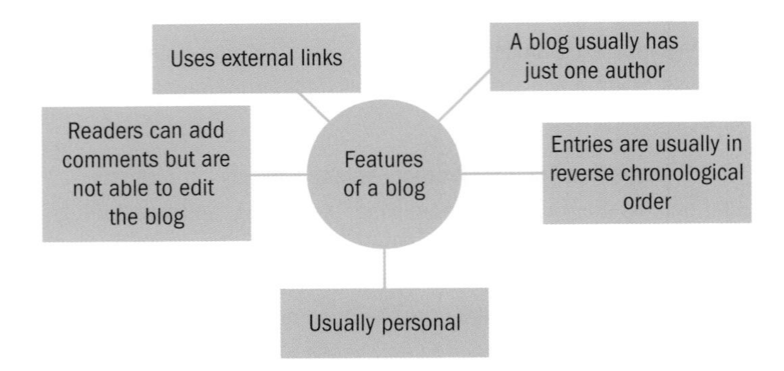

Wikis

A wiki is a web page that can be viewed and modified by anyone who has a web browser. This means that anyone with a web browser can change the content of a web page. The problem with this is that someone could alter the web page to post incorrect information or even offensive messages.

You will have seen the online encyclopaedia Wikipedia. This encyclopaedia has been created by ordinary people and anyone can add material or delete material from it. You may think this is bad idea, but if someone posts incorrect information then there are plenty of people around who will view it and correct it. This means that information from sites like Wikipedia may contain inaccurate information – you should always check the information with other more reliable sources.

Features of a wiki:

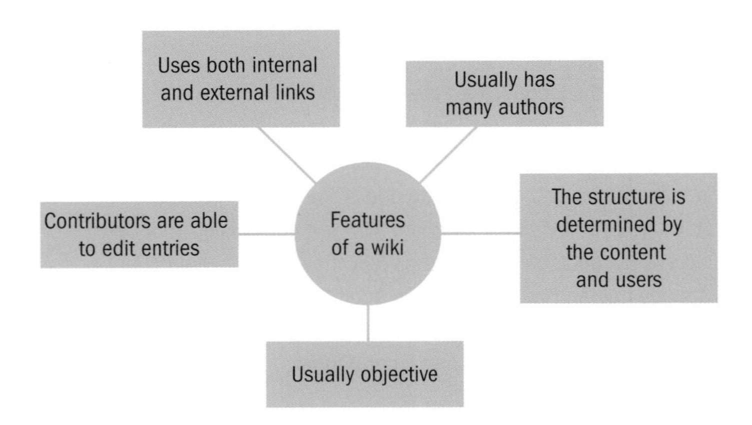

Social networking websites

Social networking websites are virtual communities of people who communicate about a particular subject or interest or just make friends with each other. Members of these sites create their own profiles with information about themselves such as hobbies, likes and dislikes, etc. You have to ensure that you do not reveal too much information about yourself as others view this information. Social networking websites enable members to communicate using instant messaging, email, a type of blog, and even voice or video-conferencing. The services allow members to invite other members and non-members into their circle of friends.

Social networking websites you might be familiar with include:

- Facebook
- MySpace
- Twitter

Reasons why the internet is so popular

The reasons why the internet is so popular include:

- There is a huge amount of information stored on every subject imaginable. It is the largest store of information in the world.
- The information can be accessed from anywhere by lots of different devices (e.g. computers, smartphones etc.) provided there is an internet connection.
- Search engines allow fast access to specific information provided the search criteria are carefully constructed.
- There are a huge range of services you can access such as email, file transfer, chat rooms etc.

Reasons why an internet search to find relevant information is not always fast

The following are reasons why an internet search to find relevant information is not always fast:

- There might be a slow internet connection so pages take time to load.
- Millions of pages could be found and you have to spend time deciding which are relevant.
- The user may not have the skills in picking good search criteria resulting in too much irrelevant information.

Issues with information on the internet

Just because information is on the internet does not necessarily make it true. Any information you get from the internet needs to be treated with caution. In this section you will be looking at how you might assess the worth of the information and how you can check its accuracy.

Unreliability of information on the internet

Many people think that because information is published on the internet it must be true. This is far from the truth, as there are many examples of sites that contain material which is completely untrue. Some sites deliberately set out to misinform or deceive.

It is important to remember that anyone is able to produce a website and publish it on the internet. They do not have to be an authority on the subject the site is about. In some cases the person creating the site has not checked the information on the site.

Advantages of using information on the internet are:

- The information is up-to-date compared to many books
- It is quick to search for the information
- It is available on your home computer or mobile device
- You can obtain information from other parts of the world easily

Disadvantages of using information on the internet are:

- The information could be unreliable
- There could be too much information to read

How can you make sure that information you use is accurate? Here are a few steps you can take:

- Check the date that the site was last updated. Bogus sites are not updated very often. Also, sites go out-of-date so you need to be sure that the site you use has been updated recently.
- Only use sites produced by organisations you have heard of (major newspapers, the BBC, etc.).
- Use several sites to get the information and check that the sites are giving similar information.
- Follow the links to see if they work. Many bogus sites have links that do not work.

Undesirability of information on the internet

It is impossible to censor what is on the internet. This is because much of the information on the internet comes from other countries. Unfortunately the internet contains pornography, violent videos, sites that promote racial hatred, and so on. Schools and parents are able to use parental controls that restrict the sites and information that can be viewed on the internet.

Test yourself

The following notes summarise this chapter, but they have missing words. Using the words below, copy out and complete sentences **A** to **I**. Each word may be used more than once.

internet **HTTP** **accessing** **address** **HTTPS**
blog **locator** **blogger** **ISP** **search engine**
encrypt **domain** **hyperlink** **web browser**

A The ______ is a network of networks and the world wide web is the means of ______ the information contained on the internet.

B ______ is the method used for transferring data over the internet when the data is not of a personal nature.

C ______ is the method used for transferring data such as banking or credit card details over the internet where it is necessary to ______ the data to keep if safe from hackers.

D URL stands for Uniform Resource ______ and it is another name for the web ______.

E A feature of a website that allows a user to jump to another web page is called a ______.

F An organisation that provides users with an internet connection is called an ______.

G A website that allows comments to be posted usually in chronological order is called a ______ and the person who keeps one is called a ______.

H The .com or .co.uk part of a web address is called the ______.

I A ______ is software that you use to access the internet and once this is loaded you can then use a ______ to find specific information.

REVISION QUESTIONS

1 Tick **True** or **False** next to each of these statements.

	True	False
An intranet is a type of blog.		
Intranets are used by people within an organisation.		
Intranets and the internet both use similar technology.		
An ISP is a type of virus.		

[2]

2 Describe **two** features of a blog. [2]

3 Describe the similarities and differences between the internet and an intranet. [4]

4 Describe what is meant by a wiki. [4]

5 Tick **True** or **False** next to each of the following statements.

	True	False
Spam is unsolicited email.		
Email groups are used so as to avoid spam.		
Email groups are useful for when the same email needs to be sent to all members of the group.		
Spam filters can remove harmful viruses.		
Storage in the cloud is useful if you need to access the data using lots of different devices.		

[5]

6 Describe the differences between a web browser and a search engine. [4]

EXAM-STYLE QUESTIONS

1 Explain the main differences between the internet and the world wide web (www). *[2]*

2 Many organisations use an intranet.

a Explain what is meant by the term intranet. *[2]*

b Describe **one** way in which an intranet is different from the internet. *[1]*

3 Place a tick in the box next to each method if it is a method of protecting copyrighted software from being copied.

	×
Use of chip and PIN.	
Registration system that makes use of a registration code.	
The use of anti-virus software.	
Encryption of the programming code used to execute the program.	
Use of data protection legislation.	
Use of activation codes which can only be used a certain number of times on different machines.	
The use of a dongle.	

[4]

4 Describe **two** methods that software manufacturers use to try to prevent their copyrighted software from being illegally copied. *[2]*

5 Many internet users make use of social networking sites.

Describe **three** features of a social networking site. *[3]*

6 A student is doing a project at school and is using the internet as a source of information.

a Describe **three** advantages of using the internet as a source of information. *[3]*

b Describe **three** disadvantages of using the internet as a source of information. *[3]*

7 Tick **True** or **False** next to each of these statements.

	True	False
The internet is a network of networks.		
Intranets can be used by anyone with internet access.		
Intranets and the internet both use web browsers.		
The internet uses HTTP.		

[3]

8 Describe **three** differences between a blog and a wiki. *[3]*

(Cambridge IGCSE Information and Communication Technology 0417/11 q7 Oct/Nov 2012)

9 Tick **Internet** or **intranet**, next to each statement, as appropriate.

Internet	intranet	
		...is a network of computer networks
		...usually exists within one organisation
		...anybody can access it
		...can be expanded to become an extranet

[4]

(Cambridge IGCSE Information and Communication Technology 0417/11 q7 Oct/Nov 2013)

10 Discuss the use of blogs and social networking sites as means of communication. *[6]*

(Cambridge IGCSE Information and Communication Technology 0417/11 q18 May/Jun 2013)

11 Many people now store their files in the cloud.

Describe how storage in the cloud works. *[3]*

11 File management

Being able to find files quickly is very important, as is being able to save files in a file format that other people can open using the software they have. In this chapter you will be looking at how to manage files effectively, so they are easy to find even if they were first saved many years ago.

Sometimes you will receive a file in an unfamiliar format which you need to convert into a format that can be opened using your existing software. Being able to convert files from one file format to another is an important part of using ICT effectively.

The key concepts covered in this chapter are:

- Managing files effectively
- Reducing file sizes for storage or transmission

Managing files effectively

File formats and their uses

Files can be in all sorts of formats depending on the software used to create and save them. You can identify a file format by looking at the file extension which is defined here.

File extension – A label after the file name, consisting of usually three or four characters, that gives information to the operating system and other software about what format the data is stored in.

Here are the file formats you will need to know about for the examination.

css (cascading style sheets) – a file used to format the contents of a web page. The file contains properties for how to display HTML elements. HTML is a special code used for making web pages. For example, a user can define the size, colour, font, line spacing, indentation, borders, and location of HTML elements. CSS files are used to create a similar look and feel across all the web pages in a website.

csv (comma separated values) – this is a list of data items that are separated from each other using commas. CSV files are mainly used for saving data from a table, spreadsheet or database where there are rows and columns of data.

gif (graphics interchange format) – a very popular way of storing graphic images on the internet as it includes compression of the file, which makes loading times shorter than uncompressed files.

htm – a popular file format used for storing web pages.

jpg (JPEG – Joint Picture Experts Group format) – a popular way of storing still images from digital cameras that uses compression to make files smaller. This enables more files to be stored on the storage medium and also faster data transmission over networks.

pdf (portable document format) – used with a particular piece of applications software for viewing documents. This file format is used by the software Adobe Acrobat. This file format is popular because the software used to view the files is widely available and free.

png (portable network graphics format) – used for storing graphic images that is an improvement on the gif format as it does not lose any pixels (i.e. the dots making up the image) when compressing the file.

txt (text format) – text files contain just text without any formatting. When these files are imported into the package you are using, you may have to edit the text in order to add formatting.

rtf (rich text format) – this file format saves the text with a limited amount of formatting. You usually must still do some editing of the data to add additional formatting.

zip – a popular archive format widely used when downloading files from websites using the internet. ZIP files are data containers since they can store one or several files in compressed form. After you download a ZIP file, you must extract its contents in order to use them.

Locating stored files

You will need to be able to locate stored files for the examination. You will also need to create folders for the various files you create. When saving files for the first time it is always best to use 'Save As' because you can see where it is that the computer intends to save the file. If the location is not suitable then you can change it.

Before you start working through the activities, ensure you know where the files you need can be found. You will also need to decide where you will save any of the files amended or produced during the activities.

Opening and importing files of different types

As part of the practical examination, you will be expected to import and open files that are not in the file format usually used by the software. For example, you may be asked to import/open a comma separated values file (i.e. a file with the file extension .csv) using word-processing or spreadsheet software. In the practical activities in the following chapters you will get plenty of experience with opening and importing files of different types.

Saving files in a planned hierarchical folder structure

Organising files into folders and subfolders is an important part of file management. Over time, the number of files you store will increase and in the future you may need to find files stored many years before. Saving files in an organised structure will help you find files quickly. A suitable filing system on a personal computer will start off with the main folder Documents. It is then up to you to decide on the hierarchy of folders/subfolders below this. For example, you might decide on two folders below Documents called Home and School. You might then decide to create folders below the School folder for all your subjects (Maths, English, ICT, Chemistry, etc.). Below the Home folder you might have folders called Sport, Hobbies, Personal, etc.

Activity 11.1

Developing a planned hierarchical folder structure

1. Open the file folder on your computer's desktop that contains all your files. Generally, the name of this folder is 'Documents'.
2. Identify the different file types you have, so you can divide them into different categories or subfolders. At this stage you should have just a few folders. Two categories/folders could be Home and School. Choose your own.
3. Create a new folder for each category you identified. There may be a 'Make a New Folder' option that you can click to use for this. Alternatively, you may need to click on 'Files', then 'New', followed by 'Folder'. Rename each folder by the category by right-clicking the album, selecting 'Rename', typing in the name you wish to use and ckicking 'Enter'.
4. Move your files into the newly created folders. Click and hold the file you want to move. While holding the mouse button down, drag the file to the appropriate folder. Release the mouse button. You can move several files at once by holding down the 'Ctrl' button on your keyboard as you select the files you wish to move. Another way to put a file into a new folder is to use cut and paste or copy and paste.

   Using cut and paste the file is simply moved from its old position to the new position so the file is removed from its old position. With copy and paste, a copy of the file is put in the new position leaving the original file in its own position.
5. Subfolders can then be created within one of the new folders. Open one of the new folders you created and identify different categories. For example, you can further arrange the categories by date, event, etc.
6. Move the files in each subfolder into one of the new internal subfolders you created.
7. Continue creating subfolders and moving files until you feel you have a logical and organised file system.

Saving using appropriate file names

In the past you were limited to the number of characters in a file name but now long filenames are acceptable. Here are some pointers about the use of filenames:

- Avoid spaces in filenames and avoid the following characters / ? < > \ : * | ' ^
- You do not need to enter file extensions; the software you are using to save the file will add the extension automatically.
- Use filenames that make it clear what file contents are.

Here are some examples of good filenames:

project_work_v1.doc
lookup-example-spreadsheet.xls
namesandaddresses.csv
index.htm
view_from_top_of_eiffel_tower.jpg

Saving and printing files in a variety of formats

Files can be saved and printed in a variety of different formats and these include:

- Draft document
- Final copy
- Screen shots
- Database reports
- Data table
- Graph/chart
- Web page in browser view
- Web page in html view.

You will be saving and printing files in all these formats in the later chapters.

Generic file formats

Generic file formats are those file formats that are able to be used by different software no matter who the manufacturer of the software is. For example, if you use one word processor to produce documents and your friend uses a different one, then you will be able to send them a document that they can read, edit, save and print, even if they save the files in different formats. To do this, you would have to save your document in a generic file format such as .rtf or .txt. All word-processors can save and load files in these generic formats. Once your friend has opened the file, they can then save into the file format used by their software. They will now be able to view and make changes to the document using their software and if necessary they can save it in the generic format and send it back to you for your comments. Being able to read and edit files using different software is an important aspect when working with others.

The following are generic file formats that can be used for saving files and then exporting them into a different software package:

.csv	.rtf	.css
.txt	.pdf	.htm

To export a file into a different piece of software when there is no common file format

Suppose a file created in application software 1 is to be loaded into different application software 2. Software 2 cannot read files created in the usual file format for software 1 so we need to use a common generic file format. The file created in application software 1 is saved in the generic file format which is then read by application software 2. Application software 2 is now able to view and edit the file.

To use a generic file format, follow these steps:

1 Find a generic format that you can save your file in and check that the other software is able to load a file in this format.

   Suppose you notice that you can save the original file in .csv format and that the new software is able to use files in this format.

   The original file is now saved into the generic file format.

2 You now import the file in its common file format into the new software.

   Suppose you have a file which you have saved as a .csv file (i.e. this file has been saved into a generic file format).

   On loading the file into the new software, the data is converted into a form that can be used by the new software.

3 You now save the file in its new file format (i.e. the format used by the new software).

   The file can now be used.

Saving and exporting data into generic file formats

Suppose you have produced a word-processed document that you would like someone else to see in order for them to make comments about it. If they did not have the same word-processing software as you, you could save the file in a generic file format. If they just needed to open the file and view it, then you could save it in .pdf file format.

In the next activity you will save a word-processed document as a .pdf file and then view it using the pdf reader.

Activity 11.2

Saving a word-processed document as a .pdf file and then using a pdf reader to view the file

In this activity you will learn how to:

- save a word-processed document as a .pdf file
- read a document stored as a .pdf file

1 Load Microsoft Word or your word-processing software and open the document called **Computer_control_in_Formula_One.rtf**

2 Click on Save As. When the Save As menu appears, use the drop-down arrow in the Save as type section and select (*.pdf), and then click Save.

3 You have now saved the file in .pdf file format.

4 The file will usually load up automatically using the PDF reader thus allowing you to view the document in .pdf file format. It is important to note that if this does not happen you may not have the PDF reader installed on your computer. A reader can be downloaded free from the website www.adobe.com.

   It is important to note that the .pdf file can be viewed but it cannot be edited; you would need to buy the software to do this. However, you can edit the file using the word-processing software and then save as a .pdf file again.

In many of the activities in the following chapters you will have to save and export data into other application packages.

Reducing file sizes for storage or transmission

It is very important to reduce file sizes when the files are used online. The reasons for this are to increase the transmission speed when transferred over a network and to reduce the memory needed to store them.

- Many email providers have a file size restriction, so if a file is too large it cannot be sent as an email attachment.
- Large files take lots of time for downloading and eat up data usage when used on mobile devices where you may have to pay for data usage above a certain amount.
- Some websites have a certain restriction on the size of image files to be uploaded to their site.
- Social networking sites often have file size restrictions on image files.
- Users of websites prefer quality images but lower file sizes as it makes browsing faster.
- Reduced file sizes mean they occupy less memory.

Activity 11.3

Compressing an image file

A plan on paper showing the outline of plots of land has been scanned in using a scanner and the image is stored as a .jpg file.

1. Locate the image file called **Land_plots.jpg**.
2. Load Microsoft Office Picture Manager (or other suitable image editing software) and open the file **Land_plots.jpg**.

   If you use Microsoft Office Picture Manager the screen should look like this. If you use other software, it should have similar icons and menus for you to follow.

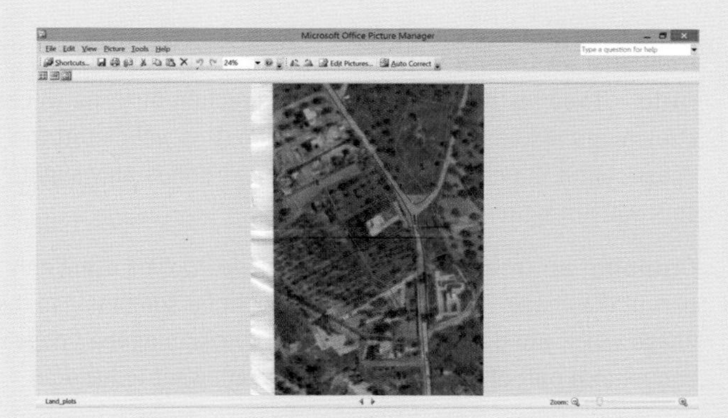

3. Move the cursor onto the image and a window pops up like this giving information about the image file.

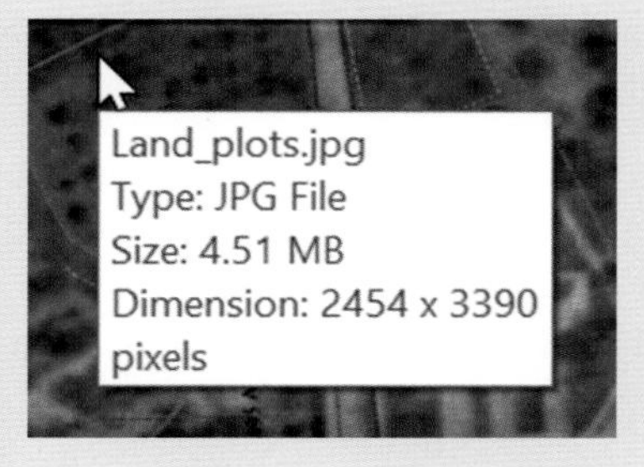

   Notice the following information in the screenshot shown here.

   The size of the file (i.e. Size 4.51 MB).

   The width and height of the image in pixels (i.e. 2454 × 3390). If you multiply these two numbers together you would obtain the total number of pixels used to create the image.

4. Notice that the plot numbers on the image cannot be read easily. The image needs to be rotated by 90° to the right. To do this, click on [icon] and when the menu appears, select Rotate right 90°.
5. Now compress this file so that it can be sent faster as an email attachment.

   Look at the left part of the screen where you should see the Compress Pictures menu. If it is not there, use the drop-down arrow to select Compress Pictures.

   Notice that you can compress for documents, web pages and email messages.

   Click on Compress for Documents.

   ◉ Documents

   Notice that the original file size was 4.51 MB and the new compressed file size is 155 KB.

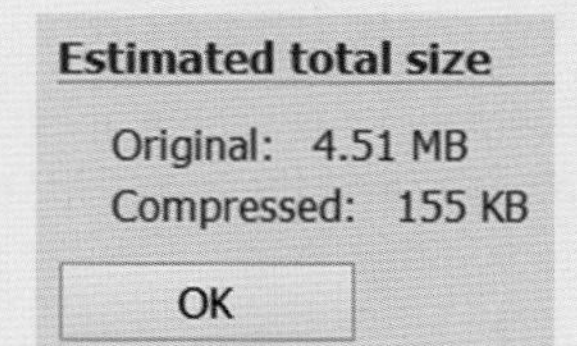

   Click on OK to confirm.

6. Click on File and then Save As to save this file in compressed form using the file name **Land_plots_compressed_file.jpg**.

12 Images

Images are used in documents and slides to communicate information. For the IGCSE you will not need to produce the images yourself as the images you will use will be supplied as a file. You will have to position the image on a document or slide as well as edit the image.

The key concepts covered in this chapter are:

- Using software tools to position images
- Using software tools to edit images

Using software tools to position images

Images, which can include clip art, photographs and screenshots, can be edited using specialist graphics/photo-editing software or alternatively you can use the software you are using to create the document or slide. In this book, we are going to use the latter method.

Word-processing, presentation and spreadsheet software allow you to insert a picture into a document, slide or spreadsheet. The cursor needs to be at the point where you want to insert the image.

The techniques for positioning images will be looked at in the activity later on in this chapter.

Using software tools to edit images

Software for creating slides and documents has simple image editing facilities and after you have positioned the image in the correct place, you can then begin to edit the image.

Aspect ratio

Aspect ratio is the ratio of the width of an image to its height, and it is expressed as two numbers separated by a colon like this 16:9. For example, if you had an image with a width of 16 cm and a height of 9 cm, its aspect ratio is 16:9. If we halved the width of the image to 8 cm then the height would need to be halved to 4.5 cm to keep the aspect ratio the same.

You will need to edit images by performing some of the following operations:

Placing an image with precision – you will normally be told where to position an image (e.g. on the next line following a certain title). You will also usually be told about the alignment of an image. For example, you may be told to align the image with the top of a paragraph of text and to the left margin.

Resizing an image, maintaining the aspect ratio – in most cases the image will need to be resized and in the examination you may be told the width or height in cm that the image will need to be resized to. If you change the width of the image the height will be automatically adjusted so that the image is not distorted. Similarly if the height is adjusted, the width will be automatically adjusted. When resizing an image in this way you are said to be maintaining the aspect ratio of the image.

Resizing an image without maintaining the aspect ratio – if you resize the image in one direction only you will be stretching the image in a single direction. This will distort the image, but if the change is small the distortion is only slight, so it will not be noticed.

Cropping an image – cropping an image involves using only part of the image. For example, there might be something in the image which you don't want to include. By cropping the image you can remove part of the image that you don't want to include.

Rotating an image – images can be obtained in portrait view (longest side is the height) or in landscape view (shortest side is the height) and you may want to turn the image around. You often have to rotate photographs taken with a digital camera so they are the right way around.

Reflecting an image – sometimes an image of an object can look better if it is facing in the opposite direction. Reflecting an image produces a mirror image of the object. Note that reflecting an image is called flipping the image in some software packages.

Adjusting the colour depth in an image – the colour depth of an image refers to how many bits of information are used to represent the colour of each pixel. The more bits the more accurate you can make the colour. The trouble is that more bits leads to greater file sizes. Using software tools you can increase the colour depth in an image.

Adjusting the brightness of an image – the brightness of images can be adjusted to make the image look more realistic.

Adjusting the contrast of an image – contrast is the difference between the light and dark areas of an image.

The implications of image resolution

An image on the screen is made up of lots of dots of light called pixels. The resolution of an image is given as two numbers like this: 640×480, where 640 is the number of pixel columns in the width of the image and 480 is the number of pixel rows in the height of the image. By multiplying these two numbers you obtain the total number of pixels in the image. In this example, $640 \times 480 = 307\,200$ pixels. As we normally use mega-pixels we divide this number by one million to give 0.3 M pixels.

A larger image taken using a smartphone might use 2560 × 1920 = 4 915 200 pixels which is approximately 5 M pixels.

The resolution of an image affects the quality of the image. The file size of a high resolution image will be higher than a low resolution image as more pixels and hence more bits are needed to store the information about each pixel. An increased file size will have the following implications:

- Transmission time will be increased – when you upload (e.g. store the image on a website, social networking site or load the image on an auction site or save the image to the cloud or attach the image to send with an email) the time taken will be greater. When you download the image (e.g. view the image on a website) the time will also be greater.
- The file size will increase – this will have implications on the number of images you can store on storage media.

Adjusting image resolution

Image resolution can be adjusted so as to reduce the file size. This is particularly important if the image is to be used on a website. There is only so long that a viewer of a web page will wait for it to load before clicking off the site and going to a different one.

Activity 12.1

Resizing images and distorting an image

In this activity you will learn how to:

- locate stored files and load a file in a generic file format
- set page size, orientation and margins
- position an image in a document
- adjust the size of an image, maintaining the aspect ratio
- adjust the size of an image, distorting the image slightly

1 Load Microsoft Word or your word-processing software and open the file **Emerging_technology.rft**.

2 Set the page size to A4. To do this click on Page layout and the Page Setup menu appears as shown here.

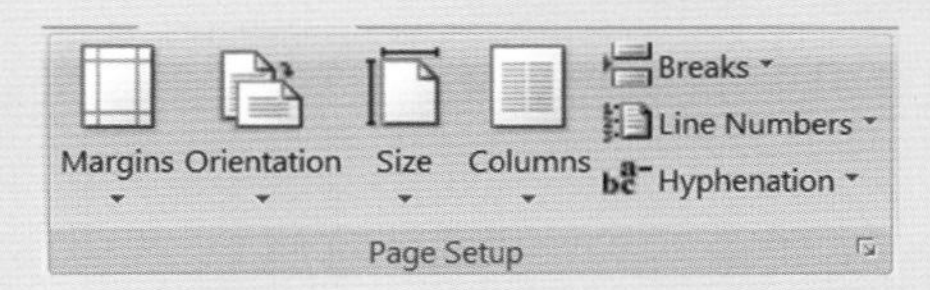

Notice all the properties you can set from this menu. As we need to set the page size, click on Size and from the list of sizes select A4.

3 Keeping on the Page Layout click on Orientation and select Portrait for the orientation.

4 Again, keeping on the Page layout click on Margins. From this menu you can select custom margins. Click on Custom Margins and the following dialog box will appear.

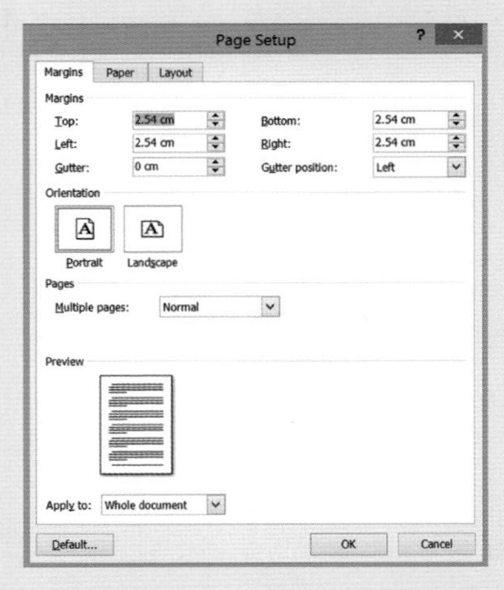

Set all the margins to 2 cm. Adjust the settings for the Top, Bottom, Right, and Left margins to 2 cm. Click on OK to confirm.

5 Save your document as Emerging Technology v2 in your work area, but do not store it as an .RTF file. Instead, save it in the file format used by your word-processing software.

6 Click on the first line after the first paragraph of the nanotechnology section.

Import the image **carbon_nanotube.jpg**.

This can be done by clicking Insert and then Picture. You will have to locate the file **carbon_nanotube.jpg**.

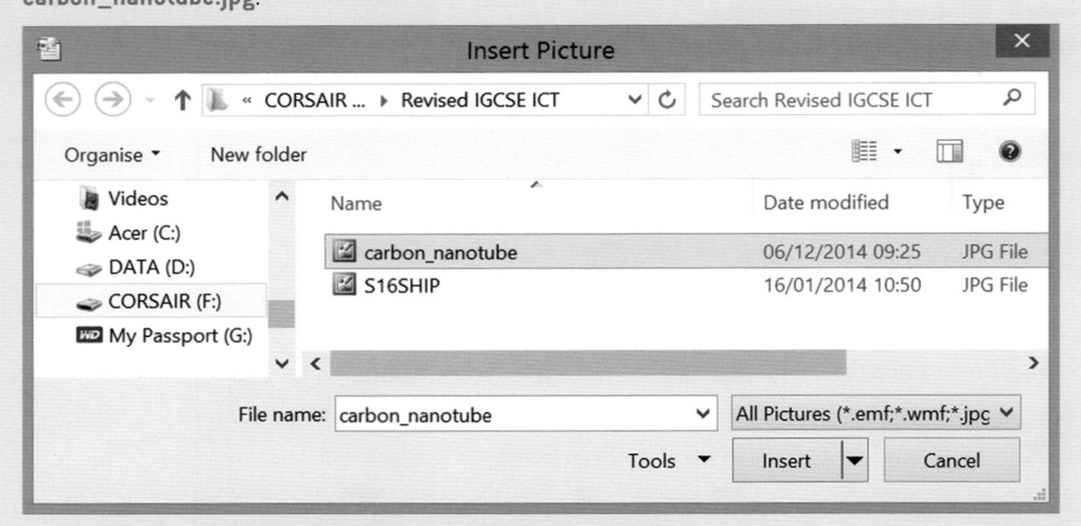

When you have located the file, select it and click on Insert.

Notice that this image has been aligned to both margins.

7 This image needs resizing. Click on the image to display format picture tools.

It is worth spending a bit of time looking at what you can do using these tools. Check the location of the tools to do the following to the image:

- Adjust the brightness
- Adjust the contrast
- Adjust the colour (Recolour)
- Adjust the way the text wraps around the object (Position)
- Crop the image
- Rotate the image
- Size the image.

Resize the image to a height of 8 cm. Notice that the width of the image changes automatically to keep it in proportion. This is called maintaining the aspect ratio.

8 Click on Position and then choose Position in Top Right with Square Text Wrapping.

The page containing the image should now look like this.

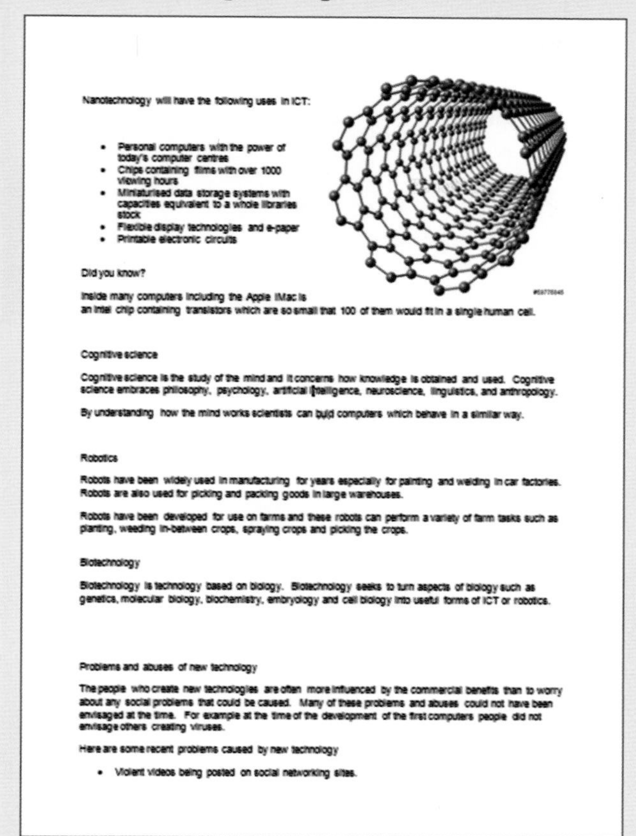

Nanotechnology will have the following uses in ICT:

- Personal computers with the power of today's computer centres
- Chips containing films with over 1000 viewing hours
- Miniaturised data storage systems with capacities equivalent to a whole libraries stock
- Flexible display technologies and e-paper
- Printable electronic circuits

Did you know?

Inside many computers including the Apple iMac is an Intel chip containing transistors which are so small that 100 of them would fit in a single human cell.

Cognitive science

Cognitive science is the study of the mind and it concerns how knowledge is obtained and used. Cognitive science embraces philosophy, psychology, artificial intelligence, neuroscience, linguistics, and anthropology.

By understanding how the mind works scientists can build computers which behave in a similar way.

Robotics

Robots have been widely used in manufacturing for years especially for painting and welding in car factories. Robots are also used for picking and packing goods in large warehouses.

Robots have been developed for use on farms and these robots can perform a variety of farm tasks such as planting, weeding in-between crops, spraying crops and picking the crops.

Biotechnology

Biotechnology is technology based on biology. Biotechnology seeks to turn aspects of biology such as genetics, molecular biology, biochemistry, embryology and cell biology into useful forms of ICT or robotics.

Problems and abuses of new technology

The people who create new technologies are often more influenced by the commercial benefits than to worry about any social problems that could be caused. Many of these problems and abuses could not have been envisaged at the time. For example at the time of the development of the first computers people did not envisage others creating viruses.

Here are some recent problems caused by new technology

- Violent videos being posted on social networking sites.

9 In the robotics section of the text, insert an extra line.

With the cursor positioned on this line click Insert and then Picture. You will have to locate the **file_robots.jpg** and select it and click on Insert.

10 The image of robots will appear between the two sentences. Insert an extra line between the two sentences.

You are now going to change the height of the image without changing the width. This means adjusting the size of the image without maintaining the aspect ratio.

Right-click on the image and the following menu appears.

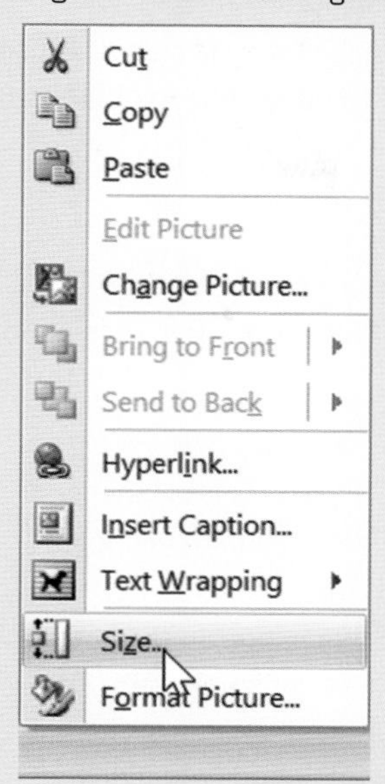

Select Size and the following menu appears:

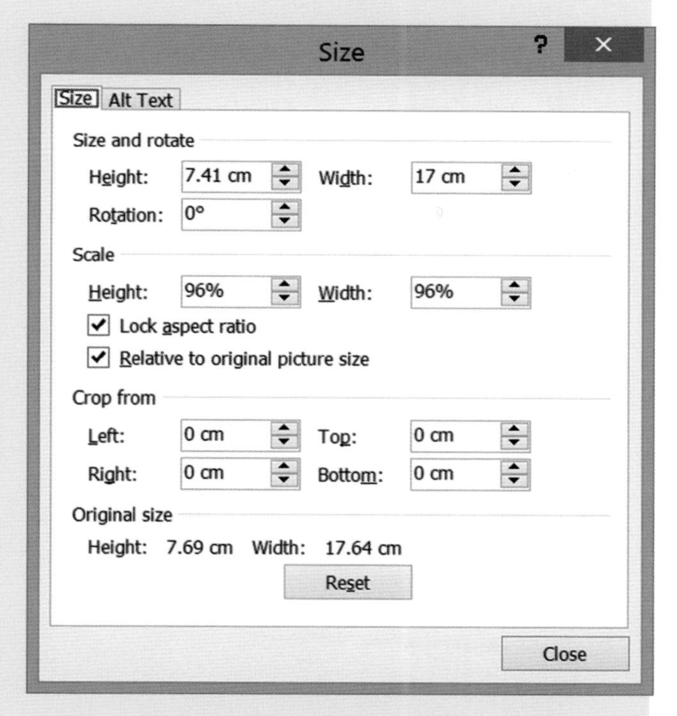

Because you are adjusting the size of the image without maintaining the aspect ratio you need to click on Lock aspect ratio to remove the tick. Now change the Height of the image to 6 cm. Notice as you are doing this the width of the image stays the same, so you are distorting the image slightly.

Click on Close.

11 Save this document in your work area using the filename **Emerging_technology_v3**. Make sure that you save using the file format used by your word-processing software.

You will be using this file in an activity in a later chapter.

Activity 12.2

Editing an image

In this activity you will edit an image using Microsoft Office Picture Manager which works in a similar way to other image editing software.

In this activity you will learn how to:

- crop an image
- change the brightness and contrast of an image
- export a file using a different file format (i.e. from a .png file to a .gif file).

1. Locate the image file called **Lindos**.
2. Load the software Microsoft Office Picture Manager or whichever image editing software you are using and open the file **Lindos**.
3. There are two problems with this image. In the bottom right of the image there is the top of someone's head in a baseball cap. At the very top of the image there is something else in the image that needs removing.

   These objects can be removed by cropping the image.

   Click on Picture and then select crop.

4. Notice the black corners. If you click on them and drag them you can move the corners.

   Now use the corners to crop the image so that only a rectangle is marked similar to the one shown below.

Then click on OK

We have now removed the unwanted parts of the image to give an image like the one shown below.

5. Now alter the brightness and contrast of the image by clicking on the drop-down arrow

then select Brightness and Contrast:

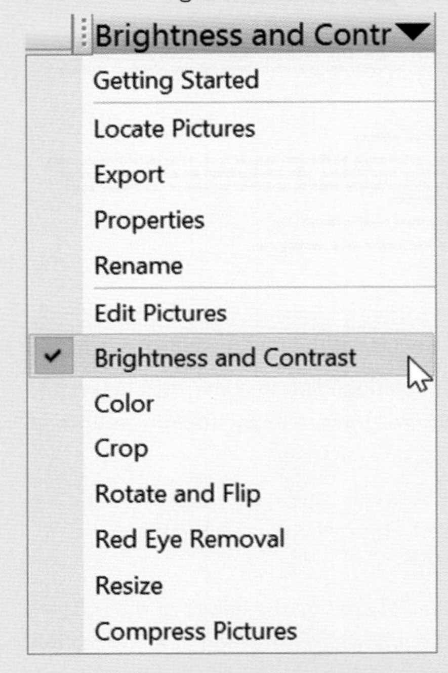

Adjust the Brightness to 10 and the Contrast to 15 using the sliders as shown below.

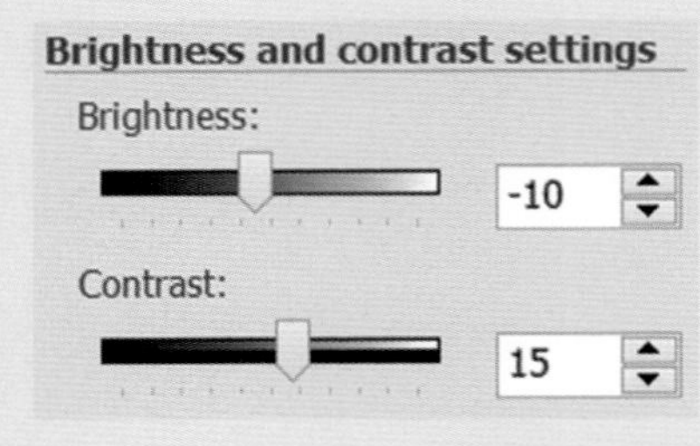

6. Click on the drop-down arrow Brightness and Contr

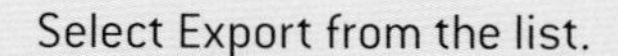

Select Export from the list.

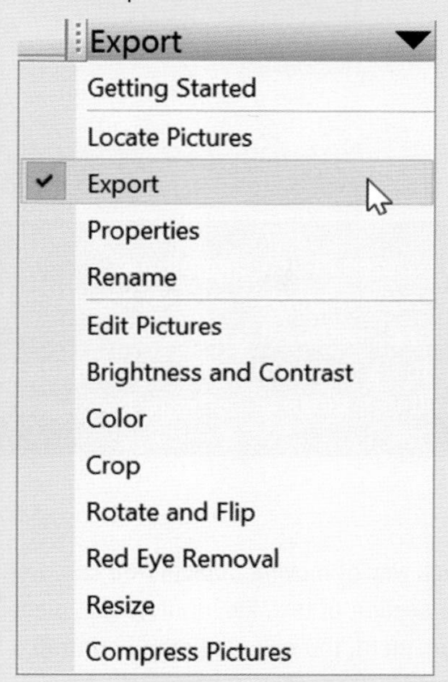

The following menu appears.

Change Export with this file name to GIF. (This changes the file format from the original .png format to .GIF format.)

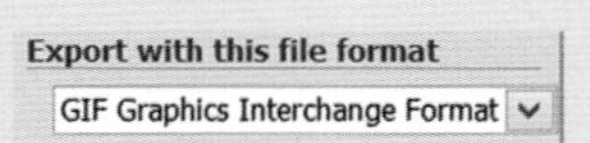

Save the file using the original name in this new file format.

Click on OK.

13 Layout

Documents and slides consist of different elements such as text, images, charts, graphs, tables, screenshots, etc. The layout of the elements on the page is just as important as the content and it needs to meet the needs of the target audience.

The key concepts covered in this chapter are:

- Using software tools to prepare a basic document to match the purpose and target audience
- Using software tools to create headers and footers appropriately within a range of software packages

Using software tools to prepare a basic document to match the purpose and target audience

Entering and editing text and numbers

Text and numbers can be entered into word-processing software in a number of ways:

- using keyboard entry
- loading/importing a file created using the same word-processing software or created by another package
- copying and pasting from another source.

Using keyboard entry

If the text is not available to load/import or copy, then it is necessary to key it in using the keyboard. Keyboard entry needs to be done carefully in order not to introduce errors. Checking your work after typing is extremely important in the examination as the text and numbers you type in must exactly match the text you are instructed to type in.

Copy and paste

To copy text in a document, select it (the selected text will be highlighted) then right-click on the mouse. A menu will appear with a number of options including the following:

- Cut
- Copy
- Paste.

If you click on Copy, the text will now be placed in temporary storage (called the clipboard) by the computer. You can now move the cursor to another part of your document, to where you want the text to be moved. Right-clicking on the position brings up the menu again where you can choose to paste the contents into position. You will now have two copies of the text – one block in the old position and one block in the new position.

Cut and paste

To cut and paste, you simply select the text and right-click, and from the resulting menu select Cut. When the text is cut, the text is taken out of the document. You then click where you want the text to go, right-click and select Paste when the text appears in the new position.

Drag and drop

Drag and drop is a quick way of moving objects. You can drag and drop objects such as a section of text, file, photograph, piece of clip art, etc., onto a document. You can also drag and drop a file into a location (e.g. a folder or the Recycle bin).

Placing objects into a document from a variety of sources

When a document is created it often involves using objects such as text, images, screen shots, spreadsheet extracts, database extracts, clip art, and graphs and charts. Sometimes you will create the objects yourself before putting them into a document and sometimes these objects will be supplied as files for you to use.

Remember that the word 'document' is used widely here so it could be a word-processed document, a slide or a web page.

Capturing a screenshot

In the practical examination you may be required to show evidence of how you did something using the software. For example, you might be asked to produce evidence of table design for a database. You might have to provide this in a report (a document outlining your findings when you are asked to do something) and will therefore have to integrate the screenshot with text in a word-processed document.

To produce a screenshot without using specialist software you can:

1. Display the screen you want to capture.
2. If you want a screenshot of everything on the screen, press the Prt Scr key once. If you have more than one window displayed on the screen, then a screenshot of the current window can be obtained by holding down the Alt key and then pressing the Prt Scr key.
3. The screenshot is copied to the temporary storage area called the clipboard. This can be pasted into position in your document, slide, web page, etc.

Using specialist screen capture programs

There are many different screen capture programs and one of the most popular ones is called Snagit. You can also use the Snipping Tool in Windows Accessories. It offers more flexibility than simple screen capture using Alt and then Prt Scr. If you want to try this software, there is a free trial version which you can download onto your computer at: **https://www.techsmith.com/snagit.html**

Capturing an image from a website

You may be asked to locate an appropriate image to be used within a document. Once you have located a suitable image, here are the steps you should take:

1 Display the web page where the image you want to capture is located.
  Move the cursor onto the image you want to capture like this:

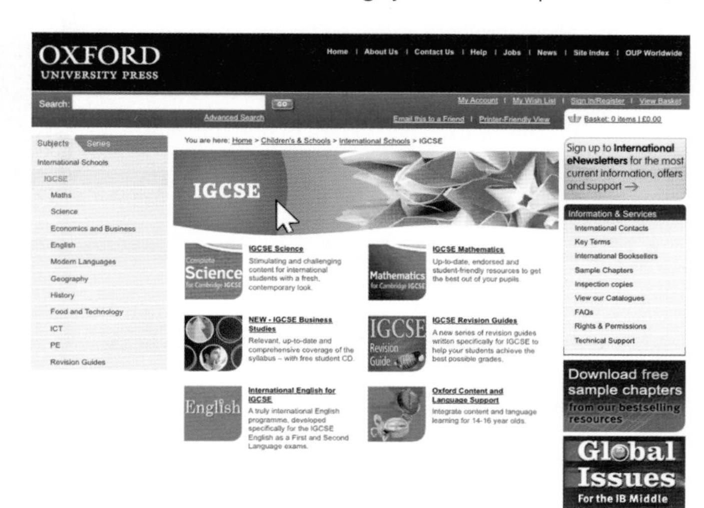

2 Right-click and the following menu appears:

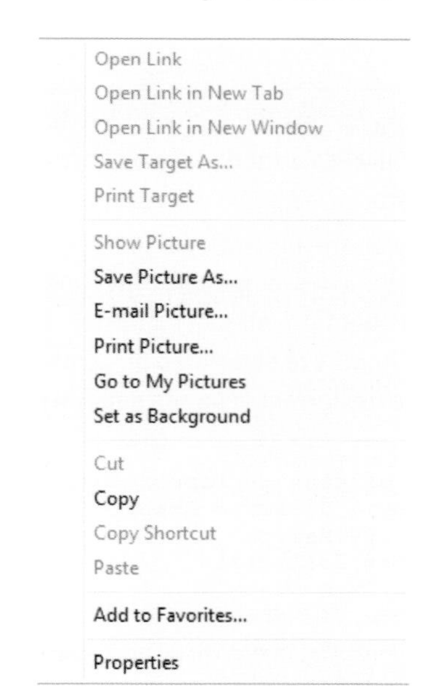

You can either save the picture as a file, or copy it to the clipboard.

- Click on Save Picture As so that file containing the picture can be saved on disk. You will have to browse to find an appropriate folder in which to store the image and also give the image a name if the one suggested is unsuitable like this:

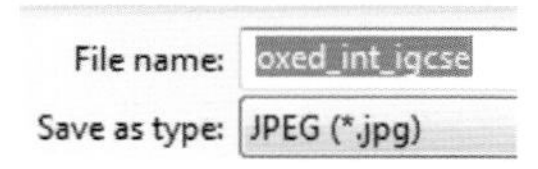

- Click on Copy to put a copy of the image into the clipboard. You can then position the cursor where you would like the image to appear in the document. You can then click on Paste to put the image into position.

Importing text from a website

You may be required to copy text from a website or email and then incorporate it into a document or presentation slide, and the steps you need to take are as follows:

1 Display the web page where the text you want to copy is located.
  Move to the start of the text and left-click and then drag the mouse until the complete section of text has been highlighted like this:

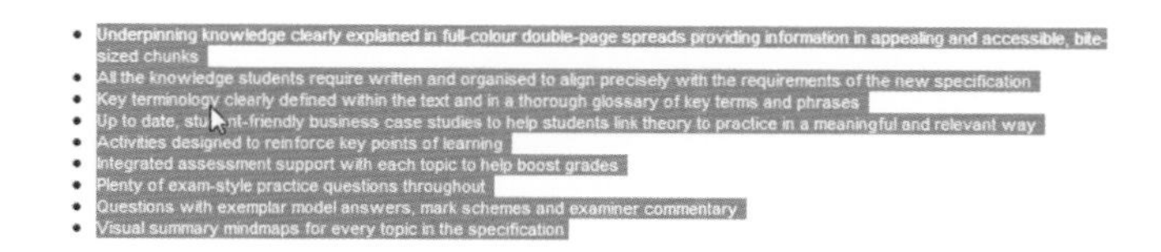

- Underpinning knowledge clearly explained in full-colour double-page spreads providing information in appealing and accessible, bite-sized chunks
- All the knowledge students require written and organised to align precisely with the requirements of the new specification
- Key terminology clearly defined within the text and in a thorough glossary of key terms and phrases
- Up to date, student-friendly business case studies to help students link theory to practice in a meaningful and relevant way
- Activities designed to reinforce key points of learning
- Integrated assessment support with each topic to help boost grades
- Plenty of exam-style practice questions throughout
- Questions with exemplar model answers, mark schemes and examiner commentary
- Visual summary mindmaps for every topic in the specification

2 With the cursor on the highlighted text, right-click and select Copy.
3 This puts the selected text into the clipboard.
4 Move to the document that you want the text copied into and click on Paste.

The text will now be copied into the required position. All that you have to do is to ensure that the text (font, font size, font style, font colour, etc.) is consistent with the rest of the document. Check other formatting such as the number of lines after a paragraph, bullets are the same as used in the same document, and so on.

Copying data from a table

It is relatively easy to take tabular data from a document, slide, website, etc. and put it into a spreadsheet. You simply select all the required data in the table by highlighting it and then copying it to the clipboard. It can then be pasted into the spreadsheet. You may have to widen columns in order to accommodate the data or headings.

Here, a table of weather data has been found on the internet. It has been selected by highlighting it. Right-clicking on the data enables the data to be copied into the clipboard.

	Maximum	Minimum
Rhodes weather in January	12°C / 54°F	5°C / 41°F
Rhodes weather in February	14°C / 57°F	6°C / 43°F
Rhodes weather in March	16°C / 61°F	7°C / 45°F
Rhodes weather in April	18°C / 64°F	10°C / 50°F
Rhodes weather in May	26°C / 79°F	15°C / 59°F
Rhodes weather in June	29°C / 84°F	17°C / 63°F
Rhodes weather in July	30°C / 86°F	20°C / 68°F
Rhodes weather in August	31°C / 88°F	21°C / 70°F
Rhodes weather in September	27°C / 81°F	18°C / 64°F
Rhodes weather in October	23°C / 73°F	13°C / 55°F
Rhodes weather in November	17°C / 63°F	10°C / [illegible]0°F
Rhodes weather in December	14°C / 57°F	7°C / 45°F

You can then open the spreadsheet software, position the cursor where you want the data to start and then paste the data into it.

	A	B	C	D
1		Maximu m	Minimum	
2	Rhodes weather in January	12°C / 54°F	5°C / 41°F	
3	Rhodes weather in February	14°C / 57°F	6°C / 43°F	
4	Rhodes weather in March	16°C / 61°F	7°C / 45°F	
5	Rhodes weather in April	18°C / 64°F	10°C / 50°F	
6	Rhodes weather in May	26°C / 79°F	15°C / 59°F	
7	Rhodes weather in June	29°C / 84°F	17°C / 63°F	
8	Rhodes weather in July	30°C / 86°F	20°C / 68°F	
9	Rhodes weather in August	31°C / 88°F	21°C / 70°F	
10	Rhodes weather in Septembe r	27°C / 81°F	18°C / 64°F	
11	Rhodes weather in October	23°C / 73°F	13°C / 55°F	
12	Rhodes weather in November	17°C / 63°F	10°C / 50°F	
13	Rhodes weather in December	14°C / 57°F	7°C / 45°F	

There is usually some tidying up to do. For example, here there are some '/' characters that need deleting and columns A and B should be made wider. If you want to draw graphs, you need to remove the units from the data and put them in the headings instead.

Activity 13.1

Creating a table using data from a csv file

In this activity you will learn how to:

- import a file in csv file format into word-processing software
- place text into a table automatically
- make text bold.

1 Load Microsoft Word and open the file **Countries** (which has been saved in csv file format) by double-clicking on the file icon.
Your teacher will tell you where to find this file.

When looking for this file in the folder, make sure that you select 'All Files' otherwise only Microsoft Word files will be shown.

2 Notice the way the file is displayed. The first line contains column headings and other lines contain the actual data about the African countries. Notice also that all the items are separated by commas, hence the name of the file format (comma separated values).

```
Name of country,Population,Capital
Ethiopia,85,Addis Ababa
Kenya,39,Nairobi
Algiers,33,Algiers
Ghana,23,Accra
Morocco,34,Rabat
Mozambique,20,Maputo
Nigeria,155,Abuja
Somalia,10,Mogadishu
South Africa,47,Pretoria
Sudan,39,Khartoum
Sengal,12,Dakar
Tanzania,38,Dodoma
Tunisia,10,Tunis
Uganda,28,Kampala
Zambia,15,Lusaka
Zimbabwe,13,Harare
```

3 Left-click on the start of the data and keeping your finger pressed down on the left mouse button drag down until all the data is selected. When it is selected it will show blue like this:

```
Name of country,Population,Capital
Ethiopia,85,Addis Ababa
Kenya,39,Nairobi
Algiers,33,Algiers
Ghana,23,Accra
Morocco,34,Rabat
Mozambique,20,Maputo
Nigeria,155,Abuja
Somalia,10,Mogadishu
South Africa,47,Pretoria
Sudan,39,Khartoum
Sengal,12,Dakar
Tanzania,38,Dodoma
Tunisia,10,Tunis
Uganda,28,Kampala
Zambia,15,Lusaka
Zimbabwe,13,Harare
```

4 You are now going to show this data in a table. You do not need to work out the number of columns and rows in the table as the software can do this automatically.

Click on 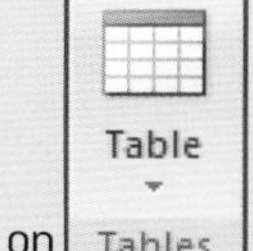

in the toolbar. The following menu appears from which you should select Convert Text to Table:

- Insert Table...
- Draw Table
- Convert Text to Table...
- Excel Spreadsheet
- Quick Tables

5 The following dialog box appears:

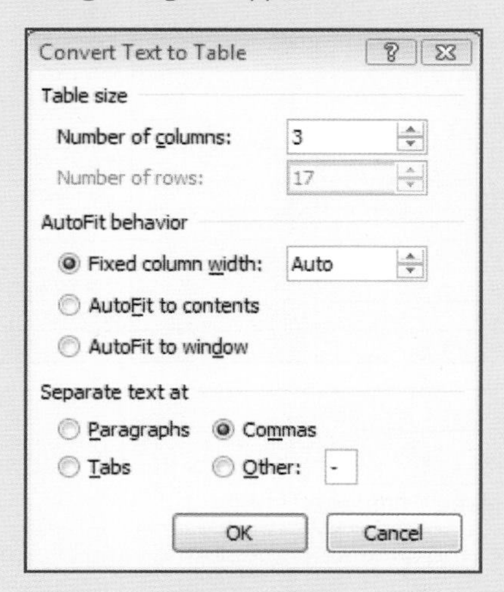

Notice that the number of columns and rows has been determined by the data itself.

Click on OK to put the data into the table. Your table should now appear like this:

Name of country	Population	Capital
Ethiopia	85	Addis Ababa
Kenya	39	Nairobi
Algiers	33	Algiers
Ghana	23	Accra
Morocco	34	Rabat
Mozambique	20	Maputo
Nigeria	155	Abuja
Somalia	10	Mogadishu
South Africa	47	Pretoria
Sudan	39	Khartoum
Senegal	12	Dakar
Tanzania	38	Dodoma
Tunisia	10	Tunis
Uganda	28	Kampala
Zambia	15	Lusaka
Zimbabwe	13	Harare

Select the first row containing the column headings and then click on **B**.

This makes the column headings bold.

6 Save this word-processed file using the filename **Table_of_countries**.

Activity 13.2

E ntering and editing text

In this activity you will learn how to:

- accurately use the keyboard to enter text
- use drag and drop to move text around
- alter the font size
- centre text
- change the line spacing in a document
- copy and paste an object (i.e. a table in this case) from one document to another
- alter the border on a table
- import an image file and position it in a document
- edit an image by cropping it.

1 Load Microsoft Word and create a new document. Type in the following text:

Africa is the world's second largest continent according to land size as well as by size of population. The African continent consists of 54 separate countries and about 14% of the world's total population live here.

The climate in Africa varies from tropical to subarctic on its highest peaks. In the northern part of Africa there are deserts and most of the land is arid. In the central and southern parts there are savannah plains and dense rainforests.

Africa boasts the greatest variety of animals and includes animals such as lions, elephants, cheetahs, deer, giraffes, camels, monkeys, etc. One ecological problem is deforestation where forests are being destroyed to make room for the growing of crops. This is destroying the habitat for many of the endangered species.

Most scientists regard the African continent to be the origin of the human species and evidence has been found of the first modern human called Homo sapiens dating to around 200,000 years old.

2 Use visual verification to check the text you have typed in by comparing it with the original text. The two should be identical. If there are any errors, these should be corrected before proceeding.

3 Select the entire third paragraph by highlighting it. Once selected, click on the highlighted text and keeping the right mouse button pressed down drag the cursor to the line between the first and second paragraphs. This technique is used to move text around.

Your text should now look like this:

Africa is the world's second largest continent according to land size as well as by size of population. The African continent consists of 54 separate countries and about 14% of the world's total population live here.

Africa boasts the greatest variety of animals and includes animals such as lions, elephants, cheetahs, deer, giraffes, camels, monkeys etc. One ecological problem is deforestation where forests are being destroyed to make room for the growing of crops. This is destroying the habitat for many of the endangered species.

The climate in Africa varies from tropical to subarctic on its highest peaks. In the northern part of Africa there are deserts and most of the land is arid. In the central and southern parts there are savannah plains and dense rainforests.

Most scientists regard the African continent to be the origin of the human species and evidence has been found of the first modern human called Homo sapiens dating to around 200,000 years old.

4 In what is now the second paragraph delete the sentence: 'This is destroying the habitat for many of the endangered species.' This should be done by highlighting the sentence and then pressing the DELete key. This is the quickest way of deleting a section of text.

Check that you have a line between each paragraph, otherwise insert one.

5 The document needs a title. Insert the title 'Africa' at the top of the page. You will need to move the existing text down two lines to make room for this title. Do this by clicking on the first letter of the first sentence and press enter twice.

Select the title by highlighting it.

Africa

Africa is the world's second largest continent according to land size as well as by s
of population. The African continent consists of 54 separate countries and about 1
of the world's total population live here.

Click on the font size and change it to 20pt like this [20 ▾].

Make the text bold by clicking on [B] and centre the text by clicking on [≡].

6 You are now going to increase the line spacing between the lines of text. Select all the text other than the title by highlighting it.

Ensure you have the [Home] tab selected and then from the paragraph section click on [↕≡].

From the resulting list, select the line spacing 1.5 like this:

✓ 1.0
1.15
1.5
2.0
2.5
3.0
Line Spacing Options...
Add Space Before Paragraph
Add Space After Paragraph

The document with the increased line spacing is now shown.

7 You are now going to add the table created in the previous activity.

Click midway between the first and the second paragraph as this is where the table is to be inserted.

Open the file **Table_of_countries** which contains the table.

Select the entire table by highlighting it like this:

Name of country	Population	Capital
Ethiopia	85	Addis Ababa
Kenya	39	Nairobi
Algiers	33	Algiers
Ghana	23	Accra
Morocco	34	Rabat
Mozambique	20	Maputo
Nigeria	155	Abuja
Somalia	10	Mogadishu
South Africa	47	Pretoria
Sudan	39	Khartoum
Senegal	12	Dakar
Tanzania	38	Dodoma
Tunisia	10	Tunis
Uganda	28	Kampala
Zambia	15	Lusaka
Zimbabwe	13	Harare

Click on [Copy] which copies the selected area to a storage place called the clipboard.

Go back to the document about Africa and click on

which puts the table into the correct position.

8 The table needs adjusting to line it up with the text.

Left-click on the left table border like this Mo Mo and keeping the mouse button pressed down drag the left border to the right so that it is lined up with the text. Now line up the right margin with the text.

Your document will now look like this:

Africa

Africa is the world's second largest continent according to land size as well as by size of population. The African continent consists of 54 separate countries and about 14% of the world's total population live here.

Name of country	Population	Capital
Ethiopia	85	Addis Ababa
Kenya	39	Nairobi
Algiers	33	Algiers
Ghana	23	Accra
Morocco	34	Rabat
Mozambique	20	Maputo
Nigeria	155	Abuja
Somalia	10	Mogadishu
South Africa	47	Pretoria
Sudan	39	Khartoum
Senegal	12	Dakar
Tanzania	38	Dodoma
Tunisia	10	Tunis
Uganda	28	Kampala
Zambia	15	Lusaka
Zimbabwe	13	Harare

Most scientists regard the African continent to be the origin of the human species and

9 You are now going to insert a digital image into the document.

Click midway between the third and fourth paragraphs.

Click on the Insert tab and then select Picture.

You now need to locate the file called **Map_of_Africa** in the area where all the files you need are stored. Double left-click on the file to insert the image into the document.

10 This image shows Europe as well as Africa and it would be best to use only part of the image (i.e. mainly Africa). You are going to cut out part of the image and discard the remainder. This is called cropping the image.

Select the image by clicking on it. You will see the picture formatting toolbar at the top of the screen.

Notice the size section where you can alter the height and width of an image as well as crop an image.

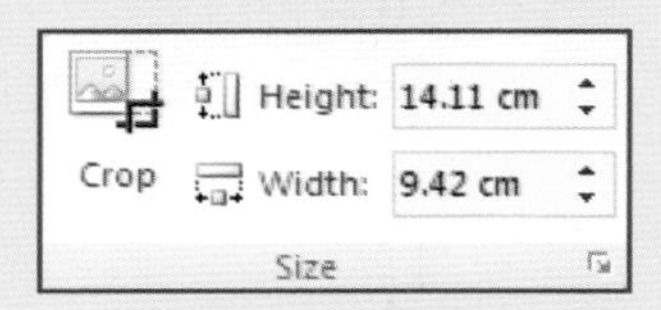

Click on the Crop icon.

The image will appear like this with the black handles that can be used to crop the image around the edges.

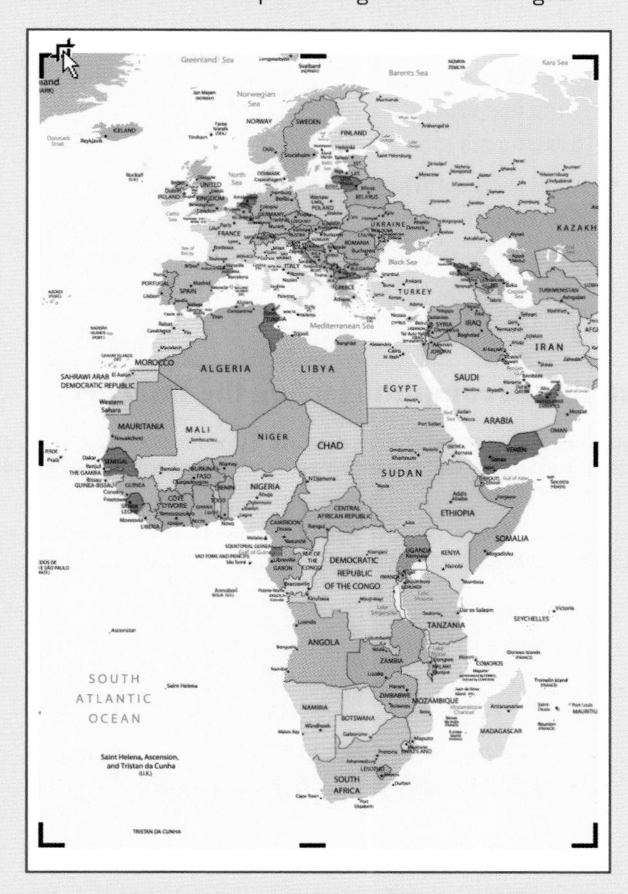

Use the handles to drag the corners and edges until the image looks a bit like this:

11 The article is now complete, so save it using the filename **Completed_Africa_article**.

Activity 13.3

Placing and manipulating images

In this activity you will learn how to:

- locate stored files
- import an image file
- size an image maintaining the aspect ratio
- wrap text around an image
- adjust a margin.

1 Load Microsoft Word and open the document called **Distributed_computing**.

Check that you have the right document displayed by comparing the start of the document with that shown here:

Distributed computing using the Internet

One main problem in collecting data is that there is so much of it to analyse. Analys
amounts of data requires a lot of computer power and this power may not be availab
because of the cost.

In many cases this problem can be solved by using distributed computing using the I
Here, instead of using a huge expensive supercomputer to do the job, it could be do

2 There are three images to be put into this text and they have been saved using these filenames:

Dist_computing_image1
Dist_computing_image2
Dist_computing_image3

All these images have been saved in the folder for this book.

Position the cursor on the line between the first and second paragraphs of the document.

Click on the Insert tab and then click on 

.

You will then have to locate the folder containing the image files that are to be imported into this document.

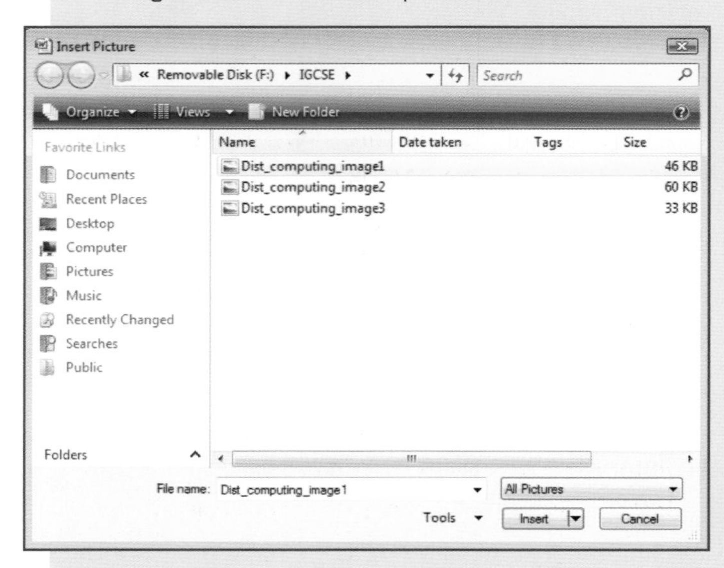

Once found, click on the file **Dist_computing_image1** and then click on the Insert button.

The image is inserted into the document like this:

Distributed computing using the Internet

One main problem in collecting data is that there is so much of it to analyse. Analysing large amounts of data requires a lot of computer power and this power may not be available because of the cost.

3 Right-click on the image and choose Size from the menu that appears. This dialog box appears:

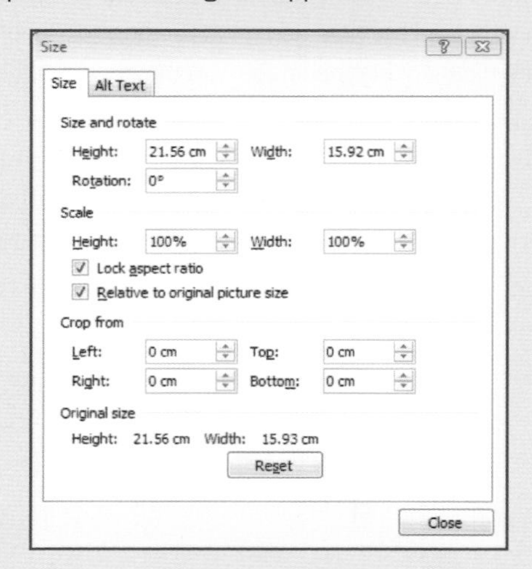

Notice the tick in the Lock aspect ratio box. This means that if either the height or width of the image is changed then the other dimension will change in order to keep the image in proportion. If you want to change the size of the image in one direction only, then this tick will need to be removed.

At the moment the image is too large, so reduce the height of the image from 21.56 cm to 13cm.

Click on Close.

4 Right-click on the image again and this time select Text Wrapping from the menu.

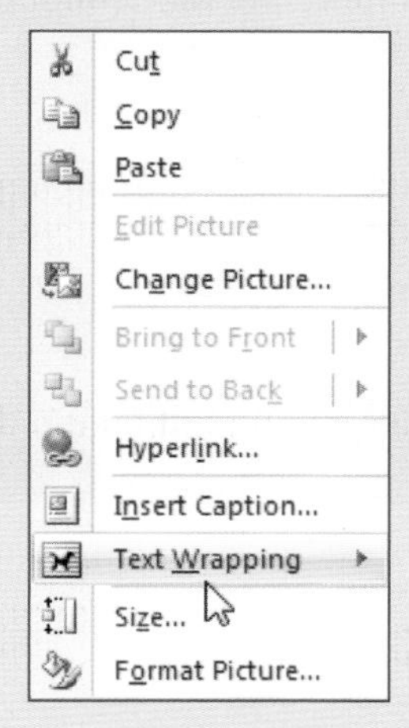

You will then be presented with a number of text wrapping options:

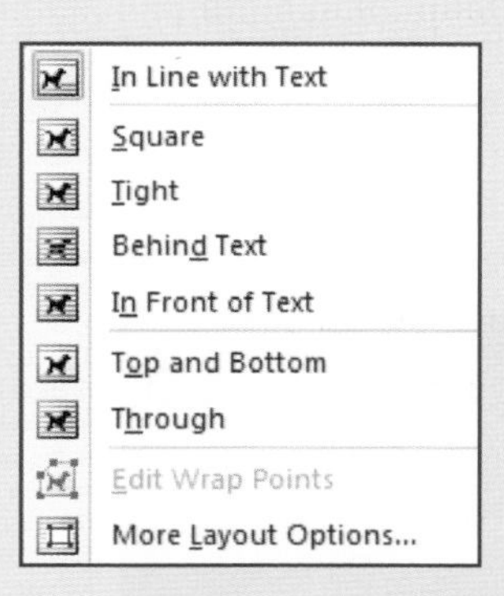

Notice these text wrapping options. You might be asked to use any of the following:

- Square – if the image is placed in the centre of text the text flows around the image leaving a square shaped image.
- Tight – here the text moves in to the shape of the image on the right- and left-hand side of the image. You can see this effect only if the image is not a regular shape.
- Top and bottom (this is called above and below in the CIE specification)—this places the image with the text above and below but with no text wrapped to the side of the image.

Click on the option 'Square' in the menu and the text will appear like this:

Distributed computing using the Internet

One main problem in collecting data is that there is so much of it to analyse. Analysing large amounts of data requires a lot of computer power and this power may not be available because of the cost.

In many cases this problem can be solved by using distributed computing using the Internet. Here, instead of using a huge expensive supercomputer to do the job, it could be done using a few less powerful computers connected together using the Internet but with each working on the same problem. In many cases computers just sit around not doing much for most of the time so why not use their processing power?

Some distributed systems ask home users to contribute some of their computing resources. For example, there is climate change research going on at the moment that requires a huge amount of data processing. Home users can contribute some of their wasted computer time to this project.

Advantages of distributed computing

- reduces cost because an expensive powerful computer such as a supercomputer is not needed

5 It is necessary to tidy up the text. Click just before *Advantages of distributed computing* and press the enter key twice to move this heading down.

6 Directly under the image, type in the following text using the font Times New Roman and a 10pt font size:

Distributed computing is a very useful tool for scientists who have huge amounts of data to analyse when trying to model systems such as the Earth's climate.

This text is a caption for the image, so it would look better if it were lined up with the edges of the image. You need to adjust the right margin to do this.

Highlight the section of text you have just typed in.

Left-click on the triangle for the right indent [14 · I · 15 · I ·△] and holding the mouse button down, drag the margin to the left until it is lined up with the right-hand edge of the image.

The caption now looks like this:

Distributed computing is a very useful tool for scientists who have huge amounts of data to analyse when trying to model systems such as the Earth's climate

It would improve the appearance if the text was fully justified. This means that the text is lined up to both the right and left margins.

With the text still highlighted click on the [≣] (Justified) button.

The text is now fully justified like this:

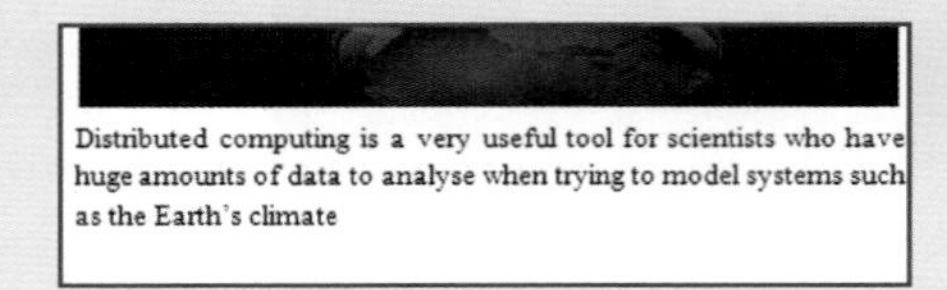

Distributed computing is a very useful tool for scientists who have huge amounts of data to analyse when trying to model systems such as the Earth's climate

7 You are now going to insert another image from a file.

Click between these two lines shown by the vertical line here, as this will be the position where the image is to be inserted.

this would provide evidence of extraterrestrial life.

|

The problem the project has is that there is a lot of background noise which includes radio

8 Click on the Insert tab and then on [Picture]. You will then have to locate the folder containing the image file **Dist_computing_image2**.

Once found, click on the file **Dist_computing_image2** and then click on the Insert button.

The image is inserted into the document like this:

signals from outer space. If the signals were confined to a narrow range of frequencies, then this would provide evidence of extraterrestrial life.

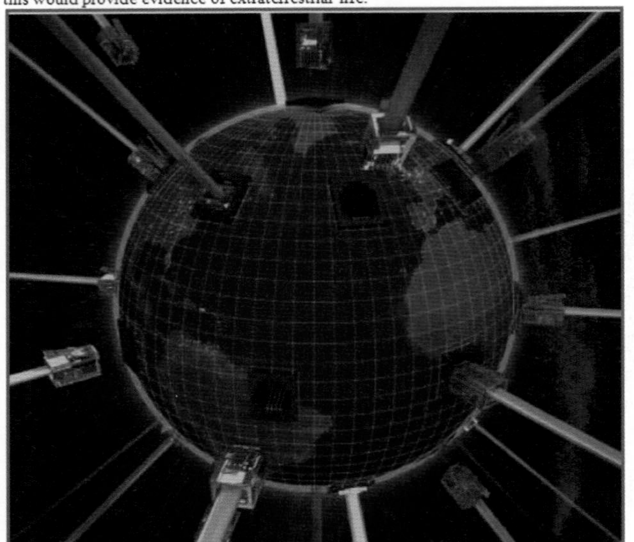

The problem the project has is that there is a lot of background noise which includes radio signals from TV stations, radar, satellites and from celestial sources. It is very difficult to analyse the data from radio telescopes and look for other signals that could indicate extraterrestrial life.

9 Resize the document so that the image is 60% of its original size and in the same proportion (i.e. make sure that the Lock aspect ratio box is ticked).

Now wrap the text so that it is square with the image.

When these actions have both been completed correctly the section containing the image will look like this:

signals from outer space. If the signals were confined to a narrow range of frequencies, then this would provide evidence of extraterrestrial life.

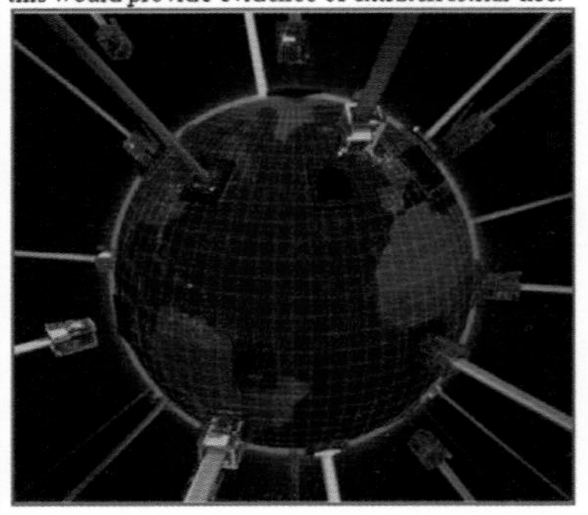

The problem the project has is that there is a lot of background noise which includes radio signals from TV stations radar, satellites and from celestial sources. It is very difficult to analyse the data from radio telescopes and look for other signals that could indicate extraterrestrial life.

In order to search for the narrow-bandwidth signals lots of computing power is needed. At first supercomputers containing parallel processors were used to process the huge amount of the data from the telescopes. Then someone came up wit the idea of using a virtual supercomputer consisting of a huge number of Internet-connected home computers. The project was then named SETI@home and it has been running since 1999.

At the time of writing the SETI@home project uses 170,000 active volunteers around the world and uses 320,000 computers but even with this power they still need more!

10 Insert the image **Dist_computing_image3** in the position shown below.

Click on the image and use the handles on the image to resize it so that the right edge of the image is lined up with the right margin for the text.

Your image should now be in the position as shown here:

home computers. The project was then named SETI@home and it has been running since 1999.

At the time of writing the SETI@home project uses 170,000 active volunteers around the world and uses 320,000 computers but even with this power they still need more!

11 Save your document using the file name **Dist_computing_final**.

Organising the page layout

The page layout is the arrangement of the items (text, images, etc.) on the page of the document. Here are some of the things that are regarded as part of the page layout. Most of these can be altered by pressing the Page Layout tab:

Once the Page Layout tab has been pressed, you are presented with many tools that help you lay out the page, and the main ones are shown here:

Size Setting the page size (this is usually determined by the size of the paper you are printing the document on). Page sizes include A4, A5 (which is half the size of A4), letter, etc.

Page orientation can be portrait or landscape. Most of the time, documents and letters should be in Portrait orientation. Save Landscape orientation for presentation, leaflets, and so on.

Margins Here you can adjust the size of the margins around the page (i.e. right, left, top and bottom). The margins define where the main text and other objects such as images can be placed. You can also set gutters which allow extra space to allow the pages to be bound without concealing any information on the page.

Using the column tool you can put the text into columns. You can also set up the width of each column and the space between the columns and even have a vertical line or lines separating the columns.

Breaks Using breaks, you can force the computer to start a new page rather than allow it to start a page only when the page is completely filled. You can insert section breaks and column breaks so that the text can run on from one section or column to the next. Breaks also allow you to prevent the occurrence of widows and orphans which you will learn about later.

Some page layout tools are accessed from the Insert tab and these include:

Here you can set headers (at the top of the page) or footers (at the bottom of the page). These can contain items such as date, page numbers, name of document author, name of the document file, etc.

Creating and formatting tables

Tables are made up of rows and columns that can be filled with text. Tables are used to organise and present information. They may also be used to align numbers in columns and once this is done you can perform calculations on them. There are many different ways of creating a table and one way in Word is to use the Table menu. To make creating tables easy there are set formats already created and you can just select the table format that suits your needs.

Merge and split table cells – if you create a table and then want to type a heading in the first cell, this is what happens:

This is the heading and you can see what happens if you type in too much				

The text is wrapped in order to keep the width of the cell the same. This is not what you want to happen with a heading, so it is necessary to merge cells. When this is done the heading fits in like this:

This is the heading when the cells have been merged to fit the heading				

It is also possible to split cells like this							

Activity 13.4

Creating a table

In this activity you will learn how to create a table with a specified number of columns and rows.

1 Click on the Insert tab.

2 Click on 

and the following menu appears:

Insert Table

Insert Table...
Draw Table
Convert Text to Table...
Excel Spreadsheet
Quick Tables

3 Tables consist of columns and rows. Suppose you wanted a table containing 5 columns and 6 rows. All you do is run the cursor so that you have 5 columns and 6 rows highlighted like this:

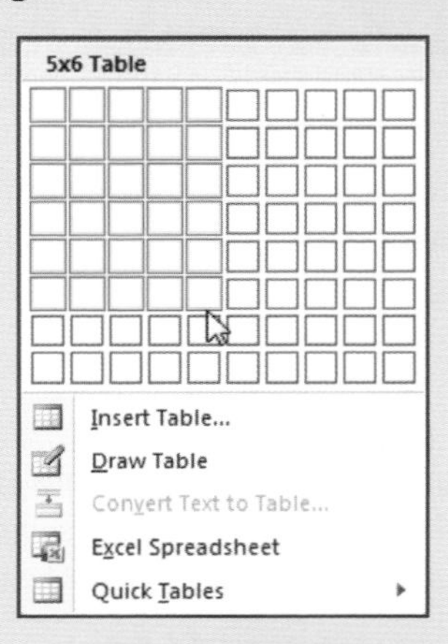

When you have the correct number highlighted, left-click and the blank table is inserted into the document like this:

4 Close the file without saving.

Activity 13.5

Formatting tables

In this activity you will learn how to:

- insert a table into a document
- adjust the column width and row height
- input data into a table.

You are going to produce a table to show two of the main differences between a wide area network (WAN) and a local area network (LAN).

1 Load Microsoft Word and start a new document.

2 Set up a table consisting of 3 columns and 3 rows. Look back at Activity 13.4 if you have forgotten how to do this.

The following table will appear at the cursor position. Notice that it spans from one margin to the other.

Adjust the width of the columns and the height of the rows in the following way. Move the pointer onto the lines and you will notice that the cursor changes to two parallel lines (either vertical or horizontal). Press the left mouse button and drag until the widths are similar to those shown below.

	LAN	WAN
Difference 1	Confined to a small area, usually a single site.	Covers a wide geographical area spanning towns, countries or even continents.
Difference 2	Usually uses simple cable links owned by the company.	Uses expensive telecommunication links not owned by the company.

3 Type the data shown in the table above. You can centre the headings LAN and WAN by typing them, highlighting them and then clicking on the centre button on the toolbar. This will centre the headings in the column. Also, embolden the words LAN, WAN, Difference 1 and Difference 2.

4 See how neat this now looks. Save your document using the filename **The_differences_between_a_LAN_and_a_WAN**. Print out a copy of your document.

Using software tools to create headers and footers appropriately within a range of software packages

Headers and footers

A header is an area between the very top of the page and the top margin. A footer is an area between the very bottom of the page and the bottom margin. Once you tell the software application you are using that you want to use headers and footers then you can insert text into one or both of these areas.

Here are some types of information that are commonly put into headers and footers:

- Page numbers
- Today's date
- The title
- A company logo (it can be a graphic image)
- The author's name
- The filename of the document.

As well as being able to place any of the above objects in a header or footer, you can also align the objects with the left margin, the right margin or to the centre of the header or footer.

Rather than adding certain text to the headers and footers manually, you can instruct the software to add it automatically. For example, you can instruct the software to add the date automatically in the header or footer on every page if required.

Activity 13.6

Including a header and footer in a document

In this activity you will learn how to:

- import a document which is in rich text format (.rtf) into your word processor
- set the margins for a document
- set the paper size for a document
- insert headers and footers in a document
- add page numbers, dates and filenames automatically
- set the alignment of text in headers and footers.

For the examination you will need to set the page layout for a document. You will need to follow the instructions given carefully.

1 Load Microsoft Word and open the document called **Whizz_Logistics**.

This file has been saved in rich text format and can be loaded into Word without alteration, as both the text and the simple formatting in this document are kept.

2 Click on the Page Layout tab on the toolbar and select Margins and from the items in the menu select Custom Margins. The following box appears:

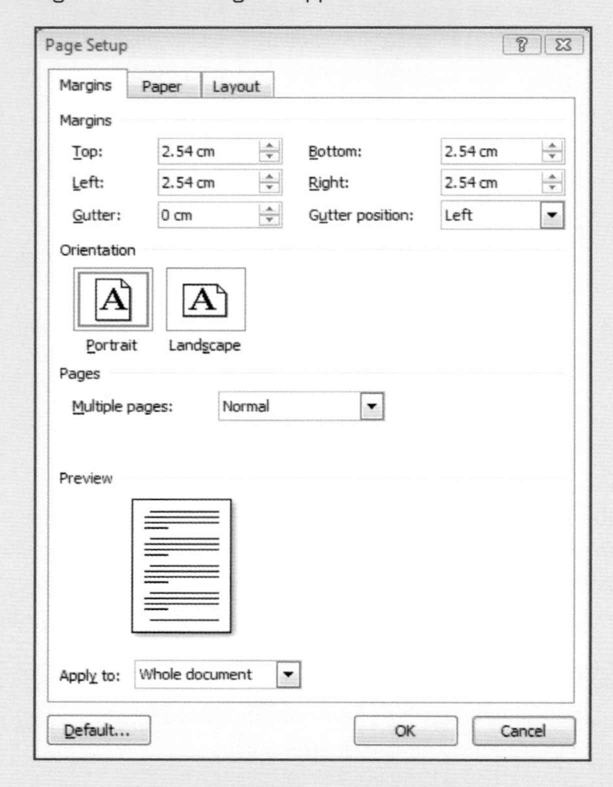

Notice the margin section where you can change the size of the margins of the page you are producing.

Set the top, bottom, left and right margins to 3 cm each. When you have done this correctly the margin section will look the same as this:

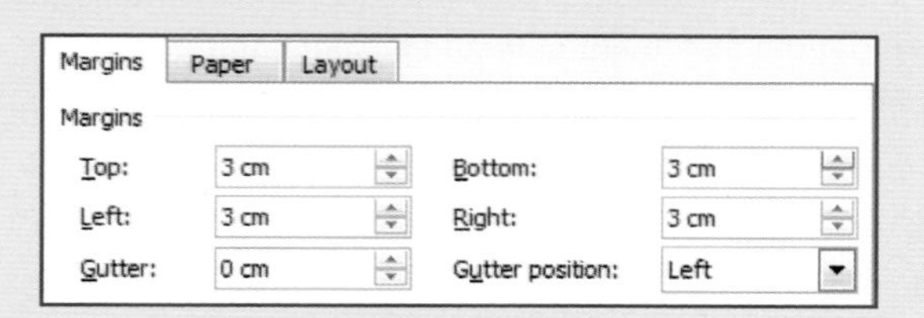

Ensure that Portrait has been selected in the orientation section.

3 Click on the Paper tab at the top. It is from here you can select the size of the paper to be used. Notice the paper sizes include A4, A5, Legal, etc. Check that A4 has been selected for this document.

Click on OK.

4 You are now going to insert headers and footers in the document. A header is an area on the page between the top margin and the top of the page. A footer is an area between the bottom margin and the bottom of the page. Headers and footers are used for such things as the name of the author of the document, the date the document was produced, page numbers, etc.

Click on the Insert tab in the toolbar, click on Header and then choose the following layout from the list.

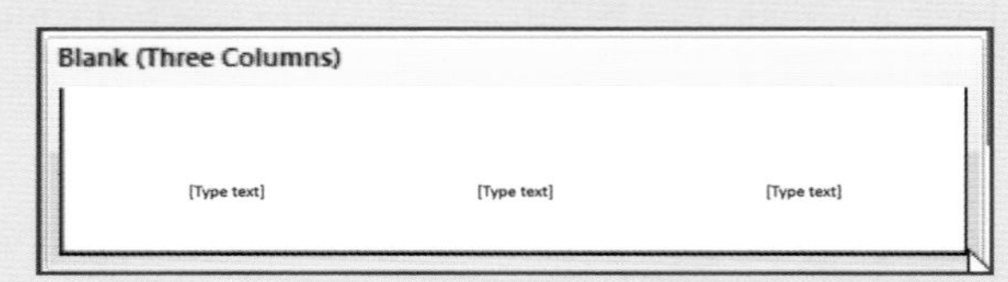

In the examination you will need to follow instructions as to what to put into the header.

Place the following information in the header:

Your own name, which is to be left aligned, and the number 121121, which is to be right aligned.

Click on the left placeholder box so that it will be highlighted like this:

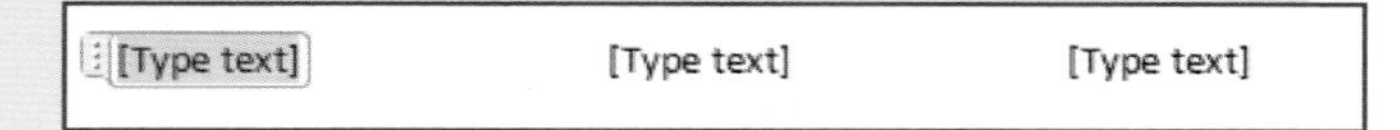

Type your name here. The text you type will replace the text already there.

Click on the right placeholder box and type in the number 121121.

Click on the centre placeholder box and click the backspace key to delete the box.

Your header should now look like this:

5 Click on the Insert tab in the toolbar and then click on Footer and choose the following from the list:

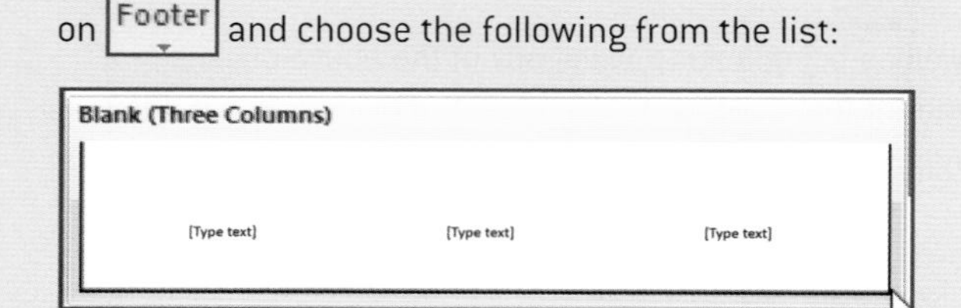

You are going to include the following in the footer:

- An automated filename, left aligned.
- An automated page number, centre aligned.
- Today's date, right aligned.

To create the automated filename click on the placeholder box on the left-hand side of the footer and click on Quick Parts and then select Field from the menu. The following dialog box appears from which you need to select FileName from the list of fields.

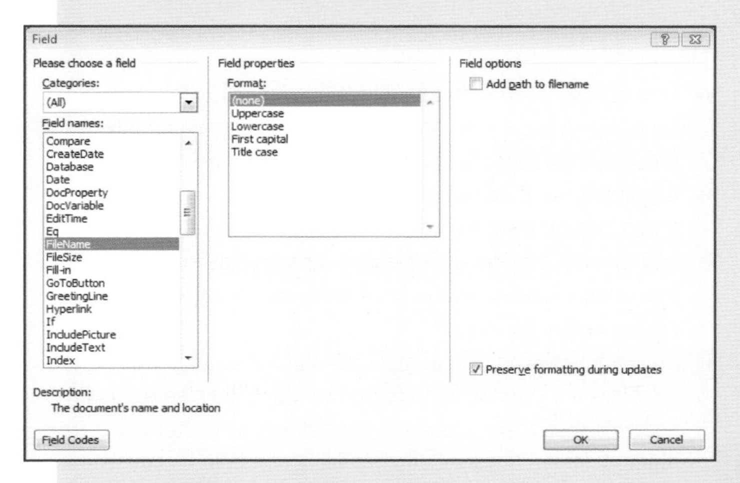

Click on OK to continue.

The filename now appears in the footer like this:

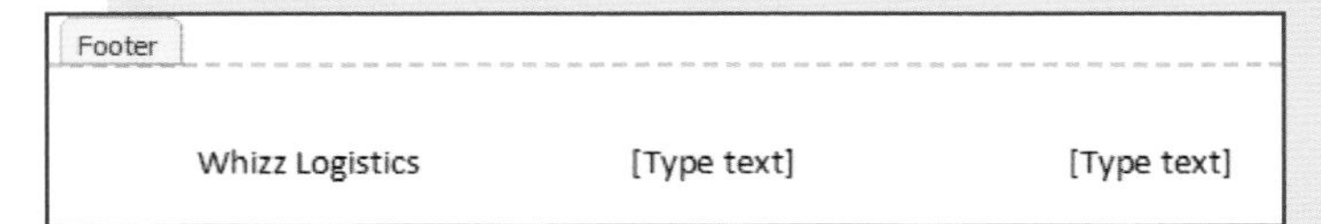

6 To insert the page number, click on the centre place holder box and then click on Page Number and pick Current Position from the menu options. You are now presented with some options and you need to choose the one shown here:

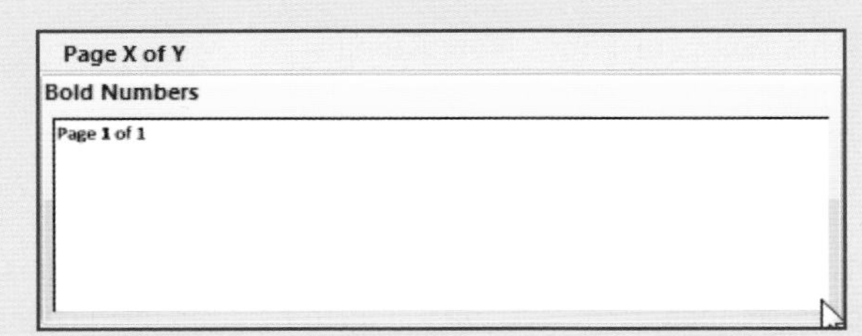

The page number now appears in the footer like this:

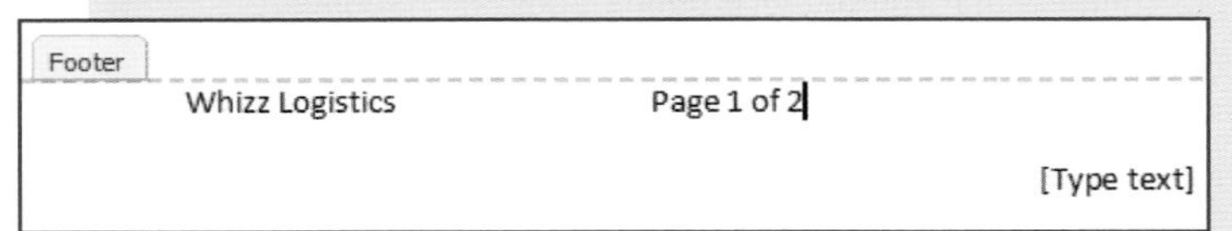

If the placeholder box has been forced onto the line below delete the place holder and then position the cursor after the text Page 1 of 2.

Click on  and then select the date format highlighted as shown:

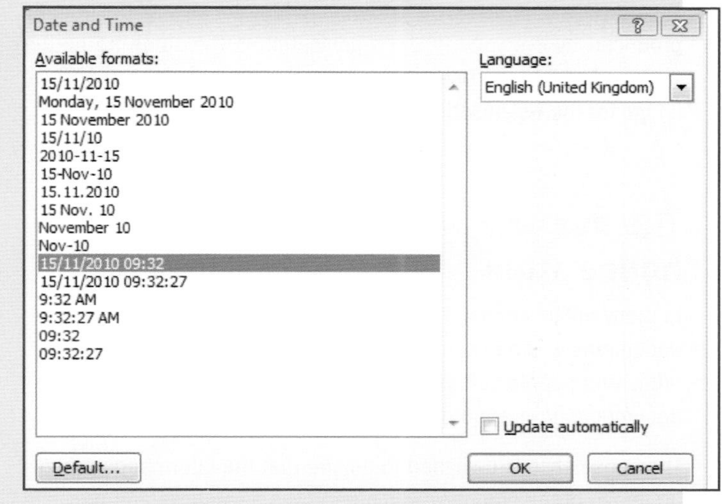

Click on OK.

Important note

Different countries write dates in different formats. When asked in the examination to insert a date, it does not matter which format you use.

The date and time is inserted next to the page number like this:

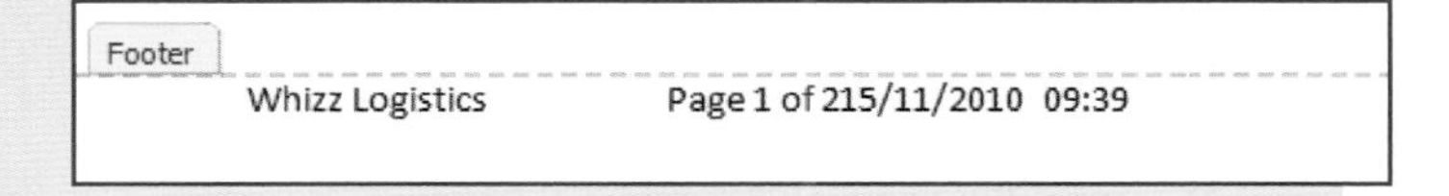

You now need to click just after the 2 of the page number and press the tab key until the end of the date-time is almost aligned with the right margin. To finish off you will need to press the space bar until the date-time is exactly aligned with the right margin.

The final footer should look like this:

7 Save the file, ensuring that it is saved still in rich text format.

8 Print a copy of the document.

Styles

When producing documents for an organisation there is usually a house style which anyone producing documents must adhere to. The purpose of house style is to ensure that the style of documents produced by everyone in the organisation is the same no matter who produces them. Use of a house style helps reinforce a positive image for the organisation.

The key concepts covered in this chapter are:
- The use and purpose of a corporate house style
- Applying styles to ensure the consistency of presentation

The use and purpose of a corporate house style

In many organisations there will be a corporate house style for documents which ensures that documents produced by different areas and people still look the same. It also ensures that all documents have a professional appearance.

The house style is designed to ensure that the information creates an image of the company which improves awareness and provides positive publicity.

In the examination there will usually be a house style specification included and this tells you what styles to apply to any documents you are producing.

Here is a sample of a house style specification sheet.

Title	Font – sans serif, 26 point, centre aligned
Subtitle	Font – sans serif, 20 point, underlined and left aligned
Subheading	Font – sans serif, 14 point with one clear line space after a subheading
Body text	Font – serif, 12 point, 1.5 line spacing with one clear line space after each paragraph

Applying styles to ensure the consistency of presentation

Ensuring that the document style is consistent

It is important to create professional-looking documents. It is therefore essential that the page layout is consistent on each page and for all the pages in a multi-page document. Page layout consistency needs to be considered when creating the document and also during the proofreading stage.

In the examination you will need to ensure that the page layout is consistent and you should ensure that each of the following styles has been used consistently.

- Font styles – ensure that headings use the same font style, subheadings have the same font style and the main body text has the same font style, unless you are instructed otherwise.
- Alignment – text can be left aligned, centred or right aligned. It can also be justified, which means the text is aligned to both the left and the right margins. Ensure that you have obeyed any specific instructions in the examination and have maintained consistency.
- Spacing between lines, paragraphs and before and after headings – if you decide to leave two blank lines after a main heading, then there should be consistency whenever there are other main headings. All other spacing needs to be consistent.

Font type and font size

Changing the font type (Arial, Times New Roman, etc.) alters the appearance of the characters. Font types are given names and you can change the font by selecting the text and then clicking on the correct part of the formatting toolbar shown below. Notice also that there is a section for altering the font size (i.e. how big the characters appear).

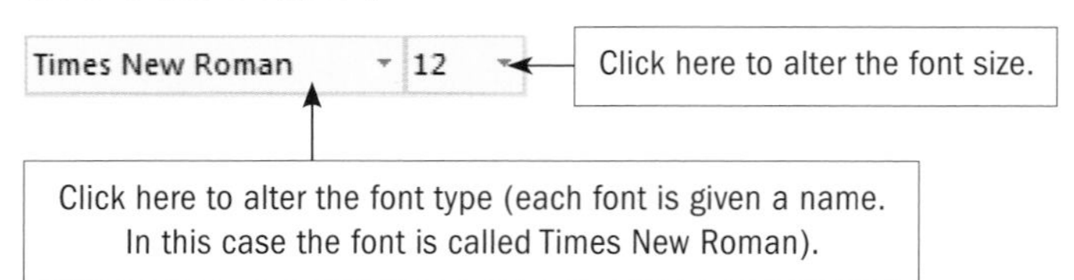

Here are some examples of different font types:

- This text is in Times New Roman
- This text is in Arial
- This text is in Century Schoolbook
- This text is in Tahoma

Font size is measured in points or pt for short – 20pt text is much larger than 12pt text. You will often be given instructions as to what point size to use for headings, sections of text, etc., in the documents that you produce during the examination.

There is another way to alter the size of selected text. You can press either of these buttons to increase or decrease the font size.

Serif and sans serif fonts

There are two main types of font: serif fonts and sans serif fonts. It is very important for the examination that you know the difference between them.

Serif fonts are those fonts that have detail attached to the strokes that make up the letters. Sans serif fonts do not have this extra detail added. Sans in French means without, so sans serif means without serifs. Serif fonts were developed so that a reader's eye is led to the next letter in a large block of text.

A

This letter is in a serif font – notice the detail at the bottom of the strokes.

A

This letter is in a sans serif font – notice that there is no detail on the strokes. Sans serif fonts are usually better for titles, headings and posters; serif fonts are better for large blocks of text.

Serif fonts	Sans serif fonts
Times New Roman	Arial
Garamond	Verdana
Century Schoolbook	Tahoma

Important note When preparing documents in the exam you will be asked to use a serif or sans serif font, so it is important to be able to spot one. Stick with the common fonts in the table.

Activity 14.1

Serif or sans serif?

In this activity you will learn how to distinguish between serif and sans serif fonts.

Here are some fonts. You decide whether each of them is a serif or sans serif font.

1	A	**6**	X
2	V	**7**	R
3	H	**8**	U
4	Y	**9**	R
5	K	**10**	F

Using an appropriate font

Most of the examination questions ask you to use specific font types (serif or sans serif) and they do not usually specify the font name. You therefore have to choose from a number of different fonts. You must make sure that whichever font you choose is easy to read.

Text enhancement

Text may be enhanced in order to make it stand out. Text enhancements include bold (text in heavy type), italic (text slanting to the right) and underlining. These can be used to draw attention to certain words or text in a document or slide.

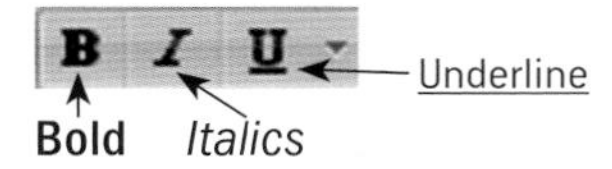

Font colour

The text/font colour can be changed by clicking on A and then choosing a colour from the menu that appears.

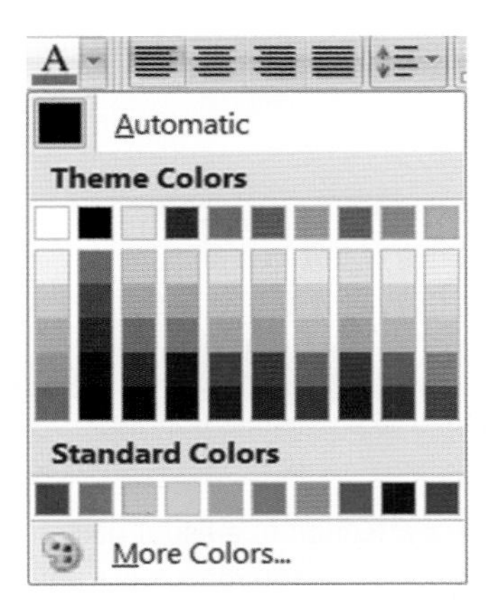

Once you have chosen a colour, all subsequent text typed will be in the chosen colour.

If you have already typed some text and want to change its colour, you need to first highlight it to select it before changing the colour.

Highlighting text

Just like you can highlight printed text using a highlighting pen (usually in a bright colour), you can produce highlighted text in a word-processed document.

To highlight text, you first select it like this:

This text is going to be highlighted.

Then click on ab.

You will then be presented with a palette of colours to choose from.

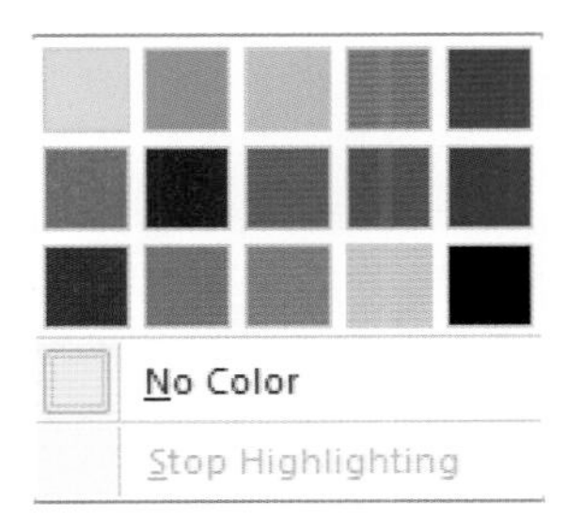

When you click on one of these colours the text is highlighted like this:

This text is going to be highlighted.

Text alignment

These buttons are used for aligning sections of text. Each button is used to align text in the following way:

Aligns text to the left.

Centres the text.

Aligns text to the right.

Aligns the text to both the left and right margins – called fully justified text.

Used to decrease or increase the indent level of a paragraph.

Line spacing

Clicking on this button brings down a menu allowing you to adjust the line spacing between lines of text in a document. There are various options such as 1.0, 1.15, 1.5, 2.0, 2.5, 3.0. You can also use this to add lines automatically before or after a paragraph.

Bullets

Bullets are useful to draw the reader's attention to a list of points you would like to make. Rather than just write an ordinary list, you can put a bullet point (a dot, arrow, diamond, square or even a small picture) at the start of the point. The text for your points is then indented a small amount from the bullet.

To produce a bulleted list:

1 Select the section of text you want bulleted by highlighting it.

2 Click on the button.

3 The list will be bulleted using the default bullet (i.e. black dots) like this:
   - Point 1
   - Point 2
   - Point 3

4 You can choose a different bullet. After highlighting the text you can click on the drop-down arrow on the bullet button. This menu appears from which you can choose a different type of bullet.

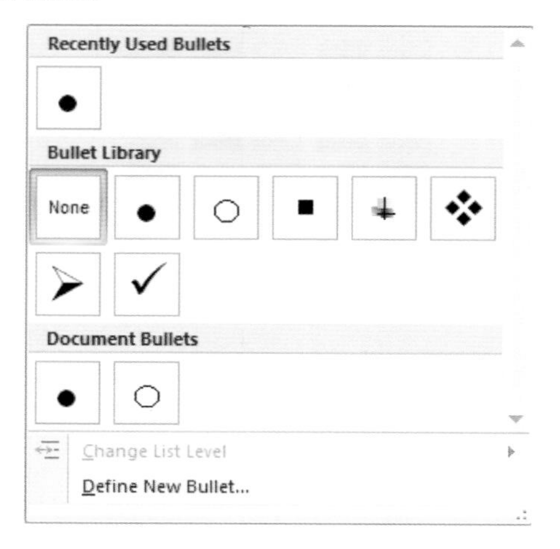

Numbered lists

Numbered lists are useful if the listed items need to be put into a particular order.

1 Select the section of text you want put into a numbered list by highlighting it.

2 Click on the button.

3 The list will be bulleted using the default numbers but you can choose a different number style. After highlighting the text you can click on the drop-down arrow on the numbered list button. This menu appears, from which you can choose a different type of number such as Roman numerals. Notice also that you can have a lettered list.

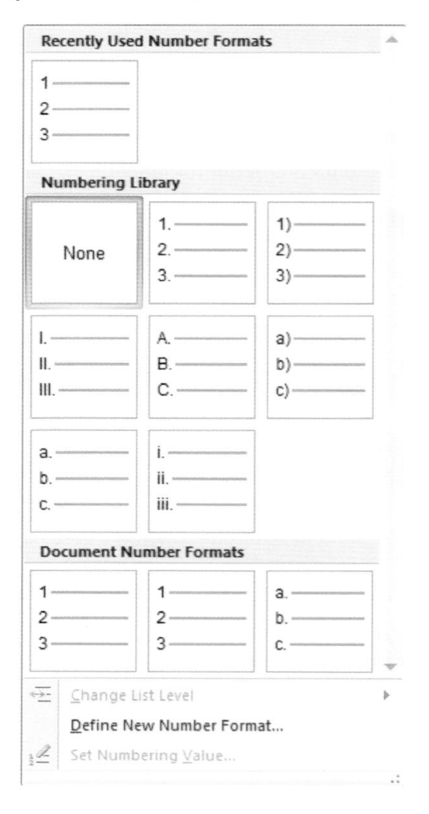

Selecting an appropriate font for a task taking into account the audience

Choosing a font is not easy, mainly because there are so many. For main body text, readability is most important.

How do you choose between serif or sans serif fonts?

According to most studies, sans serif fonts are more difficult to read. They are probably best used for headings or captions.

For more extended text, such as text in a magazine article or pages in a book, then it is best to use serif fonts.

Can I mix serif and sans serif fonts?

Yes you can and it works well. But do not use too many fonts. Limit the number of fonts or typefaces in a single document to two in a formal document and no more than three or four in a less formal document.

Activity 14.2

Choosing a font

In this activity you will be given a series of fonts and uses. You have to match the font to the activity. Remember there are no hard and fast rules about fonts, so be prepared for some lively argument at the end!

Font 1 Zapfino
Font 2 Arial
Font 3 Times New Roman
Font 4 Verdana
Font 5 Matura MT Script Capitals
Font 6 Century Gothic

Which font would be best for each of these uses?

1. The text in a children's learning-to-read book
2. The text to go on a map of Treasure Island to mark the landmarks
3. Some text at the bottom of a letter on a website to make it look as though the writer has signed it
4. The text to go with an image of a witch on a scary party invite
5. A novel

Activity 14.3

Applying styles to a document

In this activity you will learn how to:

- insert a header and footer
- apply a set of styles to the text in a document.

For the examination you will have to set the styles for the text in a document. You will need to identify the titles, subtitles, subheadings and body text and apply the styles carefully according to the instructions given in the paper.

1. Load the file **Emerging_Technology_v3**.

   This file was created in Activity 12.1.
2. Create a header and enter your own name and the number 121212 left-aligned and the number 33003 right-aligned. This header needs to appear on all the pages of the document.
3. Create a footer and enter the automated file name of the document, left-aligned and today's date right-aligned. This footer needs to appear on all the pages of the document.
4. Create or edit the following styles which should be applied to all the text in the document.

   Title

   Subtitle

   Subheading

   Body text

   The styles for the above are outlined in the house style specification sheet shown here:

Title	Font – sans-serif, 24 point, bold and centre aligned
Subtitle	Font – sans-serif, 16 point, underlined and left aligned
Subheading	Font – sans-serif, 14 point, bold and with one clear line space after a subheading
Body text	Font – serif, 12 point, 1.0 line spacing with one clear line space after each paragraph

5. Insert an extra line under the title of the document 'Emerging technologies' and apply the title style to the title.
6. Below the title, add the subtitle: **Report by:** and add your name.
7. Apply the Subtitle style to this text.
8. The subheading **Mobile working** needs a line before the body text starts. Insert a blank line after this subheading to ensure consistency with the other subheadings.
9. Identify all the subheadings in the document and apply the Subheading style to each one of them.
10. Apply the Body text style to all the remaining text in the document.
11. Save your document using the filename **Emerging_ Technology_v4**.
12. Print out a copy of your document.

15 Proofing

Proofing involves checking an ICT solution for errors. Errors can occur at various stages depending on the type of ICT solution being used. For example, incorrect data could be entered into a spreadsheet or database resulting in the information produced being wrong. When producing documents or slides it is necessary to correct spelling and grammar mistakes as well as checking for all the other possible errors that can occur. There are a variety of software tools which can help spot and correct errors. Software tools cannot spot all the errors so it will be necessary to carry out other checks, including visual verification and proofreading to spot and correct errors.

The key concepts covered in this chapter are:

- Software tools
- Proofing techniques

Software tools

The following software tools are used to ensure that all work produced contains as few errors as possible:

- Spellcheck software
- Grammar check software
- Validation checks/routines
- Double data entry.

Spellcheck software

If the spelling of your intended word is very different from the correct spelling of the word, then the suggestions for the correct spelling provided by the spellcheck software will not contain the word you require. In this situation, you are best asking someone else to see if they can spell the word correctly.

Spellchecking

Any document you produce should be spellchecked.

To check the spelling of a document you follow these steps:

1 With the document open, click on the Review tab and then on 
.

2 The following dialogue box appears.

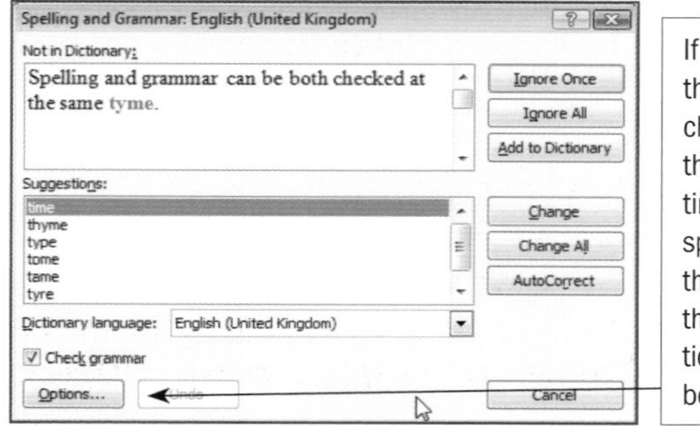

If you want the grammar checking at the same time as the spelling then ensure there is a tick in this box.

You need to look carefully at any word shown in red. This word tyme is not in the dictionary because it is spelled incorrectly or mistyped. If the correct word is in the list of suggestions you can click on the word and then click on Change.

Remember that just because a word is not in the dictionary does not make it incorrectly spelled. There are many words that are specialist words. Such words can be added to the dictionary by clicking on Add to Dictionary so that they will not be queried again in other documents.

Caution:
Do not add words to the dictionary unless they are specific to you or your college/school. Don't just add them because you think that is how they are spelled.

Using a spellchecker to remove errors as you are typing in text

When using document processing software such as word-processing and presentation software most people will have the spellchecker turned on so that any misspelled words will be marked so they can be corrected immediately.

The following word has been spelled incorrectly so the word-processing software has marked it by placing a red wavy line under the word:

Succesfull

If you right-click on the misspelled word, then a menu appears showing a list of possible spellings for the word. As the incorrect spelling was close to the correct spelling, the spellchecker has managed to find the correct word. To replace the word with its correct spelling you simply click on the correctly spelled word.

As well as finding incorrect spellings, spellcheck software is able to spot if the same word has been typed in twice.

Activity 15.1

In the text below, try to locate the mistakes a spellchecker would detect and those it wouldn't.

The internet can be described as a network of networks. All most any type of information can be found. It is posible to acess anything from a database on space sceince to train timetables. The internet is accessed though internet service providers. Through use of the internet, it is possible for data data to be rapidly transfered around the world alowing people all over the world to read it.

Grammar checkers

There are rules about the construction of sentences, and grammar checkers are software tools most often found with word-processing software that will check that these rules have been obeyed. Grammar checkers can be used to check that:

- Sentences end with only one full stop.
- There is a capital letter at the beginning of a sentence.
- Common errors like writing 'you and I' rather than 'you and me' have been avoided.

Activity 15.2

Spellchecking, proofreading and making corrections

In this activity you will learn how to:

- spellcheck a document
- proofread a document.

1 Load your word-processing software then locate and open the file called **Personal_data.txt**.

2 Insert two blank lines at the start of the document and enter the title SELLING PERSONAL DATA. The style for the title is: Font – serif, 20 point, upper case, bold and centre aligned. Apply this style to the title.

3 Other than the title, there is only body text. The style for the body text is: Font – sans serif, 12 point, 1.15 line space and fully justified. Apply this style to all the body text in the document.

4 Now spellcheck this document by right-clicking on all those words that are highlighted as being spelled incorrectly by having a wavy red line under them.
   - Enter the correct words from the list of suggested spellings for each word.
   - Note that in the section of text there is one word that has mistakenly been typed in twice, so delete one of the occurrences.

5 Carefully proofread the text looking for and correcting the following:
   - Instances where a sentence does not start with a capital letter.

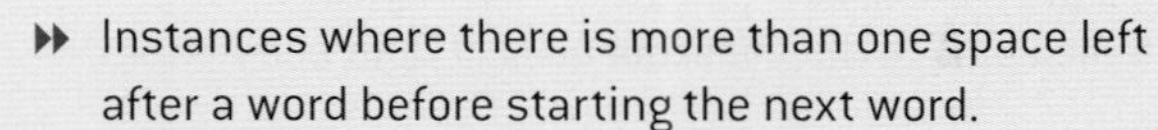

   - Instances where there is more than one space left after a word before starting the next word.
   - Other mistakes in grammar.

6 Save your document in the file format your word-processing software generally uses, giving the document the filename **Personal_data_final_version**.

Using validation routines to minimise errors

When data is being entered into a spreadsheet or database it is important that any errors are spotted by the program so that it can alert the user so they can check and then re-enter the data. When creating ICT solutions using database or spreadsheet software we need to produce validation routines that will help prevent incorrect data being accepted for processing.

When the spreadsheet or database is being designed a series of validation routines can be produced that govern what a user can and cannot enter. It is impossible to trap every type of error, so if someone's address is 4 Bankfield Drive and the user incorrectly types in 40 Bankfield Drive, we would not be able to devise a simple validation routine that would detect this.

Data type checks

Databases can automatically check to see if the data being entered into a field fits what is allowed by the data type for that field. So, for example, text cannot be entered into a field that has a numeric data type. If a field is given a data type of text, then you can enter any characters from the keyboard into this field. Therefore, you can enter numbers into a text field but you cannot perform any calculations on them. This type of check is called a data type check.

Range checks

Range checks are performed on numbers to make sure that they lie within a specified range. Range checks will only check absurd values. For example, if you typed '50' as the number of children in a household instead of '5' then a range check would spot the error; if you typed in '7' then the range check would not pick up the error.

Presence checks

You can specify that some fields must always have data entered into them. For example, everyone has a date of birth but not everyone has an email address. Checks such as this are called **presence checks.**

Validation routines will be looked at in more detail in the chapters where spreadsheets and databases are covered.

Proofing techniques

Accuracy of data entry

Computers can produce accurate output (e.g. payslips, bills, invoices, posters, business cards, web pages, presentation slides, brochures, etc.) only if the data entered is accurate. Incorrect data can have consequences of varying degrees of seriousness.

Consequences of data entry errors
Here are some of the consequences of data entry errors:

- **Embarrassment** – imagine putting up a poster in your school for a disco with a spelling mistake in it or the wrong date on it. A utilities company (e.g. water, gas, electricity or phone) could send a bill for a ridiculous amount and then have to admit their mistake.
- **Loss of money** – mistakes may mean having to compensate customers. Managers could base their business decisions on wrong information resulting in reduced profits.
- **Prosecution** – if data about a person is incorrect and it causes them loss in some way then an organisation could be prosecuted because they did not take enough care in ensuring the data was correct. For example, a person could lose a job because someone entering in data about them had recorded that they had a criminal record when they did not have one.
- **The wrong goods being sent** – mistakes in orders can result in goods being sent that were not ordered. The firm has to then pick them up and send the correct goods.

Ensuring the accuracy of information
When you create your own websites, blogs, multimedia presentations, posters, letters, etc., you need to ensure that the material is accurate. Here are some ways you can do this:

- If you present facts you should always check these using several reliable sources.
- Ask someone knowledgeable about the subject to check the material for accuracy of information.
- Use the spelling and grammar checkers provided in the software you are using. Be aware that just because text has been spell - and grammar checked doesn't mean it is correct.
- The more people who look at your work, the more mistakes are likely to be spotted. So ask friends and family to read through your material for their comments.
- Proofread your work carefully by reading through it slowly to check it makes sense and there is nothing missing.

Editing and proofreading documents
The terms editing and proofreading are often taken to mean the same thing but there are some differences between them.

Editing involves looking at the overall document to check its content and organisation. Proofreading looks more closely at the actual sentences and word combinations to make sure that they make sense and are grammatically correct.

What is the best way to edit a document?
After you have produced a document, it is a good idea to wait a couple of days before editing it because, with a fresh mind, you are more likely to spot any errors. It is common to miss words if you just read in your head. It is better to read aloud (make sure there is no one around!) because you can't skip words.

What is the best way to proofread a document?
The first thing is to read slowly. If you read too fast you may read a word that should be there but isn't. Every word must be read before moving to the next. Again by reading aloud you are more likely to read every word. Some people like to read with a cover such as a piece of paper covering the sentences below. Doing this allows the reader to concentrate on a single line at a time.

Transposed numbers and letters
It is easy, when typing quickly, to type numbers and letters in the wrong order.

Here are some examples:

- 'fro' instead of 'for'
- the account number 100065 instead of the correct account number 100056
- the flight number AB376 instead of BA376.

Transposed numbers can have a serious effect.

Suppose you transposed two numbers in your passport number when checking in for a flight online. You may not be allowed onto the aircraft because the number on your passport and booking do not match.

If you transfer money from one account to another using EFT and transposed some numbers in the account numbers, then the money could end up in somebody else's account.

Consistent line spacing
Check that you have consistently left the same space between blocks of text, paragraphs, and between text and diagrams, etc.

In the examination paper you will be given instructions about the line spacing after titles, subtitles, heading, subheadings and body text. You must be very careful that you apply the instructions throughout the text.

Consistent character spacing
When creating documents, the spacing between characters such as a comma or a full-stop and the next character should be consistent throughout the document. This is one of the things you should check for when proofreading.

Consistent case
Upper case letters are capital letters such as A, B, C, D, E... and lower case letters are small letters such as a, b, c, d, e ... The use of capital letters in such things as titles or subtitles is a matter of house style. If it is decided that all headings should be in upper case, then this should be applied consistently throughout the document and you would need to check for this when proofreading.

Removing blank pages/slides
It is common to move images and text around in documents or slides using cut and paste and this can produce blank pages. During proofreading blank pages/slides should be identified and deleted.

Widows and orphans

An orphan is the name given to the first line of a paragraph when it is separated from the rest of the paragraph by a page break. The orphan line appears as the last line at the end of one page, with the rest of the paragraph at the top of the next page. This looks untidy so we need to spot them during proofreading and adjust the text to rectify them.

A widow is when the last line of a paragraph appears printed by itself at the top of a page and again this can ruin the appearance (and readability) of a document. Again, these need to be spotted and rectified.

Verification

Verification means checking that the data being entered into the computer perfectly matches the source of the data. For example, if details from an order form were being typed in using a keyboard, then when the user has finished, the data on the form on the screen should be identical to that on the paper form (i.e. the data source). Also if data was sent over a network, the data needs to be checked when it arrives to make sure no errors are introduced during transmission.

QUESTIONS A

1 Here are some dates of birth that are to be entered into an ICT system:

a 12/01/3010
b 01/13/2000
c 30/02/1999

Assume that all the dates are in the British format dd/mm/yyyy. For each one, explain why they cannot be valid dates of birth. *(3 marks)*

2 A computer manager says, 'data can be valid yet still be incorrect'. By giving **one** suitable example, explain what this statement means. *(3 marks)*

Here are some methods of verification:

- Visual comparison of the data entered with the source of the data – involves one user carefully reading what they have typed in and comparing it with what is on the data source (order forms, application forms, invoices, etc.) for any errors, which can then be corrected.
- Double entry of data – involves using the same data source to enter the details into the computer twice and only if the two sets of data are identical will they be accepted for processing. The disadvantage of this is that the cost of data entry is doubled. Double entry of data is often used when creating accounts over the internet. They may ask you to create a password and enter it twice. This ensures there are no mistakes that would prevent you from accessing the account.

REVISION QUESTIONS

1 When students join the senior school, a form is filled in by their parents. The details on the form are then typed into a computer. The details are verified after typing. Explain briefly, how the details may be verified. [2]

2 A person's date of birth is entered into a database. State **three** things the validation program could check regarding this date as part of the validation. [3]

EXAM-STYLE QUESTIONS

1 The head teacher of a school has employed Pierre, a systems analyst, to create a new database system to keep details of all students. Each student will be allocated a unique ten-digit reference number.

Pierre has written out some of the questions that the head teacher might ask. He can then analyse these in order to design a database which may answer these questions. Some are:

Which year is Johann Schmidt in? (the school has students in years 7 to 13)

Does Anita Nash have a sibling (brother or sister) in the school?

a Complete the data dictionary table below giving the field names which would be used in the database and identifying a validation check for each field.

Field name	Validation check
Reference_number	
Year	
First_name	
	None

[6]

b Explain the differences between verification and validation giving reasons why both are needed when data is entered into a database. *[5]*

(Cambridge IGCSE Information and Communication Technology 0417/11 q14 Oct/Nov 2013)

2 Give the name of **two** different methods of verification and describe their use. *[4]*

Graphs and charts

Tables of figures can be hard to understand, so it is much easier to show these figures as pictures using graphs and charts. Using graphs and charts makes it easy to look at the relative proportion of the items and spot any inconsistencies in the data. Also, graphs and charts are helpful for spotting trends such as increasing or decreasing profits, or for seeing the biggest or smallest reading.

The key concept covered in this chapter are:
- Producing graphs or charts from given data

Producing graphs and charts from given data

There are many graphs and charts that you can produce using software, so it is important to pick the graph or chart that is most suitable.

Pie charts

These are good for displaying the proportion that each group is out of the whole. For example, you could show a class's crisp preferences using a pie chart.

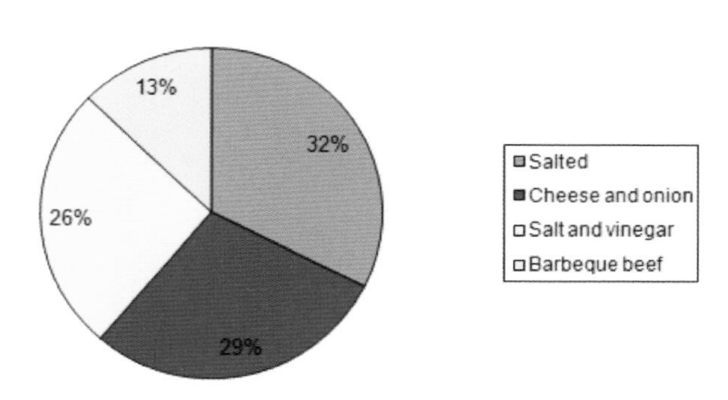

Bar charts

These are good for displaying the frequency of different categories. Here is a bar chart to investigate the types of vehicle using a certain road as a shortcut.

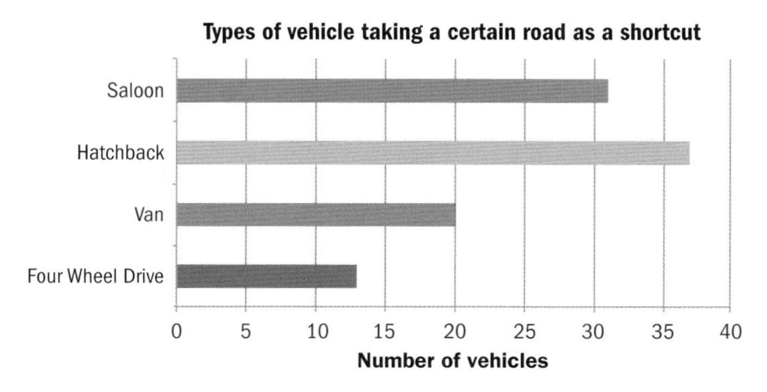

Bar charts in spreadsheet software can be:

- Vertical bar charts (or column charts as they are called in Excel) which display the bars vertically.
- Horizontal bar charts are simply called bar charts in Excel and they show the bars in a horizontal direction across the screen.

Column charts

Column charts are used to compare values across different categories. For example, you could use a column chart to compare sales of cars for the first six months of the year.

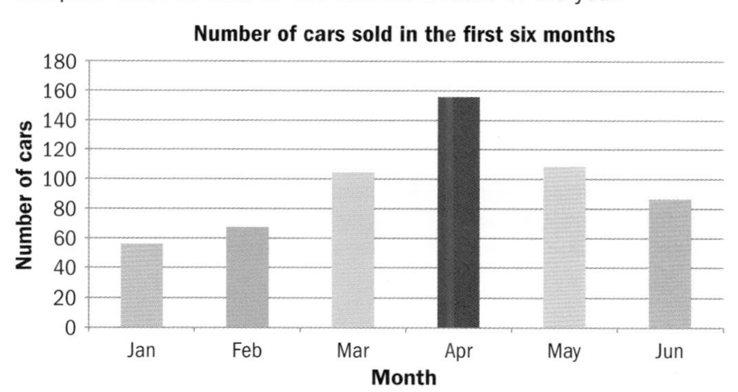

Scattergraphs

These are used to see how closely, if at all, one quantity depends on another. This is called correlation. For example, you might start with the hypothesis that someone who is tall will have big feet. You would then collect height and shoe size data and then plot the pairs of values.

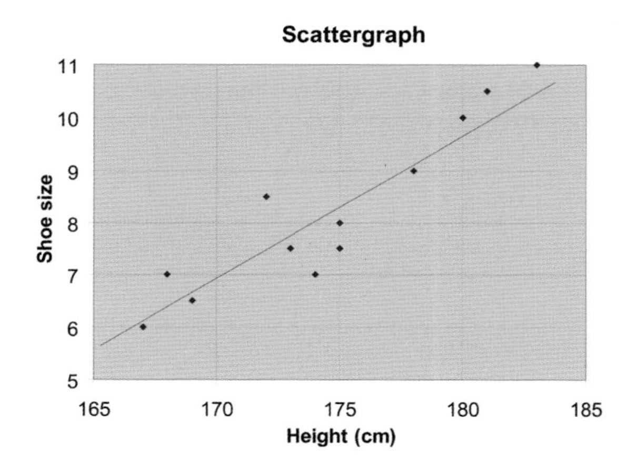

Line graphs

These can be used to show trends between two variables.

The graph below shows how the value of a car falls over the first four years from new. Here the value of the car is plotted at the end of each year.

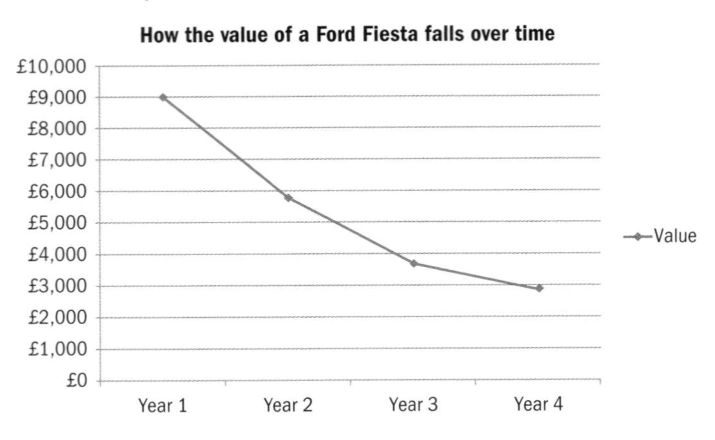

In this graph, average monthly precipitation for some location is plotted monthly over a one year period. The location is unknown, as there has been no title provided.

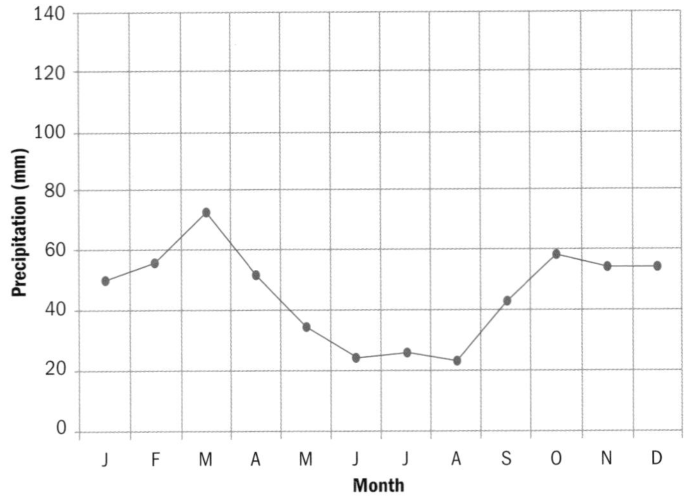

Activity 16.1

Producing a scattergraph

In this activity you will learn how to:

- select data to produce a graph or chart
- label the graph or chart with a title, legend, and axes.

A driving school advertises for pupils and they would like to answer the following question: If we spend more on advertising, will we get more pupils?

They have collected the following data and put it into a table:

Amount spent on advertising per week ($)	Number of new pupils per week from adverts
20	1
30	2
40	4
50	4
60	5
70	6
80	6
120	9
150	10
160	12

This data is quite hard to interpret, so they have decided to present it as a scattergraph. This will enable them to see the relationship (if there is one) more clearly.

1 Open your spreadsheet software and enter the data into columns A and B like this:

	A	B
1	**Amount spent on advertising per week ($)**	**Number of new pupils per week from adverts**
2	20	1
3	30	2
4	40	4
5	50	4
6	60	5
7	70	6
8	80	6
9	120	9
10	150	10
11	160	12
12		

Check that you have centred the data in both columns.

2 Select the data by clicking and dragging the mouse from cells A1 to B11. The selected area will be shaded.

3 Click Insert in the toolbar and notice the Charts section shown:

Click Scatter and from the choices of scattergraphs choose the Scatter with only markers option: 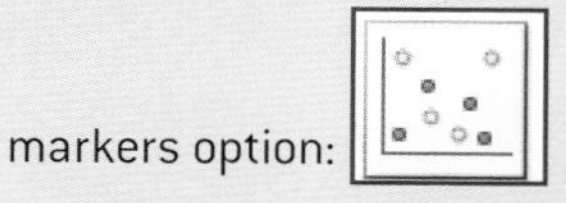

4 The scattergraph appears on the spreadsheet like this:

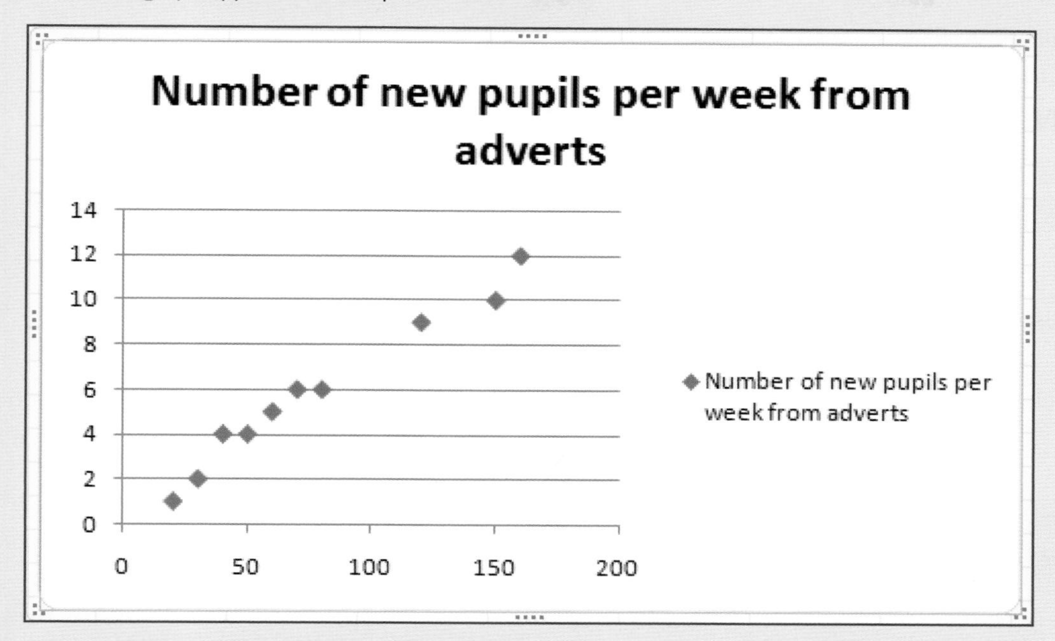

5 Single click on the border of the chart to select it.

You will now see a Chart Tools section of the toolbar. Click on Layout.

A whole series of aspects of the layout now appear.

In the Labels section, click on Axis Title:

6 You can now choose which axis on which you'd like a title, and where to place it.

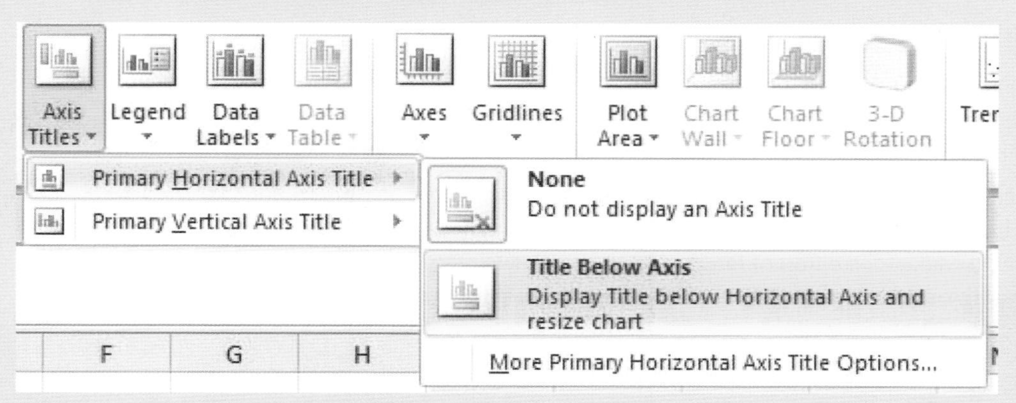

Click on Title Below Axis.

7 You will now see the Axis Title box appear on the scattergraph. Change the text in this box to read 'Amount in Dollars spent on advertising each week'.

8 Put the title 'Number of new pupils per week' on the vertical axis. From the list of options, choose to put the axis title horizontally next to the axis. Remember you will need to select the chart before you will see the layout tab on the toolbar.

Your chart should now look like this:

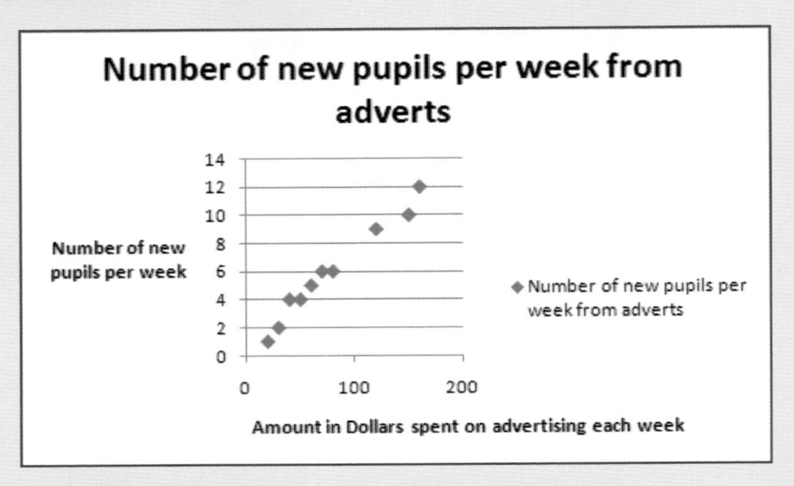

9 Save your spreadsheet using the filename **Scattergraph**.

Using contiguous and non-contiguous data

Contiguous data is data in columns or rows that are next to each other. Non-contiguous data means data in rows or columns that are not next to each other. For the examination, you will have to produce graphs and charts using both types of data.

Activity 16.2

Producing a column chart

In this activity you will learn how to:

- copy and paste data
- widen columns
- change row height
- use contiguous and non-contiguous data to produce charts
- label the chart with title, legend, category, and value axes
- scale and preview a printout.

The following table has been created using word-processing software in a document about nutrition and healthy eating. The writer of the article would like to produce a bar chart that can be used to compare the protein and fat in each of the different foods.

Product	Oven chips	Soup	Yoghurt	Salad Cream	Corn Flakes
Amount (g) of protein per 100g	2	1	5	1	9
Amount (g) of fat per 100g	5	3	1	10	6

1 Load Microsoft Word or your word-processing software and open the file called **Table_showing_fat_and_protein**.

Make sure your table looks the same as that shown above.

2 Select all the data in the table by clicking and dragging until all the data is highlighted like this:

Product	Oven chips	Soup	Yoghurt	Salad Cream	Corn Flakes
Amount (g) of protein per 100g	2	1	5	1	9
Amount (g) of fat per 100g	5	3	1	10	6

With the cursor positioned somewhere on the table, right-click the mouse and in the menu that appears select Copy. This copies the selected data to the Clipboard.

3 Load Microsoft Excel or your spreadsheet software.

When the spreadsheet grid appears, click on cell A2 as this is where we want the copied data to start.

Right-click the mouse and from the menu that appears select Paste.

Check the table has appeared like this:

	A	B	C	D	E	F
1						
2	Product	Oven chips	Soup	Yoghurt	Salad Cream	Corn Flakes
3	Amount (g) of protein per 100g	2	1	5	1	9
4	Amount (g) of fat per 100g	5	3	1	10	6

4 Adjust the column width by clicking and dragging the line between the columns like this:

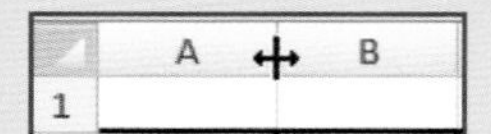

Moving to the right will widen the column, and the text will reflow accordingly.

5 Adjust the row height of the data rows by right-clicking and dragging the lines between the rows:

Amount (g) of
3 protein per 10

The spreadsheet should look similar to this:

	A	B	C	D	E	F	G
1							
2	Product	Oven chips	Soup	Yoghurt	Salad Cream	Corn Flakes	
3	Amount (g) of protein per 100g	2	1	5	1	9	
4	Amount (g) of fat per 100g	5	3	1	10	6	
5							

6 You are now going to draw a vertical bar chart (called a column chart) using some of the data in the spreadsheet (the first two rows). Because these two rows are next to each other, the data in them is called contiguous data.

Select the data by left-clicking on cell A2 and then dragging to the end of the data in cell F3. This area will now appear highlighted.

7 Click on Insert in the toolbar and then in the Charts section select .

Choose the first chart in the 2D column section .

The column chart is now drawn:

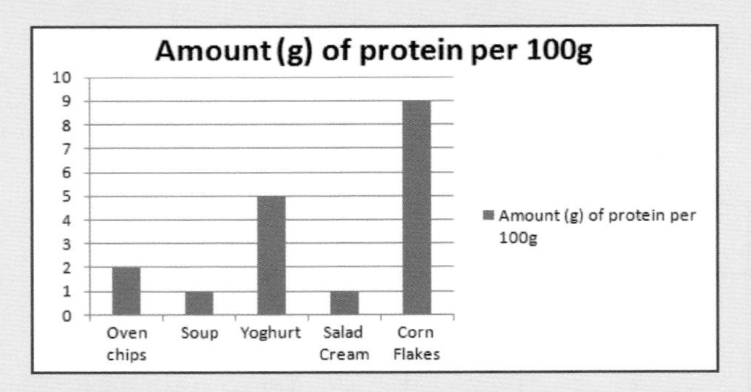

Notice the following about this vertical bar chart (or column chart):

The category axis (i.e. the horizontal axis) has the categories (i.e. the names of the items) on it.

The vertical axis, called the value axis, has the numbers relating to the category on it.

The legend is the explanation (here on the right of the chart) that explains what the height of each bar represents.

The title of the chart gives a few words to explain what the chart shows, i.e. amount (g) of protein per 100g. This is only a suggested title and you can change this.

8 You are now going to change the title of the chart.

Click on the title to select it. When the item is selected it will appear like this:

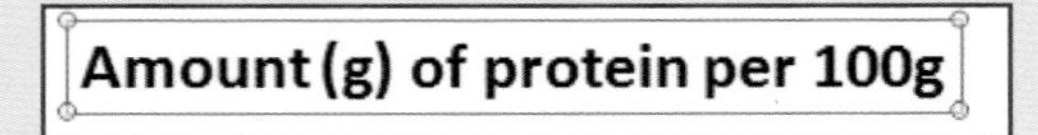

Replace the text in the title with the text shown below.

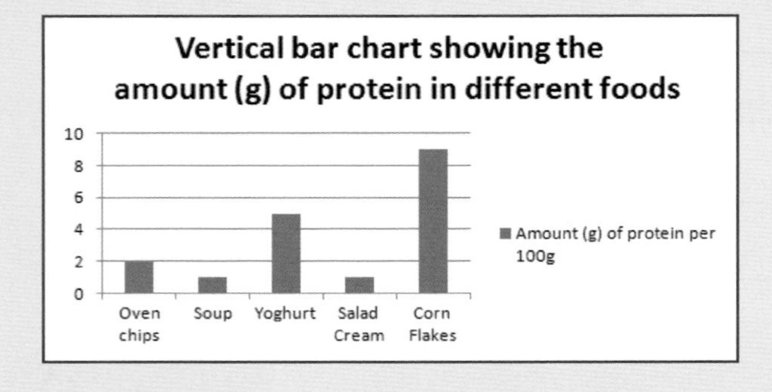

9 Rather than using a legend, you are going to label the value axis.

Select the legend by clicking on it and then and press either the backspace button or the Delete button on your keyboard to delete it.

10 You are now going to add a title to the value axis.

Click on the edge of the chart to select the entire chart.

The chart tools will appear like this:

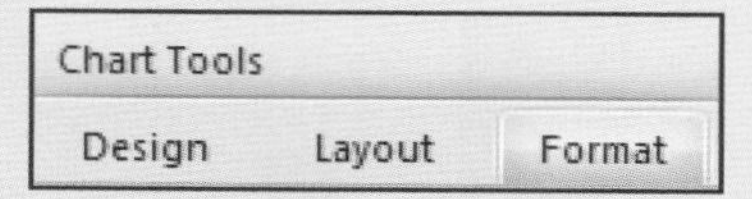

Click on Layout, then Axis Titles. Choose Primary Vertical Axis Title and finally choose Horizontal Title.

Notice the text box next to the vertical (value) axis.

Delete the text and replace it with the text 'Grams of protein'.

Your chart should now look like this:

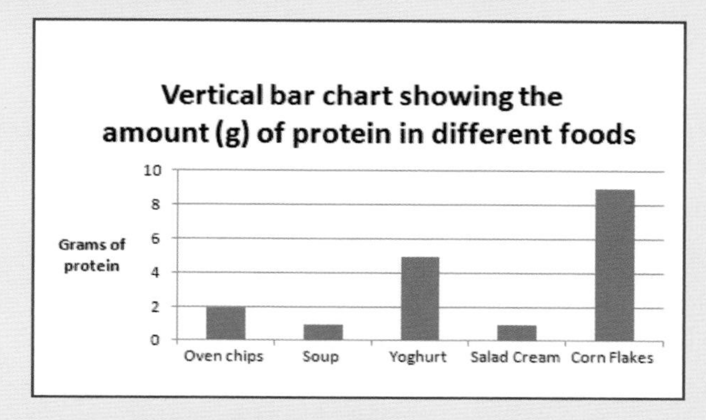

11 Save your spreadsheet using the filename **Food_bar_chart_version_1**.

12 You are now going to draw a similar vertical bar chart, this time showing the amount (g) of fat in different foods.

You need to use the first row of data (the header row) and the third row of data (the amount of fat). This data is **non-contiguous** because the two rows are not next to each other.

First select the cells from A2 to F2 by clicking and dragging. Now press the Ctrl key down and keeping it pressed down click on cell A4 and drag the mouse as far as cell F4. Rows 2 and 4 (and not 3) should now be highlighted like this:

	A	B	C	D	E	F
1						
2	Product	Oven chips	Soup	Yoghurt	Salad Cream	Corn Flakes
3	Amount (g) of protein per 100g	2	1	5	1	9
4	Amount (g) of fat per 100g	5	3	1	10	6

Now produce the vertical bar chart in a similar way to the way you did the one for the amount of protein. Make sure you alter the text so that it appears like that shown below.

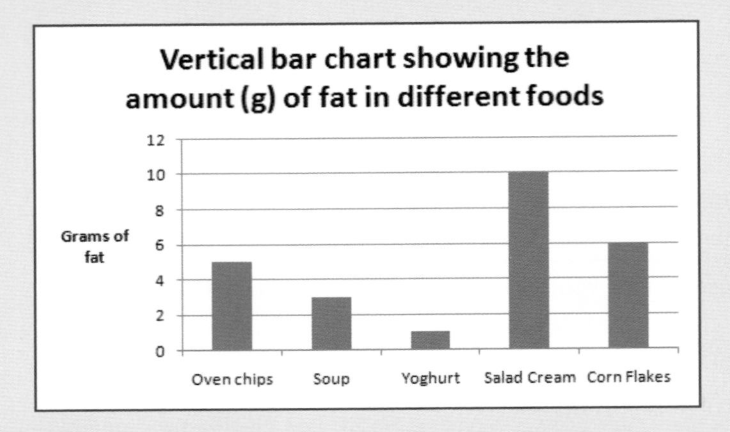

You will probably need to move the charts so they appear next to each other. To do this, click on the border of the chart; when a four-headed arrow appears, hold the left mouse button down and drag the chart into the correct position.

14 Save your spreadsheet using the filename **Food_bar_chart_version_2**.

15 You are now going to print the spreadsheet (table of data and the two charts).

Use Ctrl-P to open the Print dialog box and select your printer, if it's not already the default printer.

16 The Print window opens.

Click on Properties.

As this spreadsheet is wider than it is tall, it would be better to print it using landscape orientation. Click the radio button for Landscape.

Click on the Paper tab at the top of the window in the Scaling Printing section select Fit to page.

Click on OK and you will return to the Layout information. Click on OK again and you are taken to the main Print menu.

Click on Preview to check that the printout is on one page in landscape orientation.

Click on the Print icon and a printout is produced.

Activity 16.3

Producing a pie chart

In this activity you will learn how to:

- create a pie chart
- label pie chart segments with percentages
- add a title.

The table below shows the type of heating used in 100 houses.

Type of heating	Number of houses
Solar	16
Wood	24
Coal	12
Gas	8
Electricity	10
Oil	30

1 Load Microsoft Excel or your spreadsheet software and type in the data shown in this table, using columns A and B, in rows 1 to 7.

Visually check the data.

2 Select all the data in the table by clicking and dragging from cells A1 to B7.

3 Click on Insert and then Pie. Select the type of chart 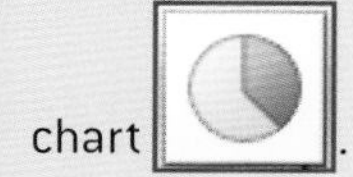.

The pie chart appears like this:

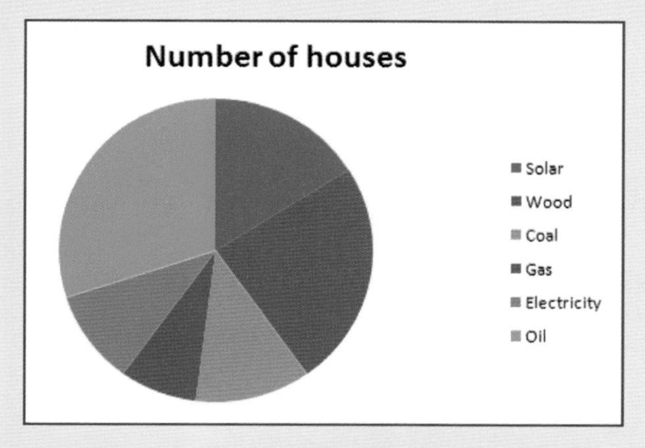

4 There are a number of problems with the chart as it is shown here:

- The title does not correctly express the purpose of the chart.
- It could do with more explanation in the legend.

Click on the chart title and change it to 'Pie chart showing the type of heating used in a sample of 100 houses'.

From the main toolbar, click on Insert and then on Text Box.

Click and drag to create a text box in a position just above the legend:

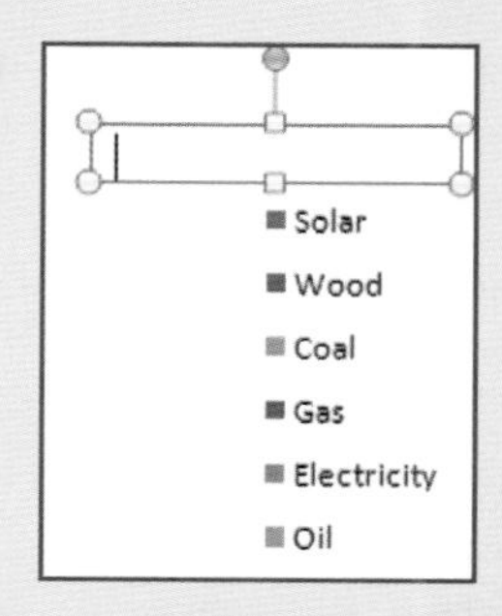

Enter the text 'Type of heating' in the text box.

Centre the text in the text box by highlighting the text and clicking on [centre icon].

5 You are now going to put labels on the segments (these are the slices of the pie). You can add the values themselves (i.e. the actual number that represents a segment) or you can add percentages.

Click on the chart area to select it. Now click on Layout | Data Labels | More Data Label Options...

Notice how the actual values are now displayed on the segments. (You may need to move the Format Data Labels window if it is over the pie chart.)

Click on the tick in the Value box to deselect it and then put a tick next to Percentage.

Notice that the percentage values have been added to the segments:

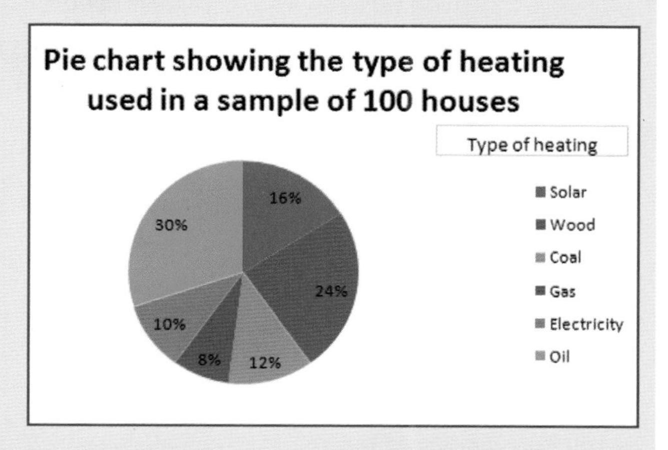

6 Save your spreadsheet using the filename **Heating_pie_chart**.

Activity 16.4

Changing the appearance of the pie chart

In this activity you will learn how to:

- change the colour scheme or patterns of a chart
- extract a pie chart segment.

1 Load your spreadsheet software and open the file created in the last activity called **Heating_pie_chart** if not already opened.

2 Click on the border of the chart to select it.

You are now going to change the colour scheme for the chart.

Click on [Design] in the Chart tools section.

Notice the following examples of chart colour schemes:

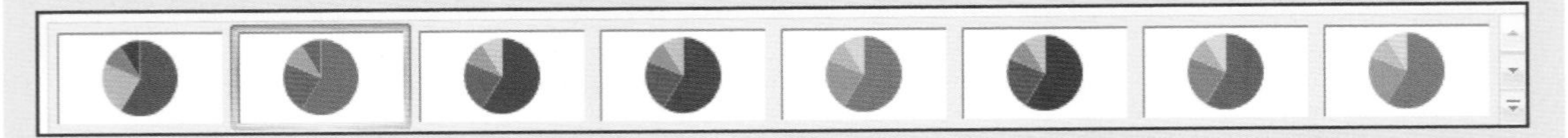

If you are going to print the pie chart in black and white, the first colour scheme should be used (i.e. shades of grey) as this will give enough contrast between the segments.

Click on the pastel green colour scheme to select it.

3 You can also add patterns to a chart.

Right-click on a segment inside the pie chart and then select Format Data Series.

On the Fill page, select Picture or texture fill, as shown here:

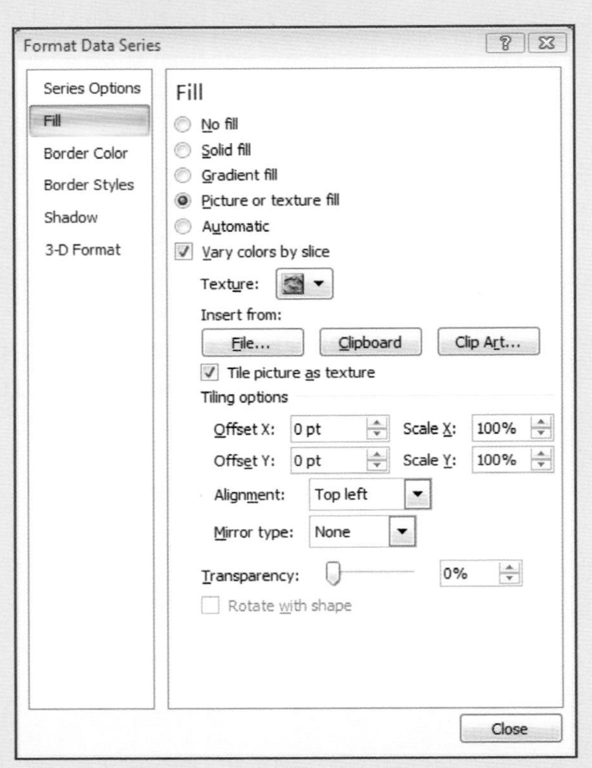

Click on the drop-down arrow for Texture [Texture: ▾] and the following textures appear.

Choose a suitable pastel colour such as 'Blue tissue'.

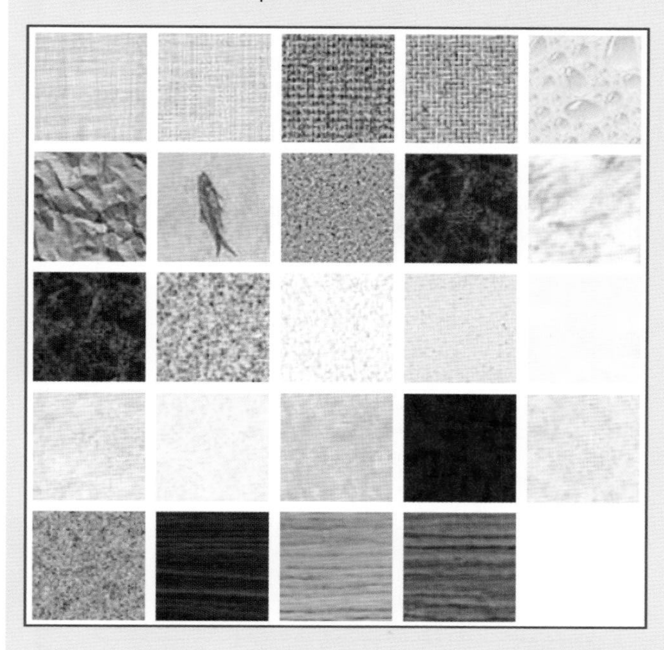

You are now going to emphasise a segment by moving it away from the rest of the pie chart.

Double left-click on the segment with the value 24% to select it. When this segment has been selected you will see the small circles appear like this:

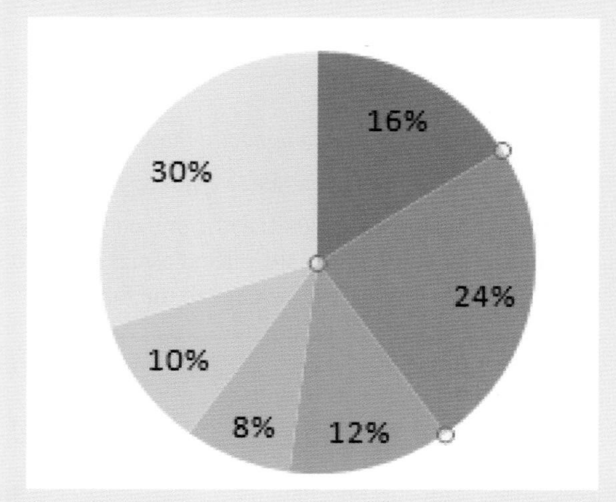

Right-click and a menu appears from which you need to select Format Data Point:

Move the slider on the Point Explosion until the value is 25%.

Now experiment by changing the Angle of first slice and the Point Explosion.

You can then save the spreadsheet using the filename **Exploded_pie_chart**.

Activity 16.5

Producing line graphs

In this activity you will learn how to:

- produce a line graph
- add a second axis
- change the axis scale maximum and minimum.

Line graphs are useful to show trends. For example, you can show the trends in hours of sunshine per day and the maximum daily temperature for each month over a year.

You are going to produce two such line graphs for Dubai, UAE.

1 Load Microsoft Excel and open the file **Dubai_weather_table**.

Check you have the following data in your spreadsheet:

	A	B	C
1			
2	**Month**	**Average hours of sunshine per day**	**Average max daily temp °C**
3	Jan	8	23
4	Feb	9	24
5	Mar	8	27
6	Apr	10	30
7	May	11	34
8	Jun	11	36
9	Jul	11	38
10	Aug	10	39
11	Sep	10	37
12	Oct	10	33
13	Nov	10	31
14	Dec	8	26

2 To create the graph, click on cell A2 and drag to cell C14. Then click on Insert and finally on Line which tells the computer to create line graphs.

3 Select the first line graph from the menu by clicking on the icon:

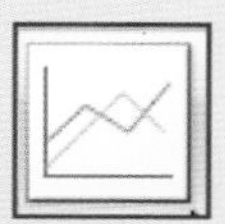

4 A pair of line graphs is drawn on the same axes like this:

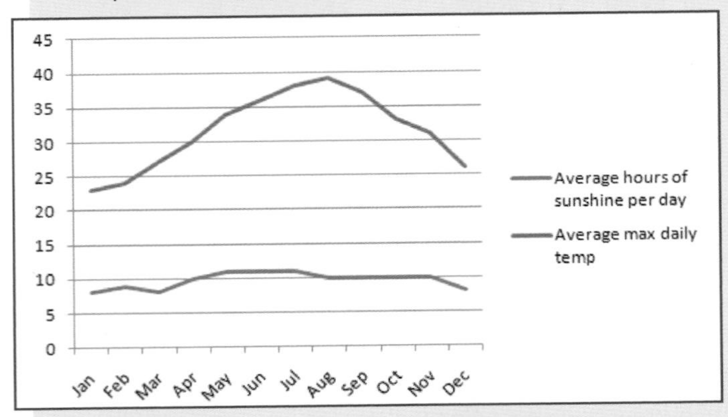

5 The problem with putting two graphs on the same set of axes is that one can appear a bit flat, rather like the one here for the average hours of sunshine per day. It is possible to have two vertical axes with each axis referring to its own line graph.

Right-click on the blue line, as this is the line for which we want to create the secondary axis. From the menu that appears, select Format Data Series.

6 On the Series Options page of the Format Data Series window, select Secondary Axis. This tells the computer that you want to create a secondary axis for the data represented by the blue line that has been selected:

Click on Close.

7 The secondary axis now appears like this:

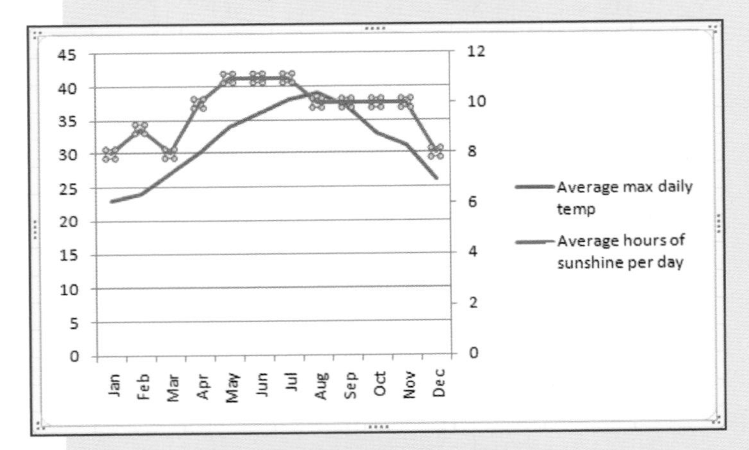

There is still a slight problem. As you can see, Excel scales the axes automatically but it is sometimes better to adjust this yourself so that the graphs are stretched out a bit more.

You are now going to scale the axes yourself.

Right-click on the secondary axis (i.e. the one on the right) and from the menu that appears select Format Axis and the following box appears:

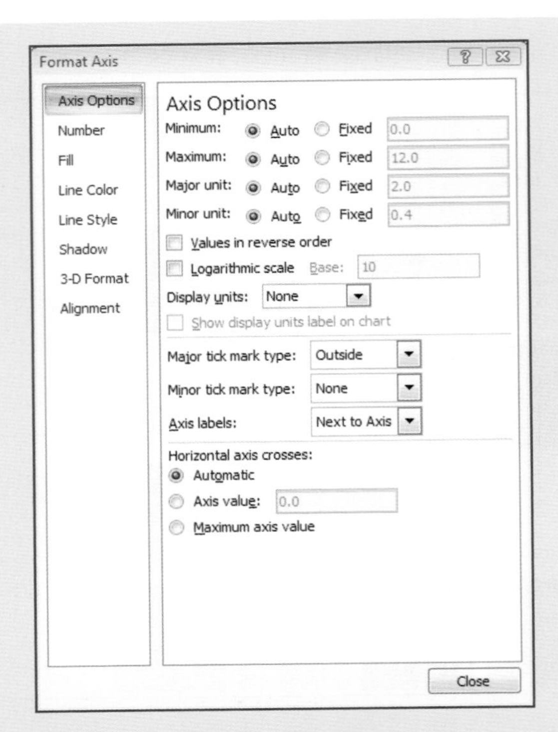

If you look at the data for the hours of sunshine either in the original table or on the graph you will see that the largest data item is 11 and the smallest is 8.

In the Axis Options section:

In the Minimum section select Fixed and change the value to 7.

In the Maximum section select Fixed and change the value to 12.

Click on Close and the changes are made.

8 The scale on the primary axis needs adjusting slightly.

Right-click on the primary axis (i.e. the one on the left) and from the menu that appears select Format Axis and you then need to change the settings to the following:

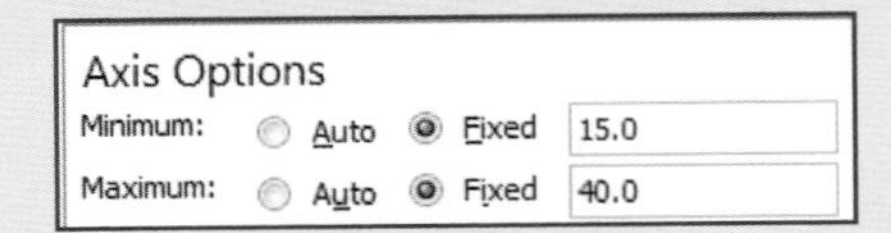

Your chart should now look like this:

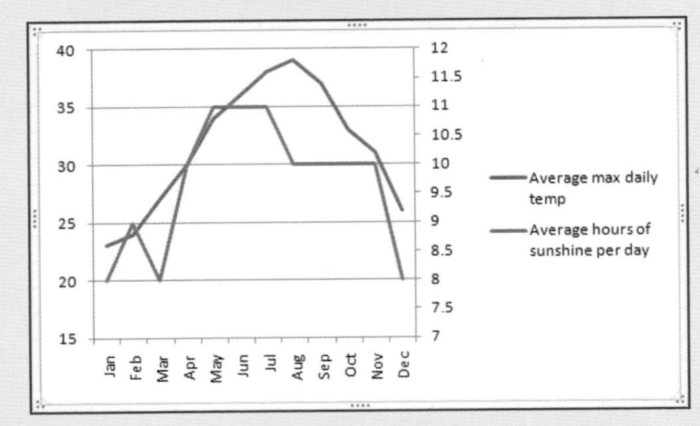

9 You now need to add a title for the entire chart and a title for each axis.

Before you do this, make sure that you click on the border of the chart to select it.

To add the chart title, click on Layout on the toolbar and then on Chart Title.

Choose 'Above Chart' from the list of options. A textbox appears above the chart into which you should enter the text **Monthly_weather_data_for_Dubai, UAE**.

Now add the axis titles as follows:

To add the chart title, click on Layout on the toolbar and then on Axis titles.

Select 'Primary Vertical Axis' Title and then select Rotated Title.

Enter the text Temperature in the text box next to the axis.

To add the chart title, click on Layout on the toolbar and then on Axis titles.

Select 'Secondary Vertical Axis Title' and then select Rotated Title.

Enter the text Hours of sunshine.

You will now need to increase the size of the chart by clicking on the corner of the border and dragging the two-headed arrow that appears until the graph is about 1.5 times its original size.

Your chart should now look like this:

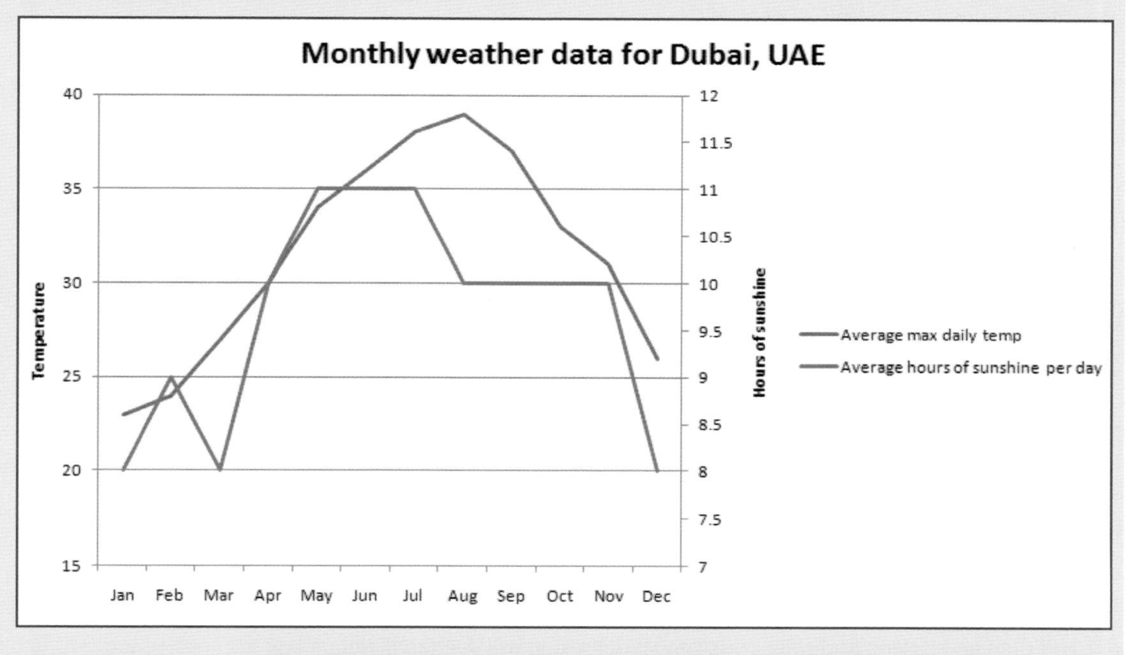

10 Save your chart using the filename **Chart_showing_monthly_weather_data_for_Dubai**.

11 Print a copy of the spreadsheet out in suitable orientation and on a single sheet of paper.

Document production

In this chapter you will be formatting and organising page layout for a variety of documents and you will be using some of the skills and techniques learnt in the previous chapters. You will be learning about how to edit tables within documents and how to format the contents of the cells. This chapter also covers mail merge, which is useful if you want to combine a list of names and addresses and some other information with a standard letter so that a series of similar letters, each addressed to a different person, can be produced.

The key concepts covered in this chapter are:

- Formatting text and organising page layout
- Using software tools to edit tables
- Mail merging a document with a data source

Formatting text and organising page layout

You looked at some formatting text and organising page layout in previous chapters. Here we will be expanding on this.

Page orientation

There are two ways in which a page can be printed onto paper. With portrait orientation the height is greater than the width; with landscape orientation the page is turned sideways so that the width of the page is greater than its height. Portrait orientation is much more common and is used for most business documents such as letters, memos and reports; this book is portrait format. However, landscape format can be useful for charts, spreadsheets and notices.

Activity 17.1

Putting text into columns

In this activity you will learn how to:

- set the right, left, top, and bottom page margins
- set the page orientation (i.e. landscape or portrait)
- set the number of columns to use
- set each column's width
- set the amount of space between columns.

1 Load Microsoft Word or your word-processing software and locate and then open the document called **Advantages_and_disadvantages_of_networking**.

2 Click on Page Layout and then on Margins and select Custom Margins; you will see the dialogue box for the page layout appear.

Set the top and bottom margins to 3 centimetres and the left and right margins to 2 centimetres. When done correctly the settings should be the same as this:

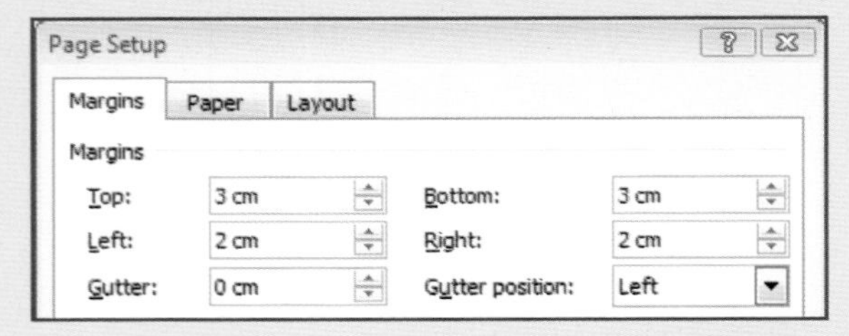

Now set the page orientation to landscape by clicking on the icon like this:

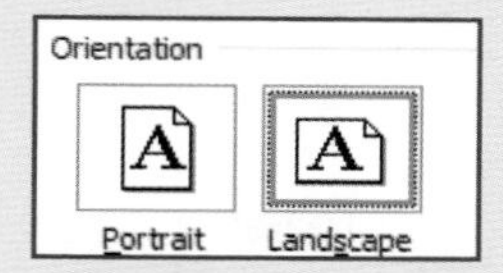

Click on OK to confirm your selections and you will see the document appear on the screen in landscape orientation.

3 Select all the text in the document by highlighting, then change it to a sans serif font and change the font size to 14 pt.

4 Select the heading and change the font size to 26 pt.

5 Select all the text other than the heading and click on Page Layout and then click on Columns.

You will be given a choice of how many columns. Choose Two from the menu like this:

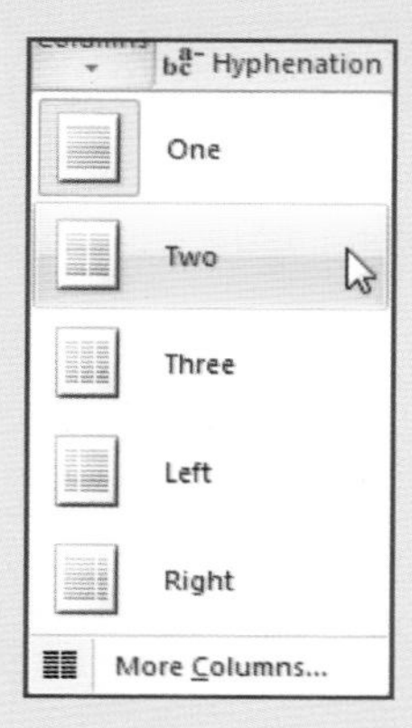

The text appears like this:

Advantages and disadvantages of networking

Advantages

The ability to share resources (e.g. printers, scanners etc.)

Only one copy of applications software needs to be installed and maintained.

The ability to share data.

Greater security because data and programs may be held in one place.

Everyone using the network is able to access the same centrally held pool of data.

Enables email to be sent between terminals thus saving time and money.

Faults with hardware, communication lines can render the network incapable of being used.

If the network is connected to external communication lines (e.g. the telephone) then there is a danger from hackers.

Viruses, if introduced onto a network, can rapidly spread and cause lots of damage to data and programs.

6 Double left-click on the gap in the centre of the ruler as shown here:

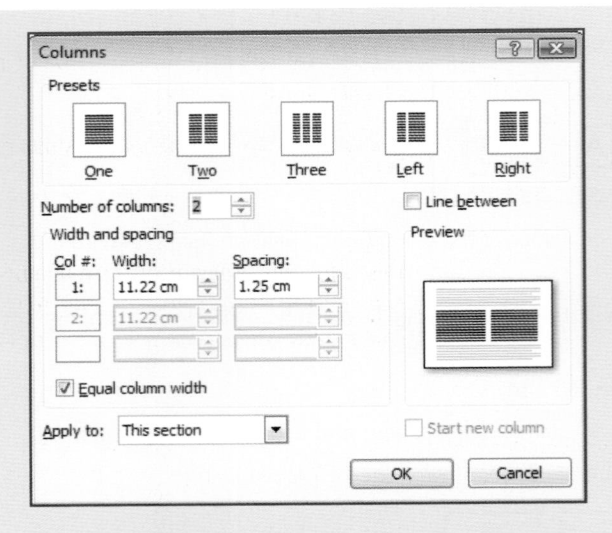

This dialogue box can be used to alter the width of each column, put a line between the columns, and alter the spacing between columns.

You are going to alter the spacing between the columns to 1 centimetre.

Alter the spacing to 1 cm like this:

Click on OK to confirm these new settings.

7 Click on the start of 'Disadvantages' and press return so that the Disadvantages heading appears at the top of the second column.

Your document should now look like this:

Advantages and disadvantages of networking

Advantages

The ability to share resources (e.g. printers, scanners etc.)

Only one copy of applications software needs to be installed and maintained.

The ability to share data.

Greater security because data and programs may be held in one place.

Everyone using the network is able to access the same centrally held pool of data.

Enables email to be sent between terminals thus saving time and money.

Disadvantages

A network manager/administrator is usually needed and it costs a lot to employ them.

Faults with hardware, communication lines can render the network incapable of being used.

If the network is connected to external communication lines (e.g. the telephone) then there is a danger from hackers.

Viruses, if introduced onto a network, can rapidly spread and cause lots of damage to data and programs.

8 Create a header and add your name left aligned and the number 121121 right aligned.

9 Create a footer and place an automated page number aligned to the right margin.

10 Save your file as a Word file in your area using the filename: **Advantages_and_disadvantages_of_networking**.

11 Print a copy of your document.

Gutter margin

A gutter margin is used when a document is to be bound. Adding a gutter margin tells the computer to add more space so that when the pages are bound, all of the content on the page can be seen.

To add a gutter margin you perform the following steps:

1 Click on Page Layout and select Margins and from the menu select Custom Margins.

2 The Page Setup dialogue box appears. Here you can apply the gutter to the left or the top of the document. Notice also that you can adjust the size of the gutter in cm. Here the gutter position is set to the left and the size of the gutter has been set to 1 cm.

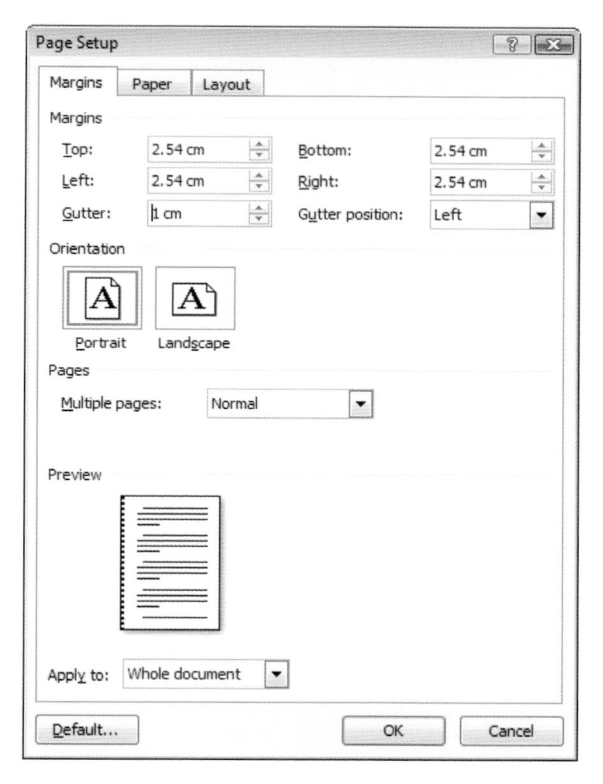

You would then need to click on OK to apply the gutter to the whole document.

Setting tabs

These buttons are used to decrease or increase the indent level of a paragraph.

The following appears no matter which tab has been selected:

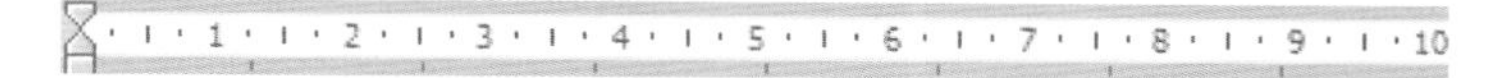

This is the ruler which contains a number of tools. On the left hand side of the ruler there are the following three sliders that can be moved along the ruler to control the way text appears on the page:

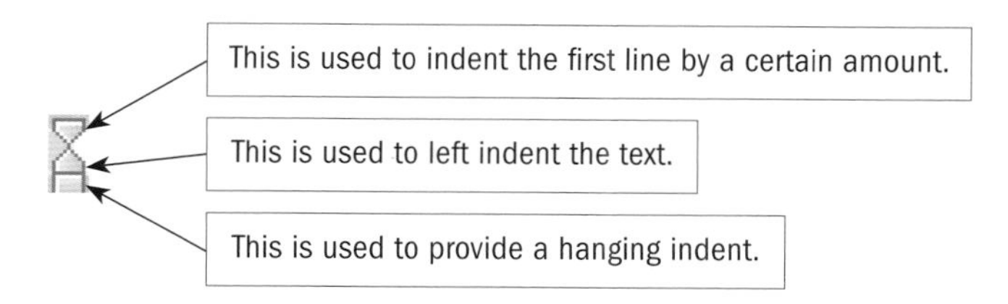

Look carefully at the following, which shows how the text appears when these indents are used.

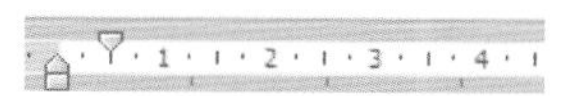

In the above, a paragraph indent of 0.5 cm has been applied to the first line of a paragraph.

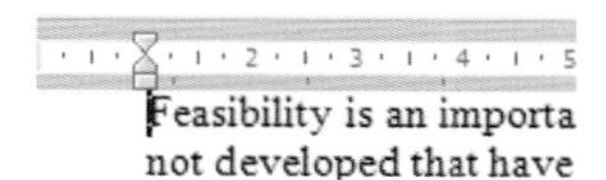

In the above, the whole section of selected text has been indented by 1 cm.

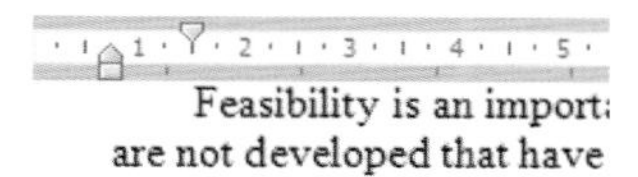

In the above, the main body of text has been indented by 1.5 cm from the left margin and the first line of the paragraph has been indented by 0.75 cm from the line of the main text.

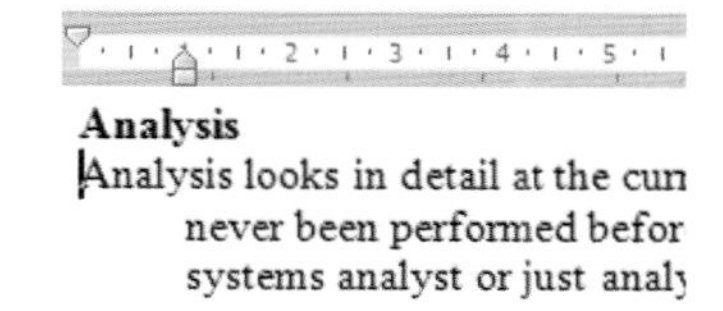

This setting produces a hanging indent of 1cm. The first line is lined up with the margin and the other lines are lined up using the hanging indent.

Using software tools to edit tables

Creating tables was discussed in Chapter 13 and here you will be looking at how to edit and format an existing table.

Activity 17.2

Editing a table

In this activity you will learn how to:

- insert rows, delete rows, insert columns and delete columns
- set horizontal cell alignment
- set vertical cell alignment.

1 Load Microsoft Word or your word-processing software and open the document called **Tours**.

Check that you have the following table loaded:

Tour	Duration (Hrs)	Adult cost($)	Child cost($)	Date
El-Alamein	6.5	45	27	12/12/11
Alexandria City	4	40	24	13/12/11
Cairo, Pyramids and Tombs	13	80	55	12/12/11
Classic Cairo	12.5	80	55	13/12/11
Caravans in the sand	13.5	110	70	12/12/11
The Pyramids and the River Nile	12	108	68	12/12/11
Cairo overland	48	250	145	12/12/11
Luxor & Valley of the Kings	14	300	175	12/12/11

2 Two extra tours have been added to the list:

- Landmarks of Alexandria for a duration of 5 hrs at a cost of $50 for adults and $35 for children on 13/12/11.
- Alexandria panorama for a duration of 7 hrs at a cost of $100 for adults and $70 for children on 13/12/11.

You need to add two blank rows to hold this data at the bottom of the list.

To do this, click on the very last cell in the table.

Press the tab key once. You will see a blank row has been inserted.

Now click on the very last cell of the row you have just inserted and press the tab key to insert another blank row.

Now add the data shown above into these two rows.

3 You are now going to insert a column between the columns for the Child cost($) and the Date.

Click on the cell with the column heading 'Child cost($)' and then right-click the mouse button and the following menu appears:

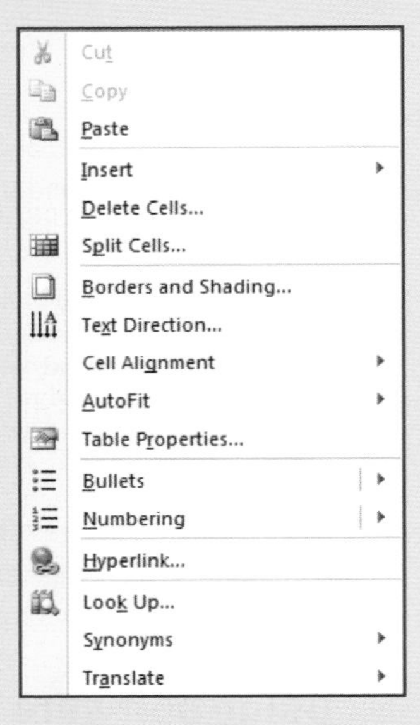

Look carefully at this menu and notice that it provides the tools for editing tables.

Click on Insert – the following choices appear:

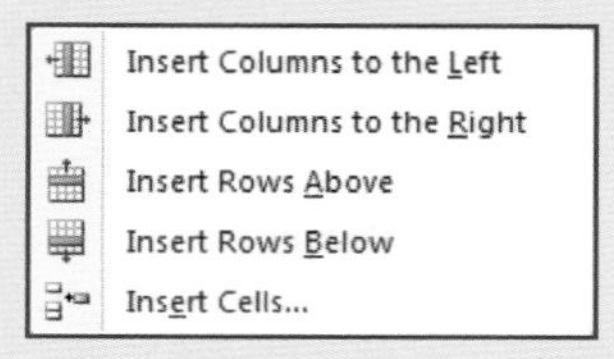

Click on Insert Columns to the Right.

The blank column now appears in the correct position.

4 Type in the data as shown here. Make sure that the column heading is made bold.

Tour	**Duration (Hrs)**	**Adult cost($)**	**Child cost($)**	**Meal provided?**	**Date**
El-Alamein	6.5	45	27	N	12/12/11
Alexandria City	4	40	24	N	13/12/11
Cairo, Pyramids and Tombs	13	80	55	Y	12/12/11
Classic Cairo	12.5	80	55	Y	13/12/11
Caravans in the sand	13.5	110	70	Y	12/12/11
The Pyramids and the River Nile	12	108	68	Y	12/12/11
Cairo overland	48	250	145	Y	12/12/11
Luxor & Valley of the Kings	14	300	175	Y	12/12/11
Landmarks of Alexandria	5	50	35	N	13/12/11
Alexandria panorama	7	100	70	N	13/12/11

5 Insert a row between the tours Classic Cairo and Caravans in the sand by clicking on the row for Classic Cairo and then right-clicking to show the menu, and then click on Insert to insert the blank row below. Then enter the data as show here:

Medieval Cairo	13	95	60	Y	12/12/11

6 It has been decided to run each tour on both dates, so the date column needs deleting. Right-click on the Date and from the menu that appears choose delete cells; the following choices appear:

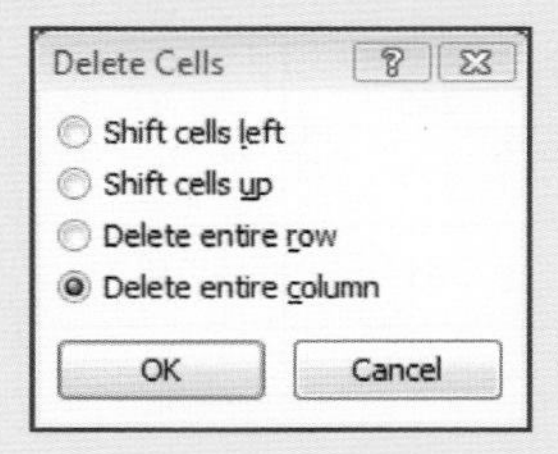

Choose Delete entire column and then click on OK to confirm the deletion.

7 The table will be improved if the contents of the cells are aligned. Highlight all the cells except the tour names like this:

Tour	**Duration (Hrs)**	**Adult cost($)**	**Child cost($)**	**Meal provided?**
El-Alamein	6.5	45	27	N
Alexandria City	4	40	24	N
Cairo, Pyramids and Tombs	13	80	55	Y
Classic Cairo	12.5	80	55	Y
Medieval Cairo	13	95	60	Y
Caravans in the sand	13.5	110	70	Y
The Pyramids and the River Nile	12	108	68	Y
Cairo overland	48	250	145	Y
Luxor & Valley of the Kings	14	300	175	Y
Landmarks of Alexandria	5	50	35	N
Alexandria panorama	7	100	70	N

Right-click on the highlighted part of the table and select Cell Alignment and the following options appear:

This is a good place to hover the cursor over each of the alternatives; for each selection, a caption will appear explaining what it does. Here is a table show how the text will appear when each button is applied to cells containing text:

Align Top Left	Align Top Centre	Align Top Right
Align Centre Left	Align Centre	Align Centre Right
Align Bottom Left	Align Bottom Centre	Align Bottom Right

Select Align Centre – this will position the data centrally (horizontally and vertically) in the cells.

Your table will now look like this:

Tour	Duration (Hrs)	Adult cost($)	Child cost($)	Meal provided?
El-Alamein	6.5	45	27	N
Alexandria City	4	40	24	N
Cairo, Pyramids and Tombs	13	80	55	Y
Classic Cairo	12.5	80	55	Y
Medieval Cairo	13	95	60	Y
Caravans in the sand	13.5	110	70	Y
The Pyramids and the River Nile	12	108	68	Y
Cairo overland	48	250	145	Y
Luxor & Valley of the Kings	14	300	175	Y
Landmarks of Alexandria	5	50	35	N
Alexandria panorama	7	100	70	N

8 You are now going to shade the cells containing the column headings to make them stand out more. Select them by highlighting them like this:

Tour	Duration (Hrs)	Adult cost($)	Child cost($)	Meal provided?

Right-click on the shaded part and the following menu appears:

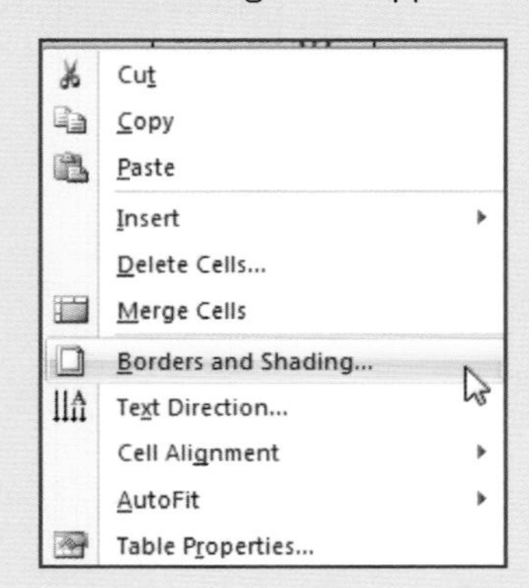

Click on Borders and Shading and the following dialogue box appears:

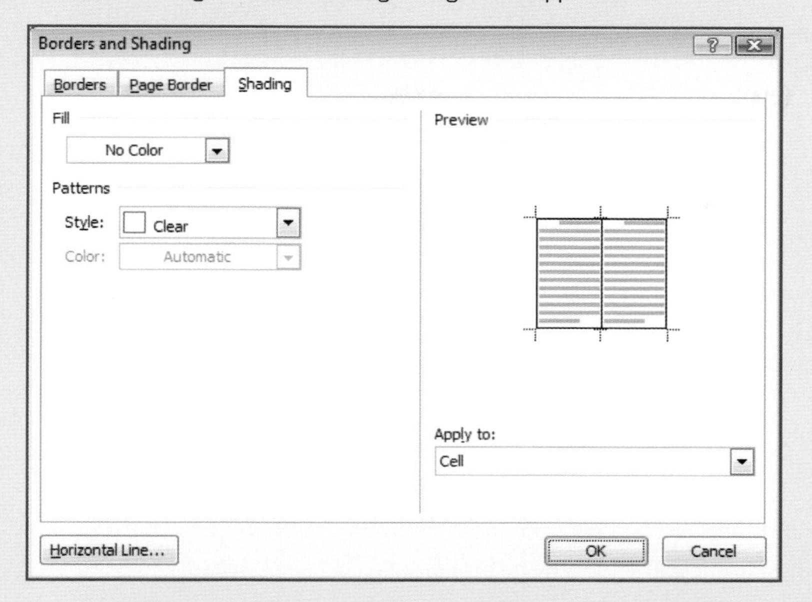

This box allows you to fill selected cells with a pattern. You can also add colour as the background to selected cells.

Click on Style and change it to the following:

This will shade the selected cells so the column headings look like this:

Tour	Duration (Hrs)	Adult cost($)	Child cost($)	Meal provided?

9 Save your new table using the filename **Final_tours_table**.

Activity 17.3

Formatting a table

In this activity you will learn how to:

- colour cells
- hide the gridlines of a table.

1 Load Microsoft Word and open the file **Stopping_distances**.

This document includes a table showing the stopping distances for a car at certain speeds.

Speed (km/h)	Thinking distance (m)	Braking distance (m)	Overall stopping distance (m)
32	6	6	12
48	9	14	23
64	12	24	36
80	15	38	53
96	18	55	73
112	21	75	96

Select the column headings by highlighting them.

2 Right-click on the highlighted area and select Borders and Shading.

Click on the Shading tab:

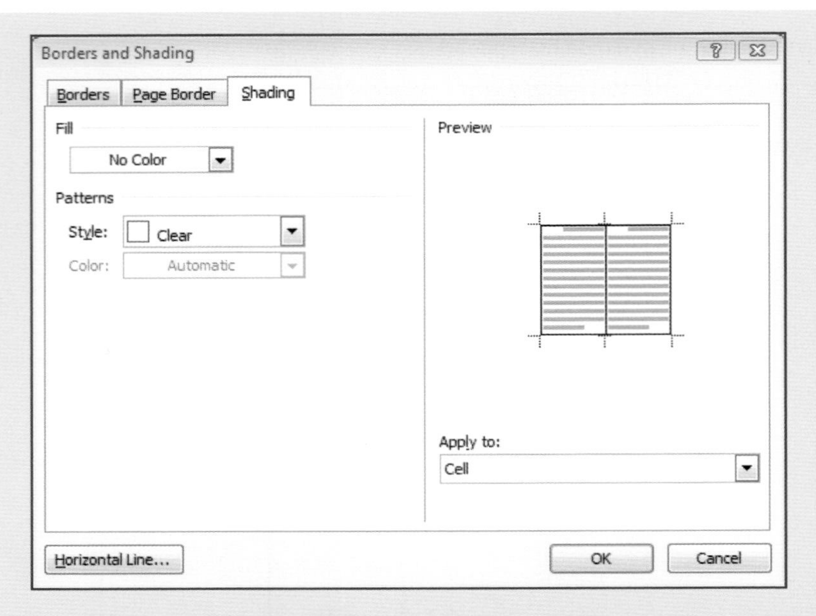

Click on the drop-down arrow on the Fill box and choose the following colour from the palette of colours by clicking on it:

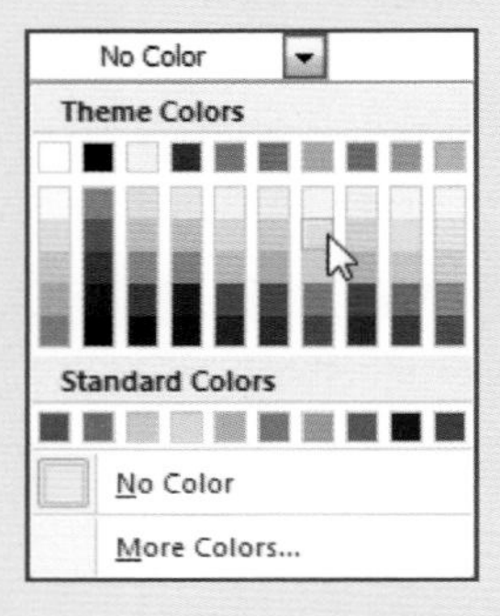

It is important to note that pastel colours are best as background colours to cells in tables if those cells also contain text.

Click on OK to confirm the colour choice and the colour is applied to the column headings like this:

Speed (km/h)	Thinking distance (m)	Braking distance (m)	Overall stopping distance (m)
32	6	6	12
48	9	14	23
64	12	24	36
80	15	38	53
96	18	55	73
112	21	75	96

3 Now shade the other cells until the table looks the same as this:

Speed (km/h)	Thinking distance (m)	Braking distance (m)	Overall stopping distance (m)
32	6	6	12
48	9	14	23
64	12	24	36
80	15	38	53
96	18	55	73
112	21	75	96

4 You are now going to remove the gridlines from the table.

Select the entire table by highlighting it.

Right-click and select Borders and Shading from the menu.

You need to change the settings on the dialogue box to the following:

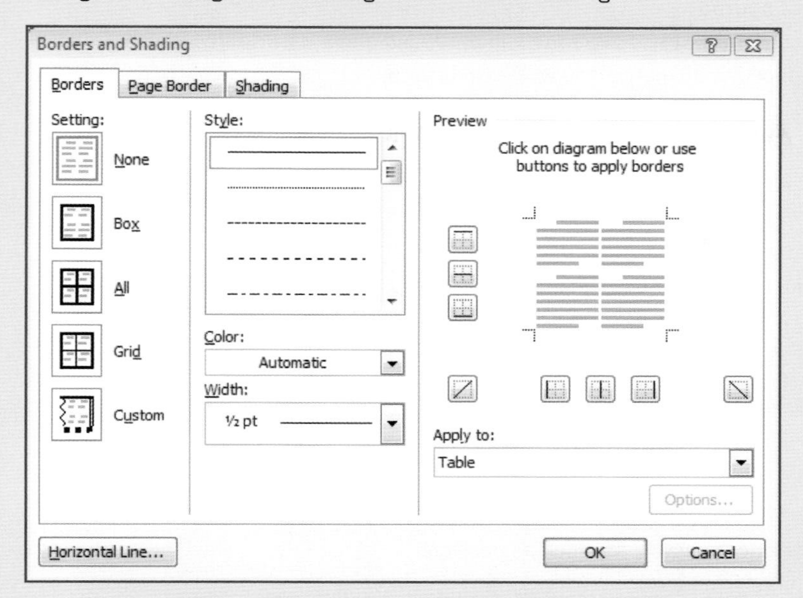

Ensure that None has been selected from the Setting and then click on OK.

Also, as you do not want any of the borders (including the borders to the cells shown) check that Table has been selected in the Apply to section. Click on OK to confirm these selections.

Notice that the table now appears like this with dotted lines showing where the borders had been.

Speed (km/h)	Thinking distance (m)	Braking distance (m)	Overall stopping distance (m)
32	6	6	12
48	9	14	23
64	12	24	36
80	15	38	53
96	18	55	73
112	21	75	96

These dotted lines do not appear when the table is printed. You can print preview the document to see how it will appear when printed.

Speed (km/h)	Thinking distance (m)	Braking distance (m)	Overall stopping distance (m)
32	6	6	12
48	9	14	23
64	12	24	36
80	15	38	53
96	18	55	73
112	21	75	96

5 Save your document using the filename **Stopping_distances_final**.

Mail merging a document with a data source

Mail merging can be used to send the same basic letter, with just slight differences, to a large number of people. The data source is used to supply the names, addresses and other details of the people to whom the letter is to be sent. The data source supplies the variable details that are merged into the letters. The data source can be produced in word-processing, spreadsheet, database or any other software that can save contact details as a file.

Performing a mail merge is worthwhile only if a large number of letters need to be sent. If you had ten letters then it may not be worth the trouble of setting up a mail merge, as you could simply create the letter with one lot of details and then change these for the next letter and so on. If you need more than ten (or you may need to send a similar letter to the same people in the future) then you should consider using mail merge.

Activity 17.4

Mail merging a document with a data source

In this activity you will be using a document that has already been produced and a data source which again has been set up for you, to produce a mail merge. You will learn how to:

- insert appropriate fields from a data source into a master document
- insert a date field into a document
- select records to merge
- merge a document with selected fields
- save and print merge master document
- save and print selected merged documents.

1 The first step in the mail merge is the creation of the letter. The letter has already been produced and is available as a file. Load the document **Green_project_letter.rtf**.

Check that you have the following text loaded.

> Dear
>
> You may have heard of The Green Project, which is a project run by your Local Authority to encourage everyone to recycle as much as possible.
>
> Our database tells us that you have a large or medium sized garden and do not make use of a compost heap. We would like to encourage as many people in your situation to make use of a compost heap to get rid of same garden and household waste by recycling. In turn you will produce rich compost that you can use on your garden.
>
> We have arranged for a free compost container for you. All you have to do is to come and pick it up from our depot at 1 Moor Lane, L12 6RE. The depot will be open from 9am to 5pm Monday to Friday. The compost container is flat-packed and comes with full instructions for assemble and use. You will be able to fit the pack into your car boot.
>
> Please take advantage of this and do your bit to help the environment.
>
> Thanking you for your time.
>
> Stephen Doyle
>
> Green Project (Compost) Manager

2 Click on Mailings and then Start Mail Merge.

3 Click on Step by Step Mail Merge Wizard.

4 Choose Letters where it says Select document type.

Click on Next: Starting document.

5 In Step 2, select Use the current document as the starting document.

Click on Next: Select recipients.

6 As you are using an existing data source to supply the name and address recipient details you can browse for the file called **Green_project_names_and_addresses**.

To find this file, follow these instructions.

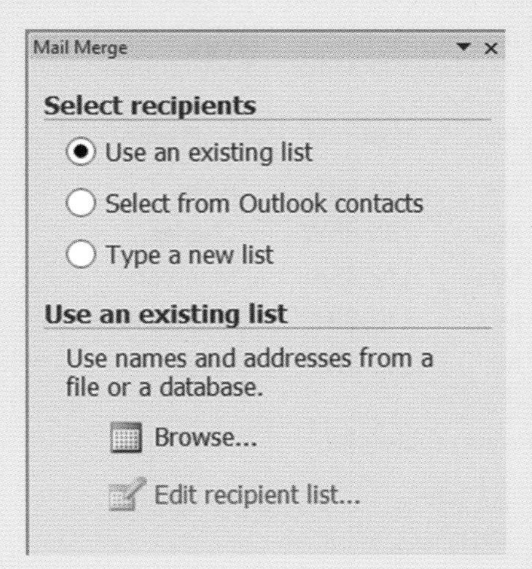

Check that Use an existing list has been selected and then click on Browse

7 Find the file called **Green_project_names_and_addresses**.

Your teacher will tell you where this file is located.

Click on the file and then on Open.

8 The data source shows the name and address details for all the recipients of the letter.

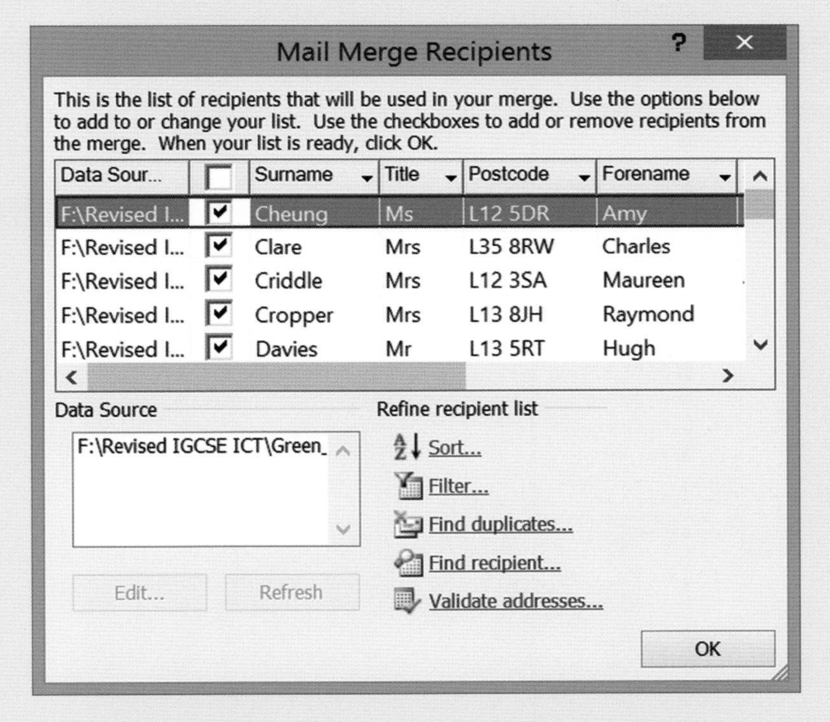

Notice that there is an option to remove ticks from the checkboxes to indicate *not* to send a letter to that person in the data source.

For this activity you are going to send the letters only to the six recipients with the surnames:

Cheung	Gant
Clare	Lee
Davies	Miles

Scroll and remove the ticks from all the recipients except these six.

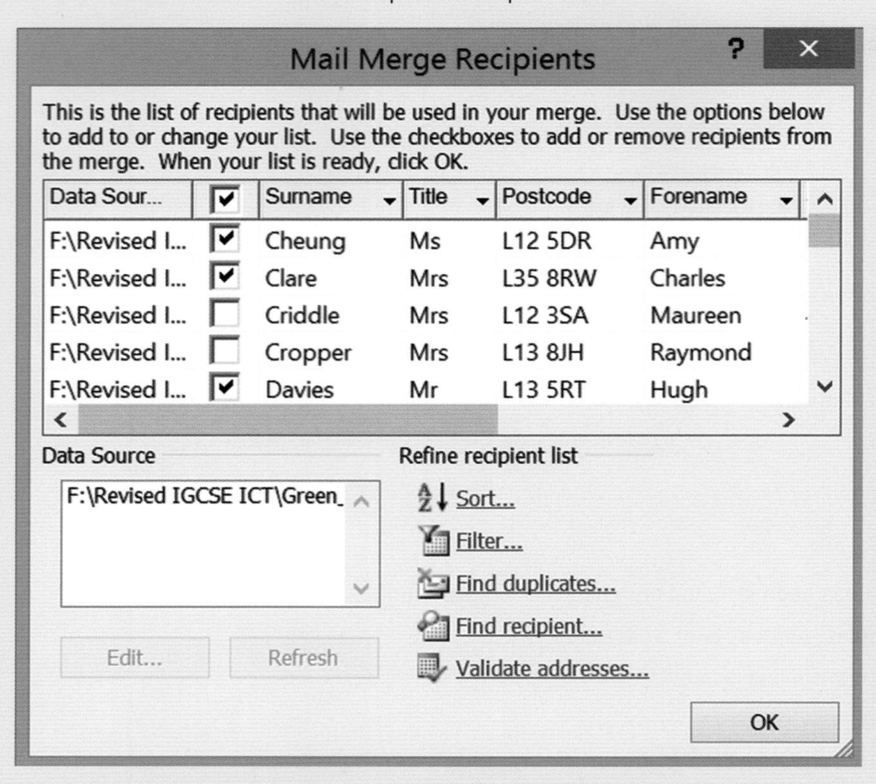

Check that you have selected the correct recipients and then click on the first row in the list (i.e. Cheung) to select it. Click OK.

9 Check that the cursor is positioned in the top left corner of the letter. To add the address details to the letter, click on Address block shown below.

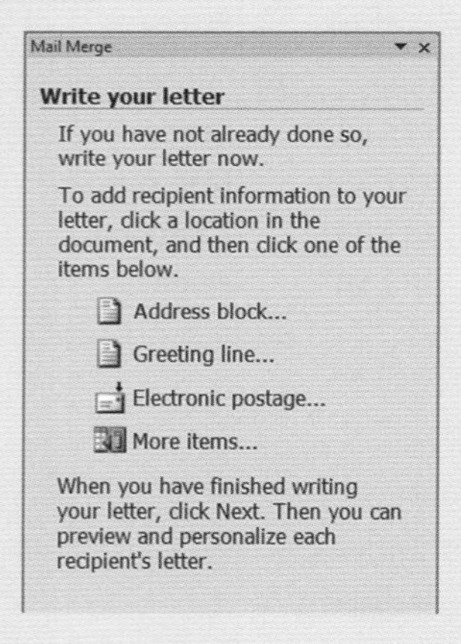

10 The following menu is displayed.

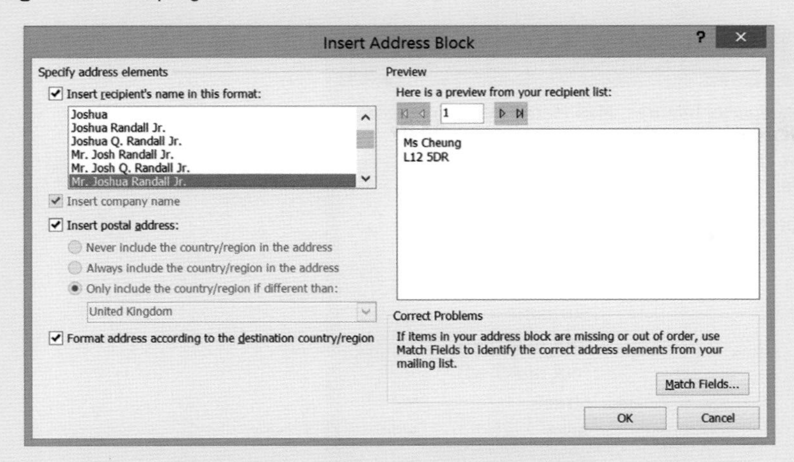

Notice that a sample address is shown. Notice that this does not contain all the address items stored in the data source. To add the items click on Match fields.

11 The following window appears.

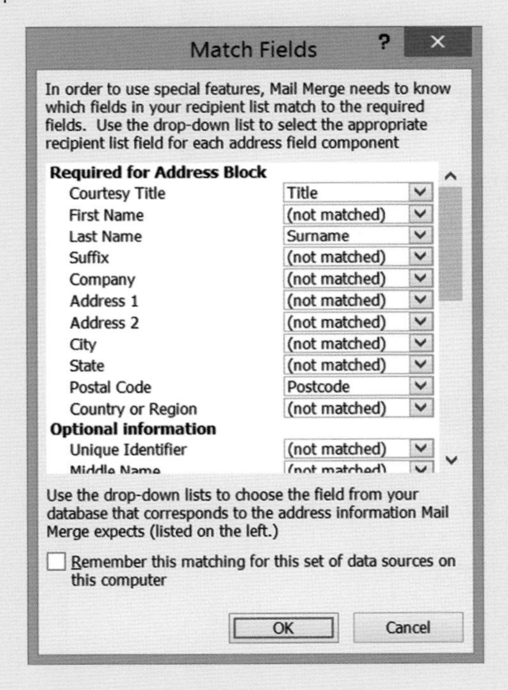

Notice that only the Title, Surname and Postcode are shown.

Click on the drop-down arrow for First Name and select Forename.

Click on the drop-down arrow for Last Name and select Surname.

Click on the drop-down arrow for Address 1 and select Street.

Click on the drop-down arrow for Address 2 and select Area.

Check your settings look the same as those shown below.

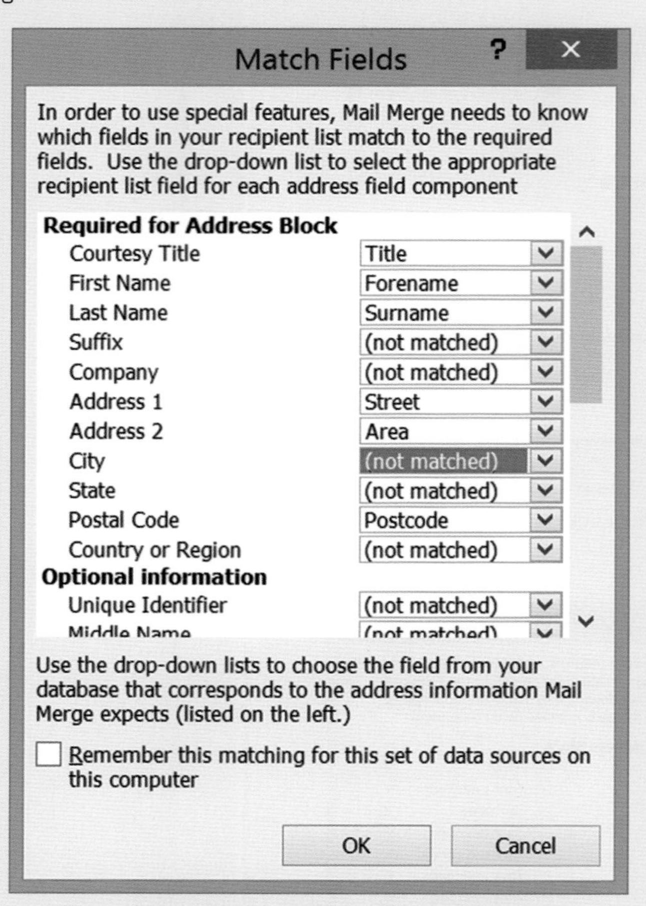

Check that the settings match and click on OK.

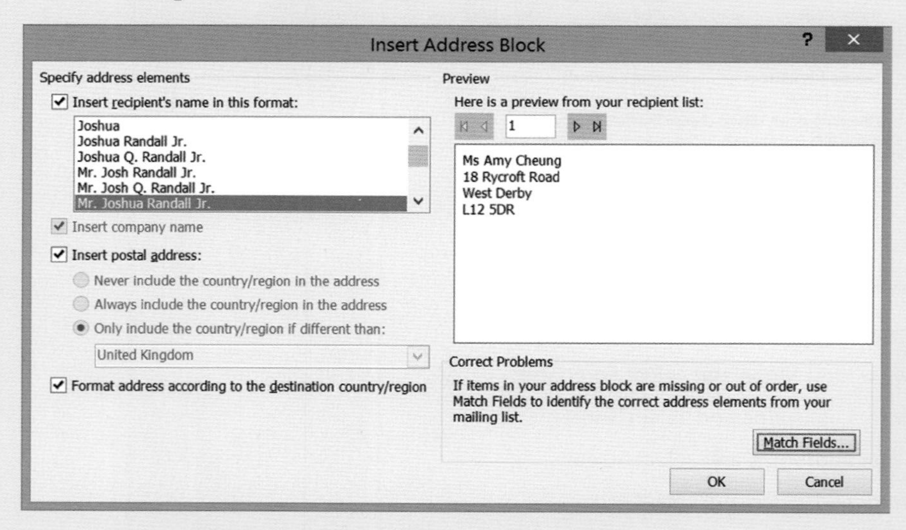

You can now see that the address is shown correctly.

Click on OK.

12 The following Address Block field is inserted at the top of the letter.

> <<AddressBlock>>
>
> Dear
>
> You may have heard of The Green Project, which is a project run by your Local Authority to encourage everyone to recycle as much as possible.

13 Move the cursor to after the 'Dear' at the start of the letter and insert a space.

Click on Greeting line and the following window appears.

In the Greeting line format: change 'Dear' to '(none)'.

Also, in the Greeting line for invalid recipient names: change 'Dear Sir or Madam', to '(none)'.

Click on OK.

14 Notice that the Greeting Line has been entered into the letter like this.

> <<AddressBlock>>
>
> Dear <<GreetingLine>>
>
> You may have heard of The Green Project, which is a project run by your Local Authority to encourage everyone to recycle as much as possible.

15 Click Next: Preview your letters.

16 The address and greeting details have now been inserted into the letter like this.

Ms Amy Cheung

18 Rycroft Road

West Derby

L12 5DR

Dear Ms Cheung,

Check that the details for the other recipients are correct by scrolling quickly through the letters using the right arrow < Recipient: 1 > .

17 A date needs to be added to each letter.

Click on the line immediately below the Address Block. This is the position where the date will be inserted.

To insert the date, click on Insert and then Date & Time.

There are various ways in which the date and time can be shown. Click on the first date in the list as shown.

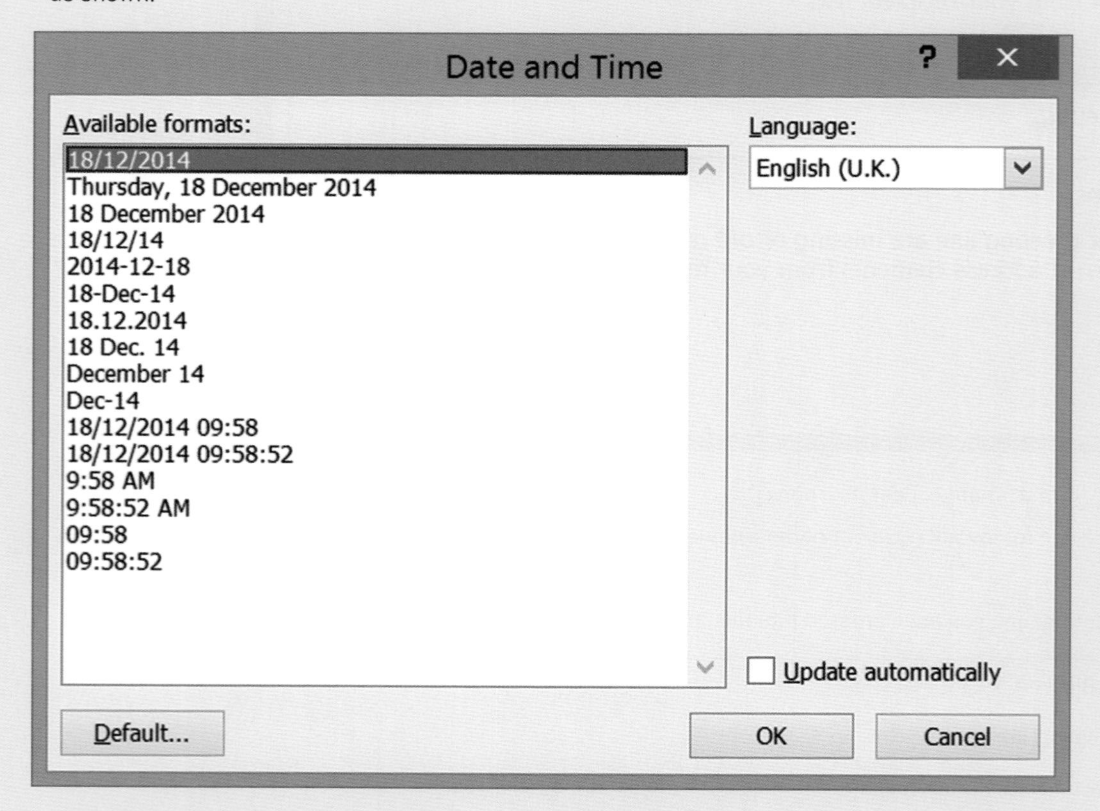

Click on OK.

Click Next: Complete the merge.

18 At this stage you can edit individual letters. You can do this to check each letter.

Click Print to print the letters.

The following menu appears.

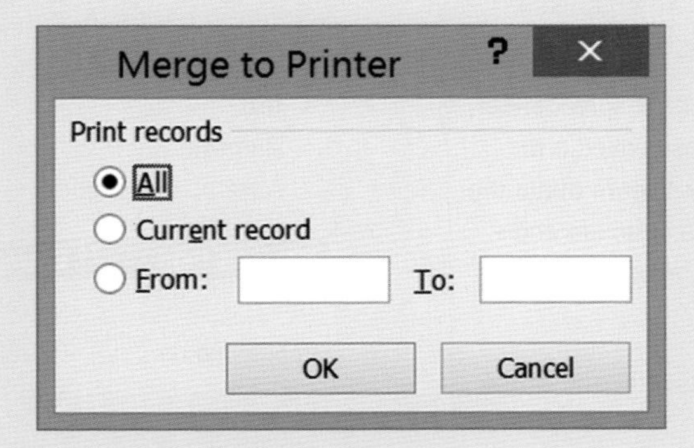

As we want all the records we have specified click on OK.

The merged letters are now printed.

Well done! You have now completed your mail merge.

18 Data manipulation

When a large amount of data needs to be stored, it first needs to be put into a structure by using spreadsheet or database software. Once the data has been stored it can then be manipulated by performing calculations, searches (i.e. queries), etc.

The key concepts covered in this chapter are:
- Creating a database structure
- Manipulating data
- Presenting data

Creating a database structure

Data can be letters, symbols, numbers, images, sound. Together these make up names, dates, exam results, photographs – as seen in your school's computer system where your information may be stored. Database, spreadsheet and many other types of software require that data is put into a structure. Once in this structure, it can be manipulated and output in lots of different ways. An organised store of data on a computer is called a database.

It is possible to create simple database structures using spreadsheet or database software. There are two types of database called a flat-file database and a relational database. Here, you will learn about the important differences between the two types of data structure and also the way data is held more efficiently using a relational database.

Data types

Data entered into a computer for processing is usually one of the following types:

- Boolean/logical
- Text
- Numeric (integer, decimal, percentage, currency)
- Date/time.

Boolean/logical

Boolean/logical data can have only one of two values: 1 or 0, yes or no, true or false, male or female, etc.

Text

Text refers to all the letters and numbers and other characters you see on the keyboard that can be typed in.

Numeric (integer, decimal, percentage, currency)

Decimal numbers are numbers containing numbers after the decimal point (e.g. 3.45) and integers are whole number (i.e. positive, negative whole numbers and zero). Percentages are numbers expressed as percentages and currency means numbers containing a currency symbol.

Date/time

There are lots of different ways of writing a date. The common ways include:

dd/mm/yy for example 12/03/16 for the date 12 March 2016

mm/dd/yy for example 03/12/16 for the date 12 March 2016

Times can also be included with the date.

QUESTIONS A

1 Data is to be stored in a structure and to do this a data type must be chosen for each item of data.
Choose a suitable data type from the following list for the data shown in the table: *(6 marks)*

Logical/Boolean
Text
Numeric
Date

Name of field	Example data	Data type
Title	(Mr, Mrs, Ms, Dr, etc.)	
Phone number	0798273232	
Sex	M or F	
Country	Botswana	
Date of birth	01/10/03	
Years at address	4	

2 Here is a list of examples of data to be put into a structure.

Some of this data can have the logical/Boolean data type and some of it cannot.

Complete the following table by placing a tick in the box next to those items of data that could use a logical/Boolean data type. *(4 marks)*

Item of data	Tick if data type is logical/Boolean
Driving licence (yes or no)	
Sex (M or F)	
Size (S, M, L, XL, XXL)	
Airport code	
Car registration number	
Date of purchase	
Car type (manual or automatic)	
Fuel type (diesel or petrol)	

Database terms

There are some database terms you will need to familiarise yourself with. They are:

Field: A field is a single item of data found in every record. A field has a 'Field Name' to identify it in the database and to help the user convert the text or number into meaningful information. An example would be Field Name: Surname, Data: Smith – means that we have information on a person whose Surname is Smith.

Record: A record is a collection of fields – the data held about one person, thing or place in a database. For example – your school record includes your name, address, date of birth, exam results and the like – all the data about you. A travel agent might have all the data about a hotel – name, address, facilities, manager's name, telephone number, room types – this makes up the **record** for this hotel.

Table: In databases a table is used to store data with each row in the table being a record and the whole table being a file. When only one table is used, it is a very simple database and it is called a flat-file database. For more complex databases created using specialist database software, lots of tables can be used and such a database is called a relational database.

Choosing the software to create a database structure

There are two types of software you could use to produce a database:

- Spreadsheet software
- Database software.

You can build a simple database by organising the data in rows and columns in a table. In the table below the columns represent each of the fields and the rows are the records.

Each column represents a field of the table.

Sex	Year	Tutor Group
M	7	Miss Hu
M	7	Mr Zade
F	8	Dr Hick
F	7	Mrs Wong
M	7	Miss Kuyt
F	8	Mr Singh

This row contains the set of the fields. Each row is a record.

Field types

There are many different data types, thus when creating a database structure you must specify the field type based on the type of data being entered. For example, a person's name would have the field type text.

In commercial databases you can have a placeholder for media that can be stored in a database. This media could be images (e.g. a photograph of a student), sound bites (i.e. small sections of sound) and video clips.

Primary key

A database is made up of tables of information that are broken down into records (rows of data) and each record has columns of information about it (fields). To distinguish a particular record, a unique data value has to be associated with it and is usually just one separate field (column of data) but can sometimes be a combination of fields.

A primary key is a field in a database that is unique to a particular record (i.e. a row in a table).

For example, in a table of children in a school, a record would be the details about a particular student. A primary key would be the student number, which would be a number set up so that each student is allocated a different number when they join the school. No two students would have the same student number.

P31 =

	A	B	C	D	E	F	G	H	I	J	K	L	M	N	O	P	Q
1	QNo	Title	Initial	Surname	Street	Postcode	No_in_house	Type	Garden	Paper	Bottles	Cans	Shoes	Carriers	Compost	Junk_mail	
2	1	Mr	A	Ahmed	18 Rycroft Road	L12 5DR	1	S	S	Y	Y	Y	Y	Y	Y	10	
3	2	Miss	R	Lee	1 Woodend Drive	L35 8RW	4	D	M	Y	Y	Y	N	N	Y	4	
4	3	Mr	W	Johnson	42 Lawson Drive	L12 3SA	2	S	S	Y	Y	Y	N	N	Y	0	
5	4	Mrs	D	Gower	12 Coronation Street	L13 8JH	3	T	Y	Y	N	N	N	N	N	9	
6	5	Dr	E	Fodder	124 Inkerman Street	L13 5RT	5	T	Y	N	N	N	N	N	N	12	
7	6	Miss	R	Fowler	109 Pagemoss Lane	L13 4ED	3	S	S	N	N	N	N	N	N	5	
8	7	Ms	V	Green	34 Austin Close	L24 8UH	2	D	S	N	N	N	N	N	N	7	
9	8	Mr	K	Power	66 Clough Road	L35 6GH	1	T	Y	Y	Y	Y	N	N	N	7	
10	9	Mrs	M	Roth	43 Fort Avenue	L12 7YH	3	S	M	N	N	Y	N	N	N	7	
11	10	Mrs	O	Crowther	111 Elmshouse Road	L24 7FT	3	S	M	Y	Y	Y	N	N	N	8	
12	11	Mrs	O	Low	93 Aspes Road	L12 6FG	1	T	Y	Y	Y	Y	Y	N	N	11	
13	12	Mrs	P	Crowley	98 Forgate Street	L12 6TY	5	T	Y	Y	Y	Y	N	N	N	15	
14	13	Mr	J	Preston	123 Edgehill Road	L12 6TH	6	T	Y	Y	Y	N	N	N	N	2	
15	14	Mr	J	Quirk	12 Leopold Drive	L24 6ER	4	S	M	Y	Y	N	N	N	Y	2	
16	15	Mr	H	Etheridge	13 Cambridge Avenue	L12 5RE	2	S	L	Y	N	Y	N	N	Y	5	
17	16	Miss	E	James	35 Speke Hall Road	L24 5VF	2	S	L	Y	N	Y	N	N	Y	5	
18	17	Mrs	W	Jones	49 Abbeyfield Drive	L13 7FR	1	D	M	N	N	N	N	N	Y	5	
19																	

A flat file uses a single table of data set up like this.

Flat files and relational databases

Computerised databases may be divided into two types: the limited flat-file database suitable for only a few applications, and the much more comprehensive and flexible relational database.

Flat-file databases

Flat-file databases are of limited use and are suitable only for very simple databases. Flat files only contain one table of data. A record is simply the complete information about a product, employee, student, etc. This is one row in the file/table. An item of information such as surname, date of birth, product number, product name, in a record is called a field. The fields are the vertical columns in a table. Because flat-file databases contain only one table, this limits their use to simple data storage and retrieval systems such as storing a list of names, addresses, phone numbers, etc. Tables consist of columns and rows organised in the following way:

- The first row contains the field names.
- The rows apart from the first row represent the records in the database.
- The columns contain the database fields.

The problems with flat-file systems

Flat files store all the data in a single table. The disadvantages of using a flat file are:

- Data redundancy. There is often a lot of duplicate data in the table. Time is wasted retyping the same data, and more data is stored than needs to be, making the whole database larger.
- When a record is deleted, a lot of data that is still useful may also be deleted.

Relational databases

In a relational database, we do not store all the data in a single file or table. Instead the data is stored in several tables with links between the tables to enable the data in the separate tables to be combined together if needed.

To understand this, look at the following example:

A tool hire business hires tools such as ladders, cement mixers, scaffolding, chain saws, etc., to tradesmen. The following would need to be stored:

- Data about the tools
- Data about the customers
- Data about the rentals.

Three tables are needed to store this data and these can be called:

- Tools
- Customers
- Rentals.

If the above were stored in a single table, (in other words using a flat file), there would be a problem. As all the details of tools, customers and rentals are stored together there would be no record of a tool unless it had been hired by a customer. There would be no record of a customer unless they had hired a tool at the time.

In the flat file there would be data redundancy because customer address details are stored many times for each time they hire a tool. This means the same data appears more than once in the one table. Hence there are serious limitations in using flat files and this is why data is best stored in a relational database where the data is held in several tables with links between the tables.

KEY WORDS

Data redundancy where the same data is stored more than once in a table or where the same data is stored in more than one table.

Primary key this is a field that is unique for a particular record in a database.

Creating databases

Before you create a structure for a database it is important to look at a sample of the data that needs to be stored.

A school keeps details of all its pupils in a database. As well as personal details (name, address, etc.), the school also holds details of the tutor group and tutors.

The person who is developing the database asks the principal for a sample of the data. This sample of the data is shown below:

Description of data stored	Sample data
Pupil number	76434
Surname	Harris
Forename	Amy
Date of birth	15/03/98
Street	323 Leeward Road
Town	Waterloo
Postcode	L22 3PP
Contact phone number	0151-002-8899
Home phone number	0151-002-1410
Tutor Group	7G
Tutor number	112
Tutor teacher title	Mr
Tutor surname	Harrison
Tutor initial	K

Three tables are used with the names: Pupils, Tutor Groups and Tutors.

KEY WORDS

Relational database a database where the data is held in two or more tables with relationships (links) established between them. The software is used to set up and hold the data as well as to extract and manipulate the stored data.

Relationship the way tables are related or linked to each other.

The fields in each table are as follows:

Pupils

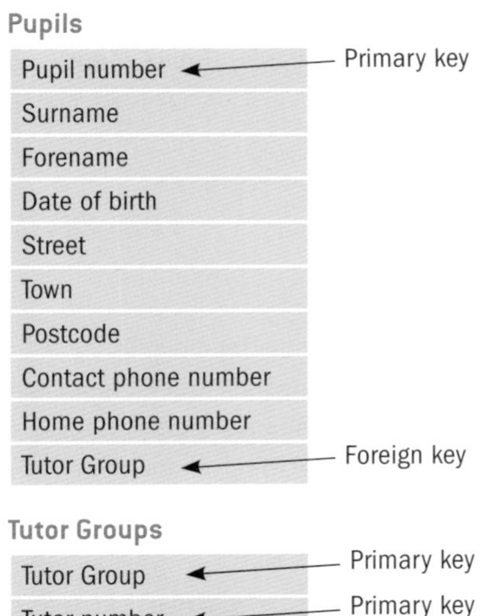

Tutor Groups

- Tutor Group ← Primary key
- Tutor number ← Primary key

Tutors

- Tutor number ← Primary key
- Tutor title
- Tutor surname
- Tutor initial

The data is put into three tables rather than one because it saves time having to type the same details over and over about the tutor for each pupil. In other words it reduces data redundancy. If there are 25 pupils in each form, the tutor's details (Tutor number, Tutor title, Tutor surname, etc.) would need to be entered 25 times. If instead we put these details in their own table, we can access them from the Tutor Group field and we then need type in the Tutor Group details only once.

QUESTIONS B

1 A luxury car rental firm keeps the details of the cars it rents out in a table. The structure and contents of this table are shown below.

Reg-number	Make	Model	Year
DB51 AML	Aston Martin	DB7	2009
CAB 360M	Ferrari	360 Modena	2008
GT X34 FER	Ferrari	355 Spider	2000
MAS 12	Maserati	3200 GTA	2001
FG09 FRT	Porsche	911 Turbo	2009
M3 MMM	BMW	M3 Conv	2010
T433 YTH	Jaguar	XK8	2009

a Give the field names of **two** fields shown in the table. *(2 marks)*

b Give the field name of the field that should be chosen as the primary key. *(1 mark)*

c Explain why the field you have chosen for your answer to part **b** should be chosen as the primary key. *(1 mark)*

d How many records are there in the table? *(1 mark)*

2 Most schools use databases to store details about each pupil. The table shows some of the field names and data types stored in one pupil database.

Field name	Data Type
UniquePupilNumber	Number
Firstname	
Surname	Text/Alphanumeric
FirstLineAddress	Text/Alphanumeric
SecondLineAddress	Text/Alphanumeric
Postcode	
LandlineNo	Text/Alphanumeric
DateOfBirth	Date
FreeSchoolMeals(Y/N)	

a Give the most appropriate data types for the following fields: *(3 marks)*

i Firstname

ii Postcode

iii FreeSchoolMeals (Y/N)

b Give the names of **three** other fields that are likely to be used in this database. *(3 marks)*

c Explain which field is used as the primary key in the database and why such a field is necessary. *(2 marks)*

d It is important that the data contained in this database is accurate.
Describe how **two** different errors could occur when data is entered into this database. *(2 marks)*

e Explain how the errors you have mentioned in part **d** could be detected or prevented. *(2 marks)*

3 **a** Explain what is meant by a flat-file database. *(1 mark)*

b Explain what is meant by a relational database. *(1 mark)*

c Describe an application where a flat-file database would be suitable. *(2 marks)*

4 Databases are of two types: flat file and relational.

a Describe **two** differences between a flat-file database and a relational database. *(2 marks)*

b A dress hire company needs to store details of dresses, customers, and rentals. They want to store these details in a database.
Which type of database do you suggest and give **two** reasons for your answer. *(3 marks)*

Linking files or tables (i.e. forming relationships)

To link two tables together there needs to be the same field in each table. For example, to link the Pupils table to the Tutor Groups table we can use the Tutor Group field as it is in both tables. Similarly, the Tutor Groups table and the Tutors table can be linked through the Tutor number field. Links between tables are often called relationships, and they are one of the main features of relational databases.

Pupils

Pupils
Pupil number
Surname
Forename
Date of birth
Street
Town
Postcode
Contact phone number
Home phone number
Tutor Group

Tutor Groups

Tutor Groups
Tutor Group
Tutor number

Tutors

Tutors
Tutor number
Tutor title
Tutor surname
Tutor initial

Foreign keys

A foreign key is a field of one table which is also the key field of another. Foreign keys are used to establish relationships between tables. In the above example the field Tutor Group would be the key field in the Tutor Groups table and a foreign key in the Pupils table.

The advantages of relational databases

Using a relational database means that you don't have to type in all the data for each pupil when you create the Tutor Groups table.

This has the following advantages:

- It saves time typing
- It reduces typing errors
- It therefore reduces redundancy.

Choosing field names

Field names should be meaningful to identify the field. Try to avoid using spaces in fields; instead use dashes or a combination of upper- and lower-case letters. Try not to make the fieldnames too long since these will be the column headings in the tables and if they are too long it will make the columns wide and not as much data can be displayed on the screen at the same time.

Using the database to extract information

The Activities for this chapter on the website require the use of database software. The suggested database software is MS Access 2007. MS Access is a brand name so you have to make sure that you do not write MS Access in an examination question when the answer is 'database software'. MS Access is popular database software but there are other examples of databases that can be chosen.

On creating the tables, linking them and adding data, the data in any of the tables can be combined together.

A query is used to extract specific information from a database, so queries are used to ask questions of databases. For example, a query could be used to extract the names of pupils in a school who are aged 16 years or over. Queries are usually displayed on the screen but they can be printed out if needed. If a printout is needed it is better to produce a report. A report is a printout of the results from a database. Reports can be printed out in a way that is controlled by the user.

Activities

Activites for Chapter 18 can be found on the website that accompanies this book. There are thirteen activities covering:

Locating, opening, and importing data from an existing file

Importing data from an existing file and creating a relationship between two tables

Entering data with 100% accuracy, showing subsets of data and sorting data

Performing searches using queries

Performing wildcard searches

Putting formulae in queries

Performing searches on the Employees database

Producing reports

Producing calculations at run time

Creating the report based on the query

Showing and hiding data and labels within a report

Database labels

Creating a data entry form

Presentations

You will probably have created many presentations in different subject areas but you may not have developed skills in all the areas needed for the examination. For the examination you need to have the skills to be able to create and control an interactive presentation.

In this topic you will be learning how to create presentations using the software MS PowerPoint 2007. There is other presentation software available having similar features that could be used. MS PowerPoint is a brand name so you would get no marks for giving brand names in the examination. So rather than say 'use MS PowerPoint' you would say use 'presentation software'.

The key concepts covered in this chapter are:

- Using a master slide to approximately place objects and set suitable styles to meet the needs of the audience
- Using suitable software tools to create presentation slides to meet the needs of the audience
- Using suitable software tools to display the presentation in a variety of formats

Using a master slide to approximately place objects and set styles

The need for consistency in a presentation

All the slides in a presentation should look as if they belong together. This means they should all have a similar design. It is acceptable if the first slide is different from the others because it usually shows just the title of the presentation and sometimes the presenter's name.

Consistency needs to be considered before you start your presentation. Here are some of the things you will need to consider:

- Styles should be the same for each slide in the presentation – this ensures that all the slides look as though they belong together.
- Fonts – you need to choose a set of fonts that work well together. Fonts should be consistent from one slide to the next.
- Point sizes – the point size is the size of the font. You will need to choose the point size for headings, sub-headings, and body text (i.e. the main text).
- Colour schemes – all the slides need to use the same combination of colours for components such as banners, borders, etc.
- Transitions – these are the movements from one slide to the next. There are lots of eye catching transitions but you need to make sure you do not use a different one for each slide transition.
- Animations – these are movements on the actual slides. For example, you can animate the way bullet points appear on each slide. Make sure that animations are used consistently.

In the examination you will usually be given instructions about these features, so you need to understand how to create/change them using the presentation software.

What is a master slide and why are they important?

Master slides are used to help ensure consistency from slide to slide in a presentation. Master slides (also called slide masters) are used to place objects and set styles on each slide. Using master slides you can format titles, backgrounds, colour schemes, dates, slide numbers, etc.

In the examination you will be given details of designs, colour schemes, fonts, etc., that you must add to a master slide.

The footer, date and slide number areas on the master slide

There is an area at the bottom of the master slide which is used to add certain information at the bottom of every slide in the presentation, such as a footer, page date, or slide number. The footer can be used for text such as a copyright message (e.g. ©Stephen Doyle 2015).

If you want to add text in the footer area placeholder (i.e. the rectangle with footer in it) you click in the box and type in the text.

The date and slide number (i.e. #) are already on the master slide but they will not be shown on the actual slides unless you make them active. To make these details appear on each page you must follow the following steps:

1 Click on Insert and then select Header & Footer and the following dialogue box appears where you can enter dates, slide numbers, and a footer.

Header and Footer

Slide | Notes and Handouts

Include on slide
- Date and time
  - Update automatically: 30/11/2010
    - Language: English (United Kingdom)
    - Calendar type: Western
  - Fixed: 30/11/2010
- Slide number
- Footer
- Don't show on title slide

Apply to All | Apply | Cancel

Preview

If you want a date and time put on each slide make sure that you put a tick in the box. You are then presented with a list of formats for the date to choose from:

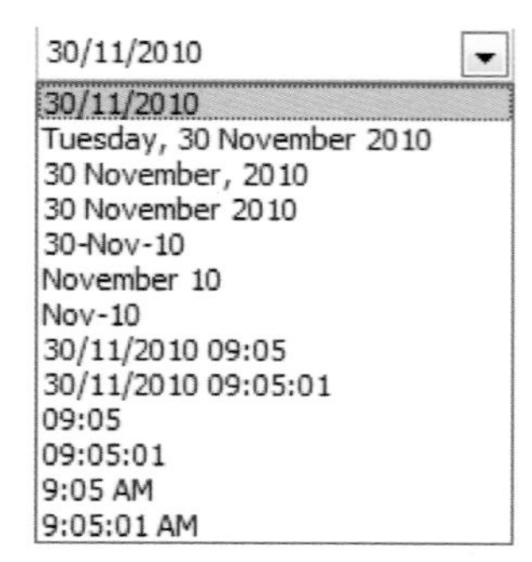

If you want the slide to always show the current date, then you should select Update automatically like this:

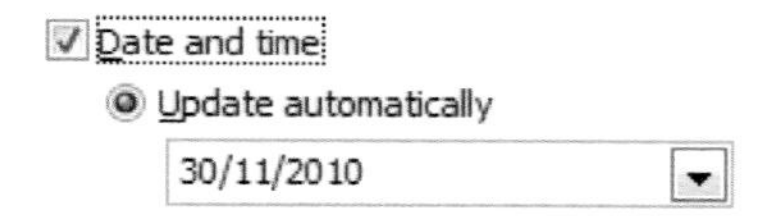

Sometimes you will want to show the same date on each slide, such as the date you produced the presentation.

Selecting ◉ Fixed will mean that a certain date will always appear on each slide.

2 To put a slide number on each slide, ensure that Slide number has been selected like this:

☑ Slide number

3 You can enter the footer text into the placeholder area marked 'Footer' on the master slide or you can enter the text for the footer into the Header and Footer dialogue box like this:

Note the (c) next to the name will automatically turn into a copyright symbol on the slides. Remember you can put any text into the footer and will be asked to do so for the examination.

Important information about headers and footers

You will see the footer placeholder areas at the bottom of the slide. If you do not require any of these, you do not have to delete them. They will only appear if you set them to appear using the method shown above.

If a footer placeholder is not in the position asked for in the examination then you can move it. You can also enlarge it so that it occupies the entire width of the slide.

Activity 19.1

Creating a master slide

In this activity you will learn how to:

- create a master slide which can be used to ensure the design consistency in all the other slides in the presentation
- change the background colour of a slide
- add a text box to a slide
- add an image to a slide
- add text to the header and footer.

1 Load Microsoft PowerPoint or your presentation software.

2 Click on the View tab and select 

.

3 The following screen appears:

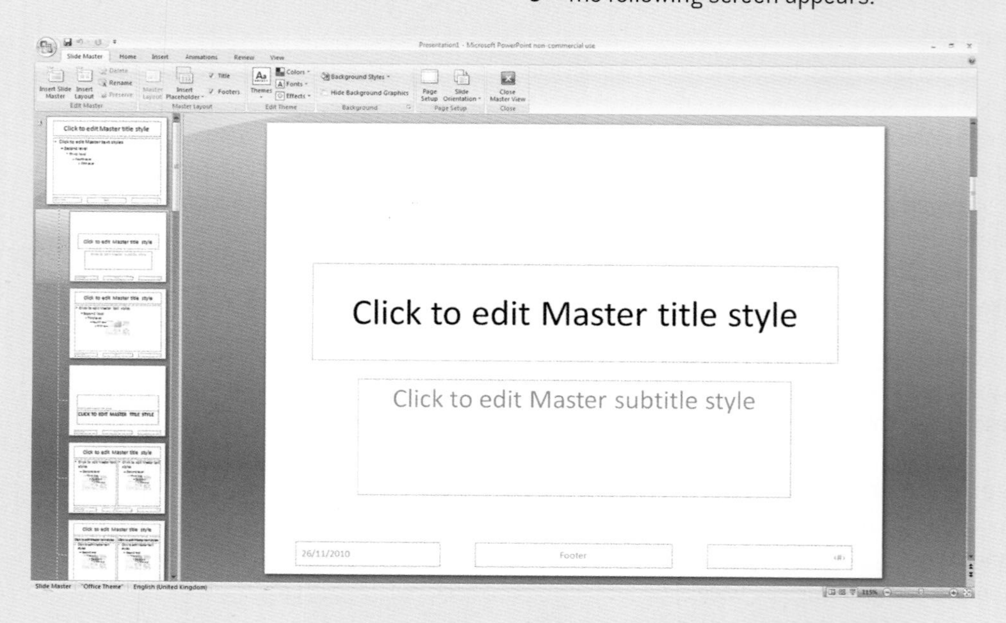

4 You now have to consider the medium for your presentation. In this case it is a presentation on a screen in landscape orientation with the audience notes (notes accompanying the presentation given to the audience) and the presenter notes (notes giving hints when giving the presentation).

Click on  Page Setup .

This allows you to set the size of the screen you are displaying the presentation on and the orientation of slides and notes.

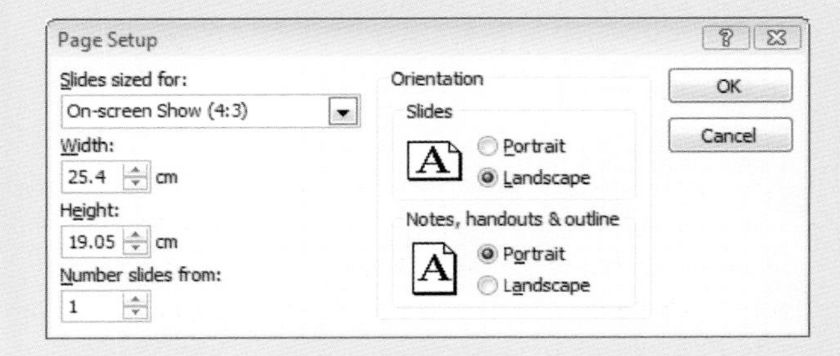

The default settings (settings that will be used unless you change them) are used for this presentation. Confirm by clicking on OK.

5 Click on the first slide at the top of the list of slides. This is the master slide and you can use this one slide to lay out the design of your slides as objects such as text and images (clip art, logos, lines, shapes, etc.) will be applied to all the slides in the presentation. This means that if you inserted a logo on this slide, it will appear at the same position on all the slides in the presentation.

The other slides you see under this one contain different style slides and these can be chosen according to the content you want to put on the slide.

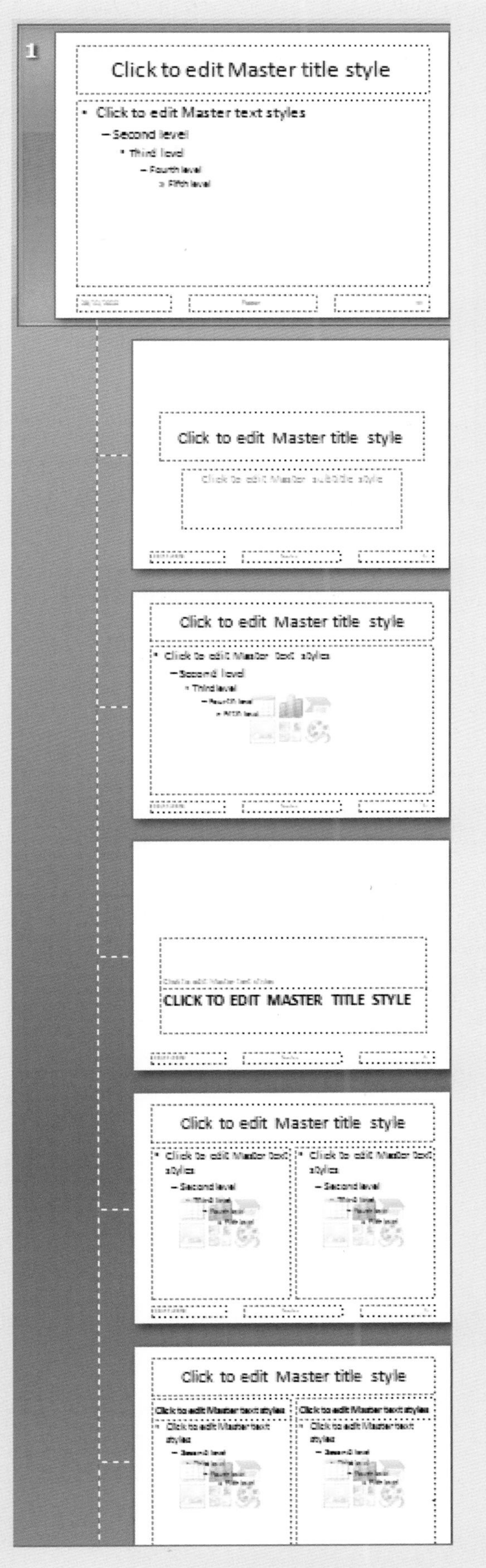

6 You are going to set up a master slide:

Ensure you have selected the master slide shown here:

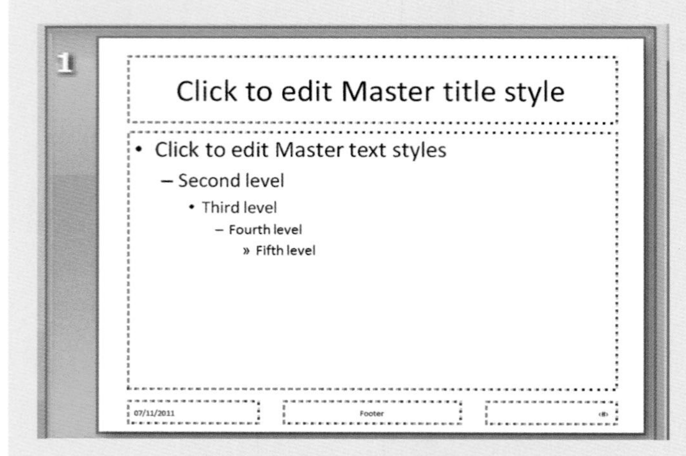

The following master slide is shown.

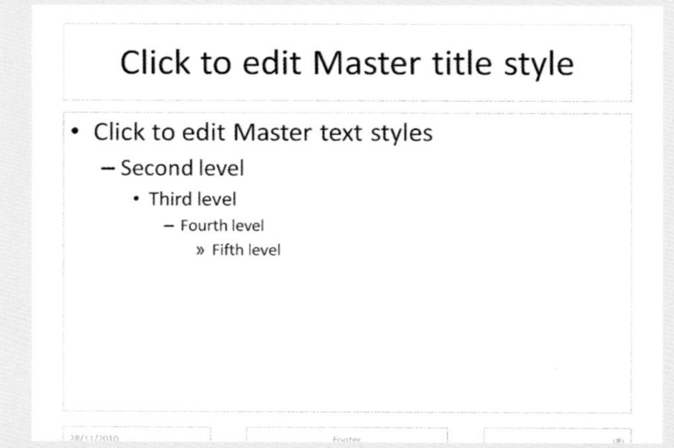

You are going to change the background colour of all the slides from white to a pale yellow.

Right-click anywhere on the white part of the slide and the following menu is displayed:

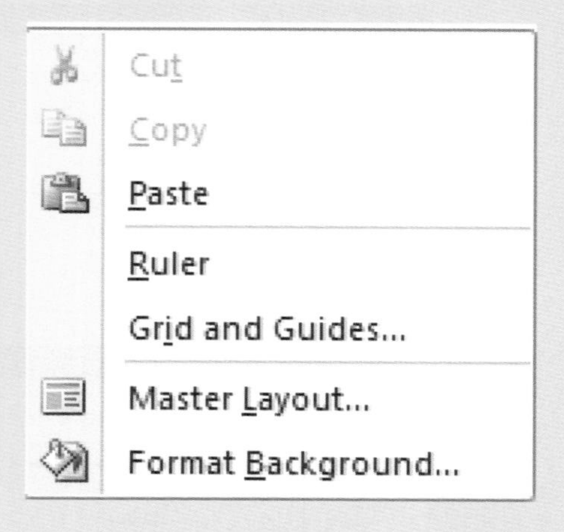

Left-click on Format Background and the following window is displayed:

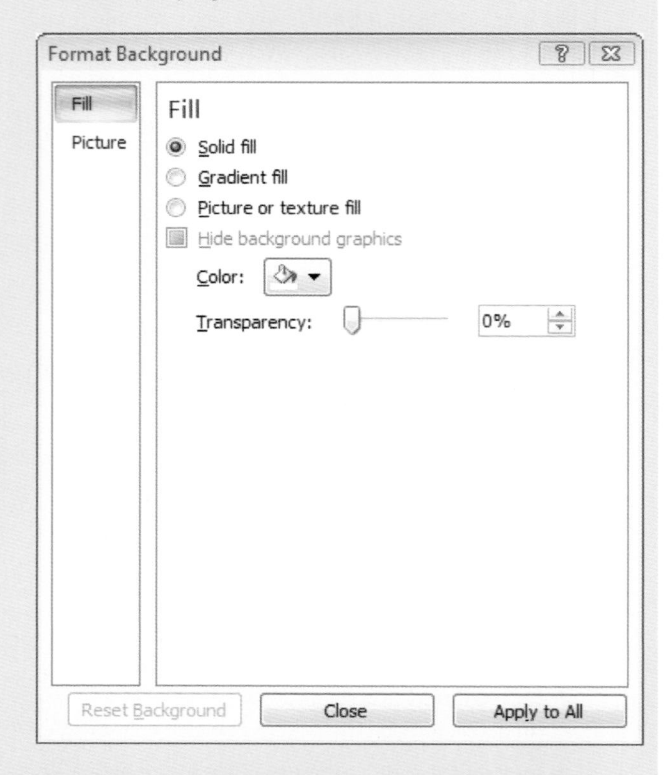

Check that Solid fill has been selected and then click on the drop-down arrow for the colour and the following palette appears:

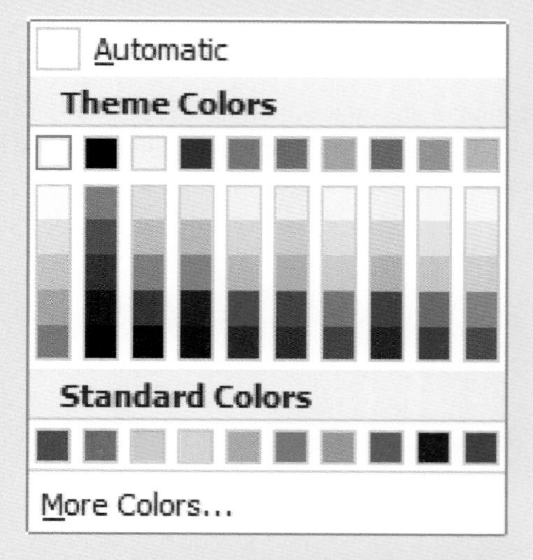

Click on the bright yellow colour in the Standard Colours section. Notice that all the slides now have this colour as the background. Now adjust the transparency of the colour by setting it to 80%.

Click on Close.

⇨

7 You are now going to add a text box to the master slide containing the following details:

- Your full name
- Your centre number (use 12122 if you don't have this information)
- Your candidate number (use 123123 if you don't have this information).

These details need to be in a sans serif font, have a point size of 12 pt and be positioned in the top left-hand corner of the slide.

There is already an object (i.e., a box containing the text 'Click to edit Master title style') and this needs to be resized to make room for the text box we want to insert.

Click on the corner like this and use the handle and drag the corner down and to the right:

Click to edit Master title style

The box should be moved into a position like this:

Click to edit Master title style

Click on the insert tab Insert and then select Text Box Text Box .

Now draw the text box in a similar position to that shown here:

Click to edit Master title style

Change the font to Arial (which is a sans serif font) and the font size to 12 Pt by altering the settings like this:

Arial 12

Type the following information into the text box:

Your name 12122 123123

Your text box containing the information will now look like this:

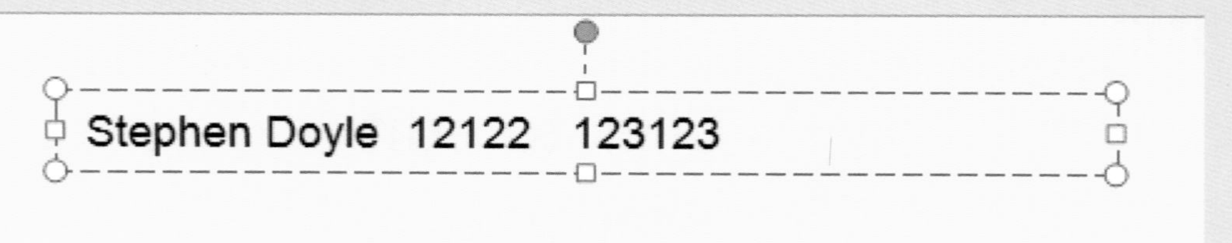

Click on a blank area of the slide to deselect the text box.

As the text in the text box is on the master slide, it will appear in the same position on all the slides.

8 You are now going to insert a piece of clip art on the master slide positioned in the bottom right-hand corner of the slide.

The piece of clip art needs to be a picture of a car (any car will do).

Click on Insert and then on  Clip Art and the following clip art pane appears:

Click on 'Car' in the search box and then click on Go.

Find a suitable picture of a car and left-click on it and then, keeping the left mouse button pressed down, drag onto the position on the slide and, when in the correct position, release the mouse button. The piece of clip art will then appear in position like this:

Stephen Doyle 12122 123123

Click to edit Master title style

- Click to edit Master text styles
  - Second level
    - Third level
      - Fourth level
        - » Fifth level

28/11/2010 Footer ‹#›

Again this image will appear in an identical position on all the slides in the presentation.

If any of these boxes are needed but not in the position shown, you can select the box and drag it into a new position.

9 You are now going to re-position the box containing the slide number so that it is on the left. You will need to move the date out of the way while you do this by left-clicking on it and drag it to another position. Do not worry about putting it into the final position as you can do this later.

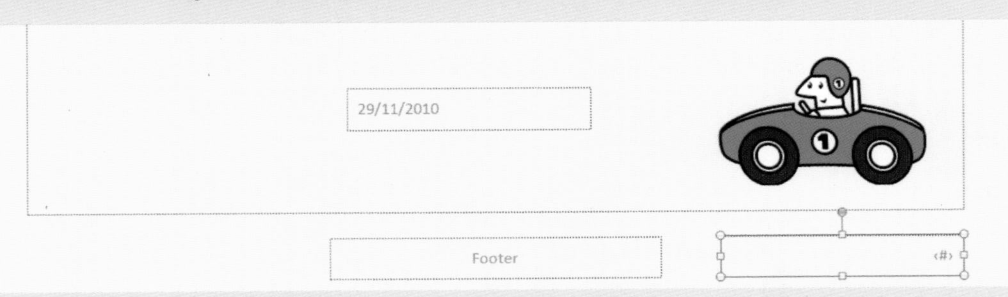

Left-click on the box with the slide number in it (i.e., shown as #) and keeping the mouse button pressed down, drag the box over to the left-hand side of the screen. Now in a similar way move the box containing the date into the position shown below:

‹#› Footer 29/11/2010

The slide number is right aligned in the box. ⇨

You are required to change this to left alignment and to do this click on Home and then on .

The slide number will move to its new position like this:

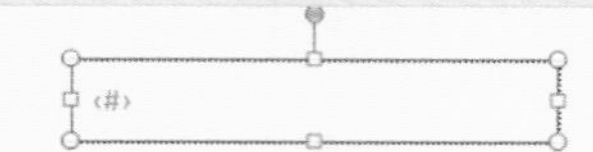

The slide number will now appear on each slide positioned in the bottom left-hand corner.

10 Details such as slide number, the footer, and the date need to be turned on before they will be shown on the slides.

Click on Insert and then select Header & Footer and the following dialogue box appears:

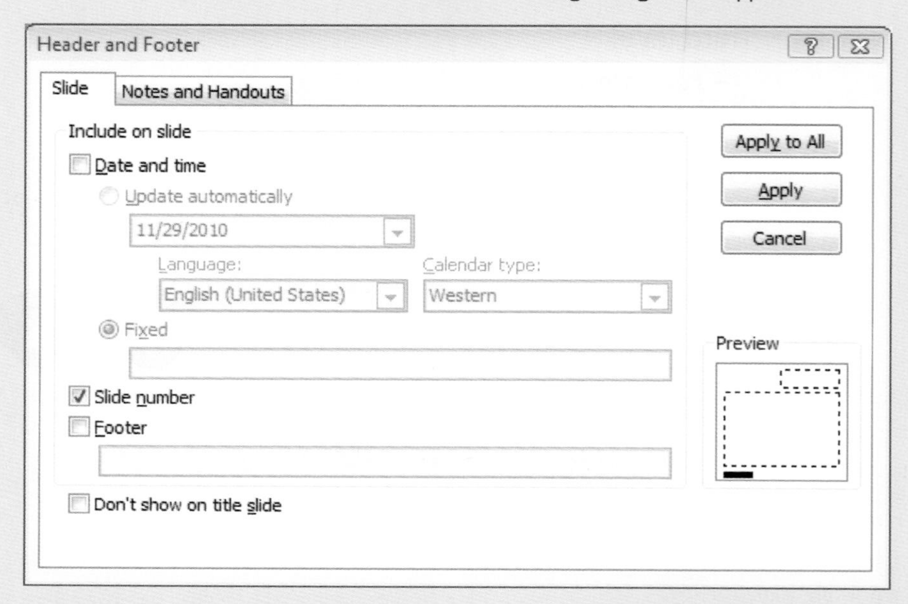

Click on Apply to All.

Ensure that there is a tick on the Slide number box. Notice also that you could turn the date on and put some text in the footer if asked to. Notice the preview shows that only the slide number will be shown.

11 You now need to check that the objects you have placed on the master slide do not interfere with any of the boxes into which you type the content when adding the material to the slides.

Click on the second slide in the list on the left.

⇨

You can see that the clip art overlaps the box to hold the subtitle:

Stephen Doyle 12122 123123

Click to edit Master title style

Click to edit Master subtitle style

‹#›

Footer

29/11/2010

To solve this problem we can re-position the two boxes by moving them up slightly. To move them you need to left-click on the border of the box and then keeping the left mouse button pressed down, drag them into a position similar to that shown here:

Stephen Doyle 12122 123123

Click to edit Master title style

Click to edit Master subtitle style

⇨

12 You now have to set up the following styles for the presentation:

- Heading: dark blue, 48 point, right aligned serif font.
- Subheading: black, centre aligned, 24 point serif font.
- Bulleted list: black, left aligned 18 point sans serif font and you are able to choose the style of the bullet used.

Click on the text 'Click to edit Master title style' as this is the heading for the slide.

Choose any serif font (e.g., Times New Roman) and change the font size to 48 like this:

Times New Roma ▾ 48 ▾

Click on A ▾ and select dark blue (it does not matter which dark blue as there are several) from the colour chart:

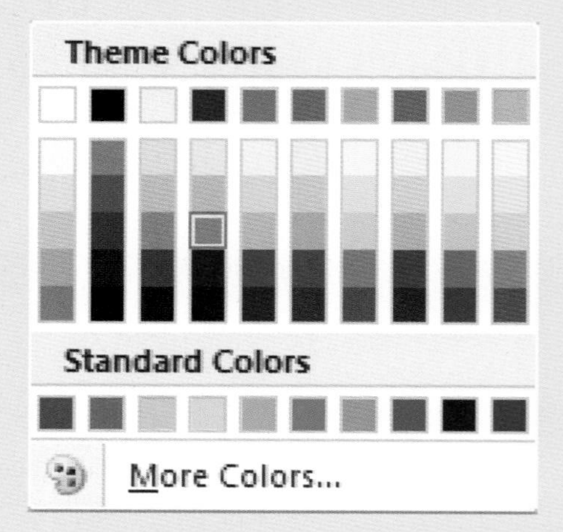

Click on ≡ to right align the heading.

Click on the text 'Click to edit Master subtitle style' as this is the subheading for the slide.

Choose any serif font (e.g., Times New Roman) and change the font size to 24 like this:

Times New Roma ▾ 24 ▾

Click on A ▾ and from the colour chart select the colour black.

Notice that there are no bullet points on this first slide.

Click on one of the slides whose layout contains bullet points and then click on the first level of bullet points (i.e., the first bullet point on the master slide):

• Click to edit Master text styles

Click on the text next to the bullet point and choose a sans serif font (e.g., Courier, Calibri, etc.) and set the font size to 18. Set the colour of the font to black and set the text to be left aligned.

Change the shape of the bullet by clicking on the drop-down arrow ☰ ▾ and choosing a bullet shape of your choice by clicking on it like this:

⇨

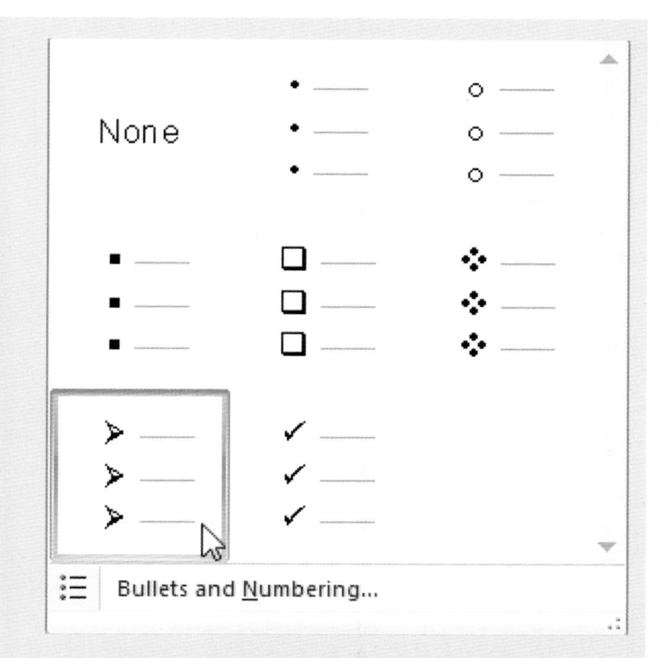

You have now made your first master slide.

Save this master slide using the filename **Master**.

Activity 19.2

Creating a master slide that includes shaded areas and lines

In this activity you will learn how to:

- use a master slide to place objects and set styles
- insert a shape onto a slide and change its colour
- draw lines on the master slide
- insert a clip art image on the master slide
- insert information in a footer
- include text on the slides including headings, subheadings, and bulleted lists
- format text (i.e., font type (serif or sans serif), point size, text colour, text alignment, and enhancements (bold, italic, and underscore).

1 Load the presentation software Microsoft PowerPoint and create a new blank presentation.

2 Click on View and then on  Slide Master to bring up the slide master.

3 Click on the slightly larger first top slide in the list down the left-hand side of the screen. You will now see the following on your screen:

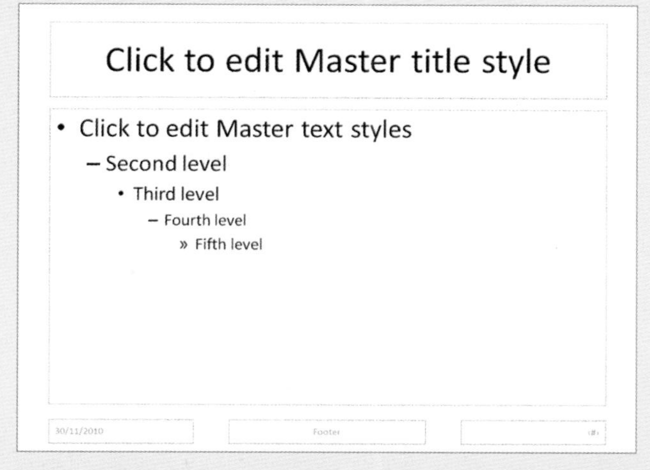

4 You are now going to place a pale grey rectangle that occupies about one quarter of the width of the slide. There are some placeholder boxes in the way of where the rectangle needs to go. You can simply click on them and drag and resize them so that they are out of the way like this:

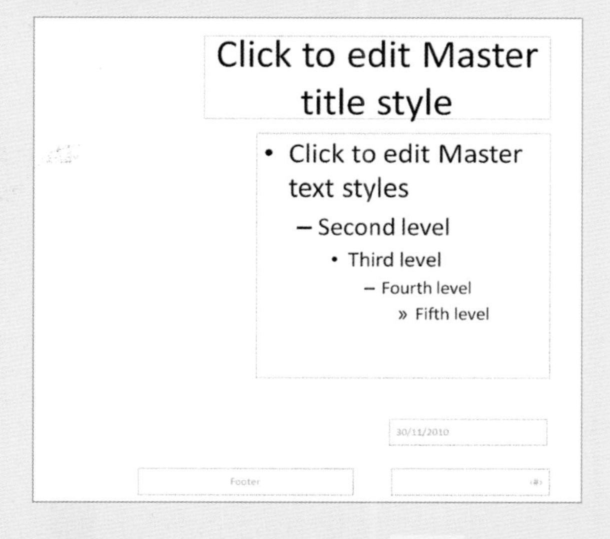

Click on Insert and then select Shapes and then choose the rectangle shape from the list of shapes by left-clicking on it. You will now see the cursor change to a cross. Position this cursor on the top left-hand corner of the slide and, keeping the left mouse button pressed down, drag the cursor until you have drawn a rectangle like this:

Click to edit Master title style

- Click to edit Master text styles
  - Second level
    - Third level
      - Fourth level
        - Fifth level

30/11/2010

Footer

Notice the handles on the rectangle to show that it is still selected. You now need to change the colour of this rectangle to light grey.

Click on Home tab.

Click on Shape Fill and the following colour palette is displayed:

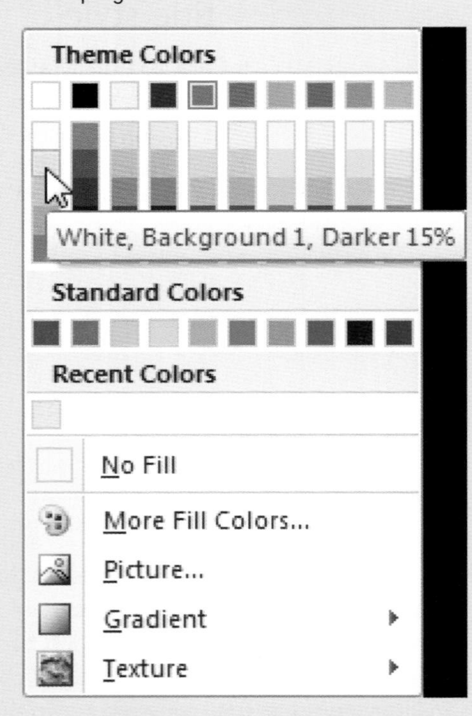

Left-click on the grey colour as shown above and the rectangular area on all the slides will be filled with this choice.

You are now going to put a red border around the rectangular area.

Click on Shape Outline and select the red colour from the standard colours by clicking on it.

The outline of the rectangular area is now in red.

The next step is to select the thickness of the border which is expressed in points. Click on Shape Outline again and this time select Weight from the menu. Now choose 1½ pt from the list of weights.

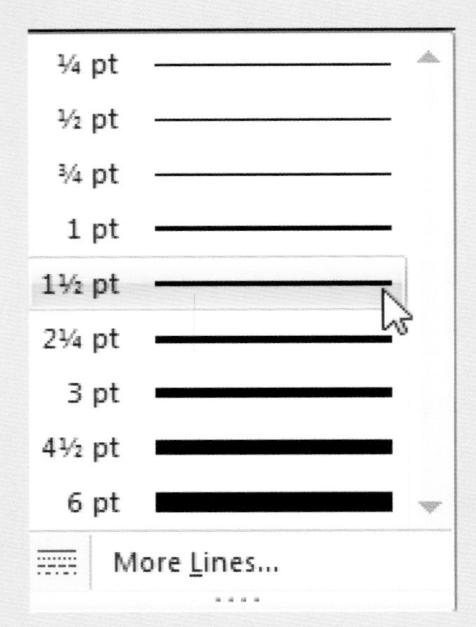

5 You are now going to insert two horizontal red lines near the top of the page.

Move the placeholder area out of the way to a position similar to that shown here:

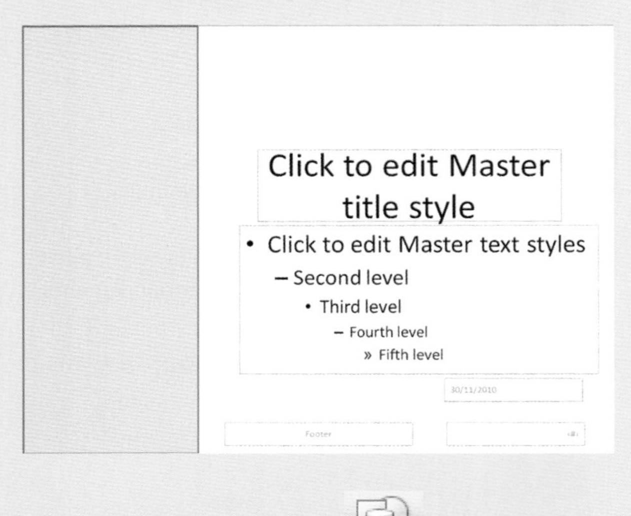

Click on Insert and then on Shapes and select the line from the list of shapes by clicking on it like this:

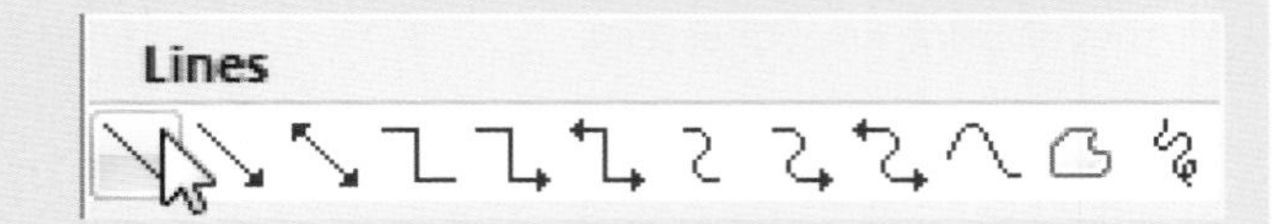

The cursor changes to the cross-wires. Position this cursor on the left at the start of the line and then drag the line until it meets the other side of the slide. If you press the shift key, it will keep the line in a horizontal position so you do not need to spend time trying to adjust the line yourself.

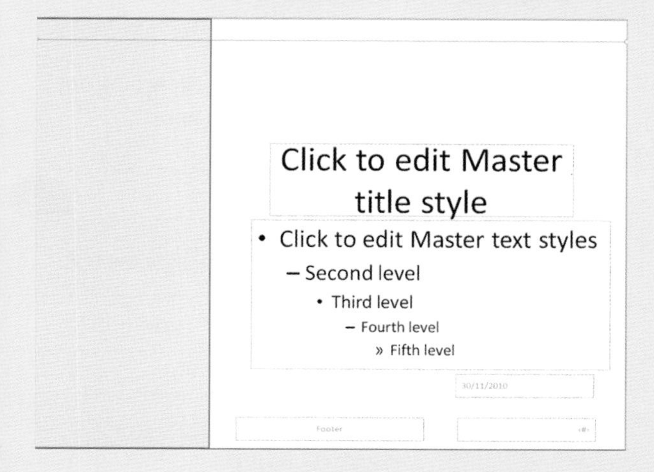

With the line still selected, click on Format and then click on Shape Outline and select red from the list of colours. Click on Shape Outline again and this time select Weight and then select a weight of 1½ pt.

The red line changes to the same width as that for the border of the rectangle. Repeat this by placing a line in red with weight 1.5 pt in the position shown here:

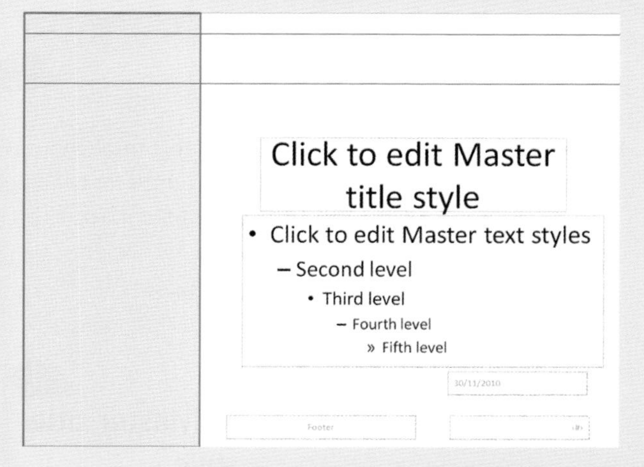

6 You are now going to insert a piece of clip art on the right of the slide and between the two red lines you have just inserted.

Click on Insert and then on  Clip Art and then enter the search word 'Computer' into the search box and click on Go. Then select an image of a computer (any will do).

Right-click on the image and drag it into the approximate position. You will need to resize the image by right-clicking on a corner handle and dragging until it is of a size that will fit between the two lines. You will probably also have to move the image into the correct position.

Your image should now be in a position similar to this:

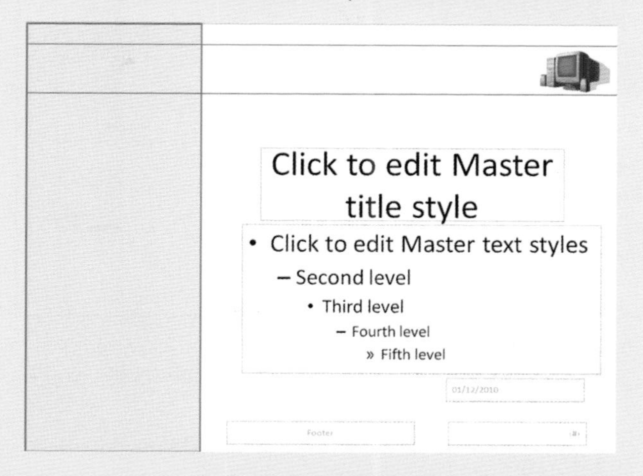

7 Move the footer to the left slightly so that it is nearer the red vertical line:

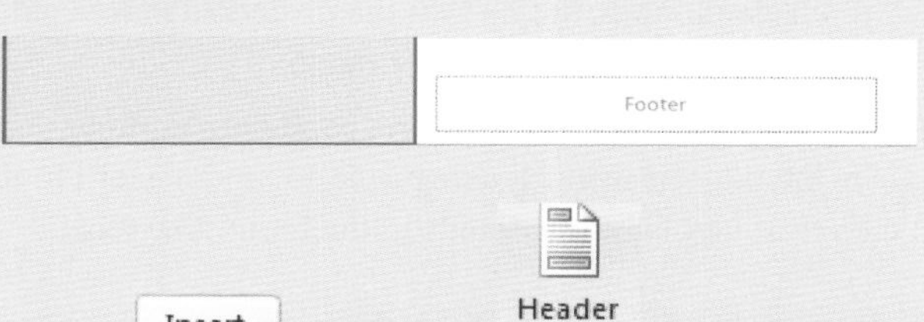

Click on Insert and then on Header & Footer when the header and footer dialogue box appears like this:

Header and Footer

Slide | Notes and Handouts

Include on slide

Date and time

Update automatically

01/12/2010

Language: English (United Kingdom) Calendar type: Western

Fixed

01/12/2010

Slide number

Footer

Don't show on title slide

Apply to All | Apply | Cancel

Preview

Click on the box for footer to indicate you want to include a footer.

Now enter the following footer details: your name, Centre number, and candidate number. If you do not have the Centre number use the number 1234 and if you do not have your candidate number, enter 56789.

Your dialogue box will now contain the footer details you entered:

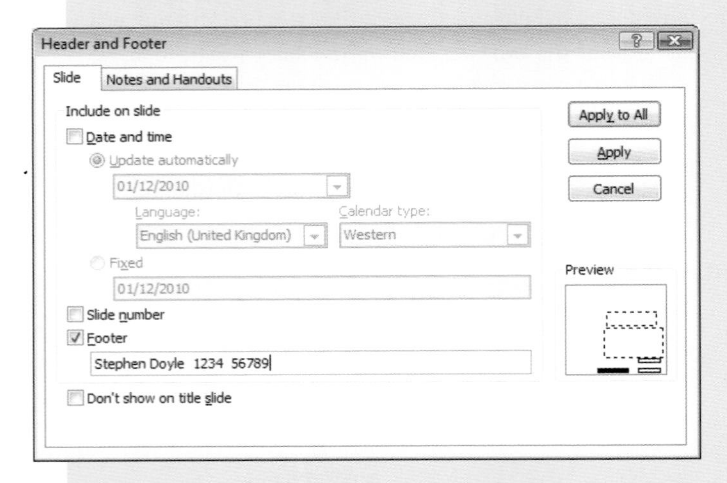

Click on Apply to All to apply the footer details to all the slides.

Note that only the footer details and not the date and slide number details will appear on the actual slides. This is because we have not selected to show the date and the slide number. Notice that the placeholder areas still appear on the master slide for these.

8 Your name, Centre number, and candidate number should be left aligned in black using a 12 point sans serif font. To do this, highlight the details in the footer like this:

Click on Home and then select any sans serif font (Arial has been chosen here) and set the size to 12 points like this:

Set the font colour to black by clicking on the font colour icon A and choosing black from the palette.

To align the text to the left, click on left align.

Check that the text in the footer appears like this:

Stephen Doyle 1234 56789

9 You are going to add the following text at the top of the master slide.

Compusolve

This text needs to be positioned at the top of the slide between the two red horizontal lines in a black bold 44 point sans serif font. The text also needs to be left aligned.

To put the text in this position you need to create a text box.

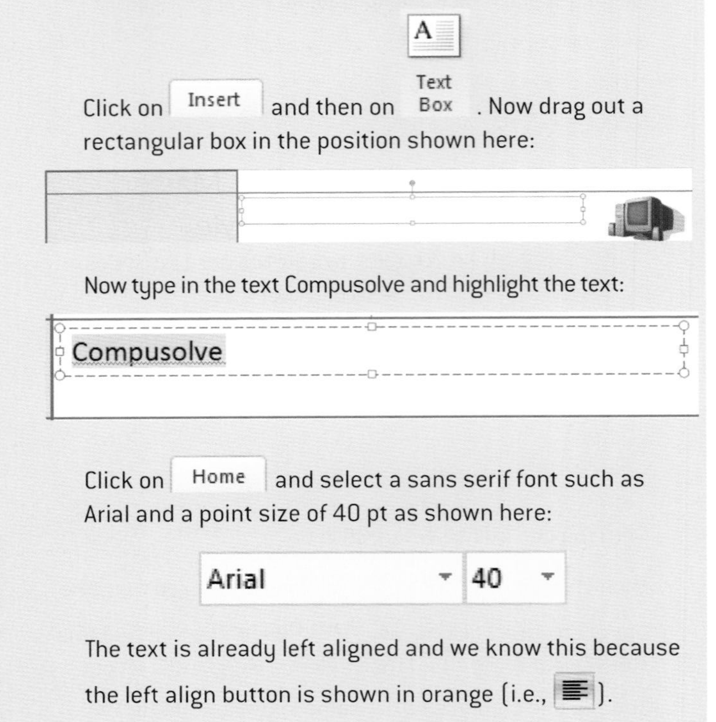

Click on Insert and then on Text Box. Now drag out a rectangular box in the position shown here:

Now type in the text Compusolve and highlight the text:

Click on Home and select a sans serif font such as Arial and a point size of 40 pt as shown here:

The text is already left aligned and we know this because the left align button is shown in orange (i.e.,).

Make the text bold by clicking on **B**.

Your text in the text box should now look like this:

10 You are now going to set the styles of text throughout the presentation. This will ensure that text used for headings, subheadings, and bulleted lists is consistent across all the slides used in the presentation.

Align the placeholder areas as shown here:

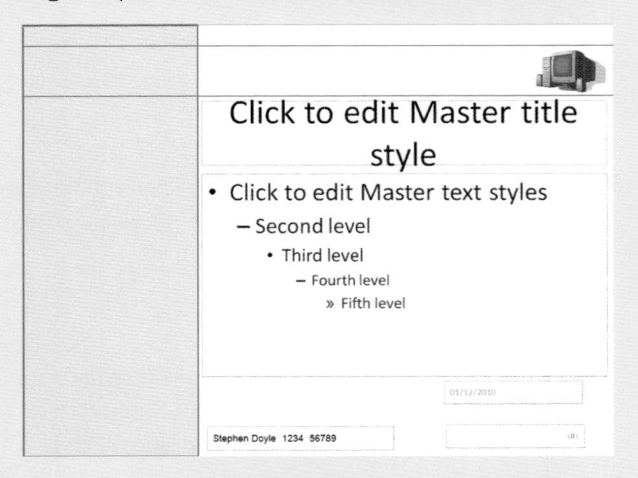

You have been asked to set the following styles of text throughout the presentation:

- Heading: sans serif font, red, left aligned, and 40 point
- Subheading: sans serif font, centre aligned, and 32 point
- Bulleted list: sans serif font, black, left aligned, and 24 point. You can choose your own bullet.

Highlight the text in the 'Master title style' placeholder area.

You are now going to format the text to that described for the heading.

Click on Home and then on change the font to a sans serif font such as Arial and change the size to 40 pt

The text is already centre aligned but you need to change the colour to red by clicking on A and choosing red from the palette of colours.

Notice that this slide does not have a subheading on it but it does contain a bulleted list, so we will change the bulleted list.

Highlight the text in the line containing the bullet and the text 'Click to edit Mast text styles' like this:

• Click to edit Master text styles

Click on Home and then on change the font to a sans serif font such as Arial and change the size to 24 pt

Arial 24

The text is already left aligned because the left align button is shown in orange (i.e.,).

To change the bullet to another shape click on the drop-down arrow on the bullet icon and the following choice of bullets appears:

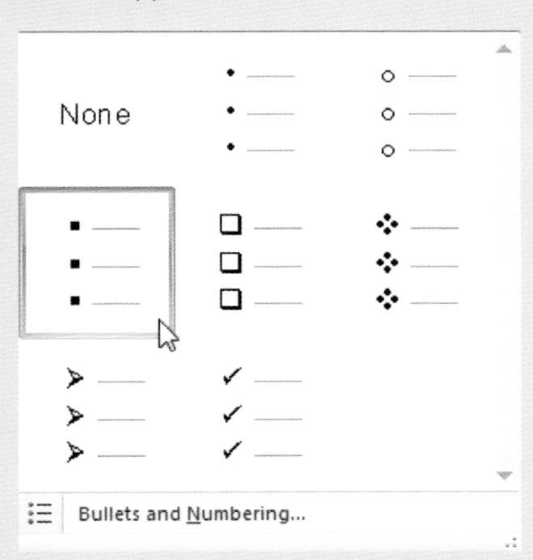

Click on the Filled Square Bullets to select that shape of bullet.

The bullet now changes to this:

▪ Click to edit Master text styles

11 You now have to change the subheading. To do this you need to have one of the slides that contain a subheading placeholder area displayed.

Choose the second slide in the list by clicking on it.

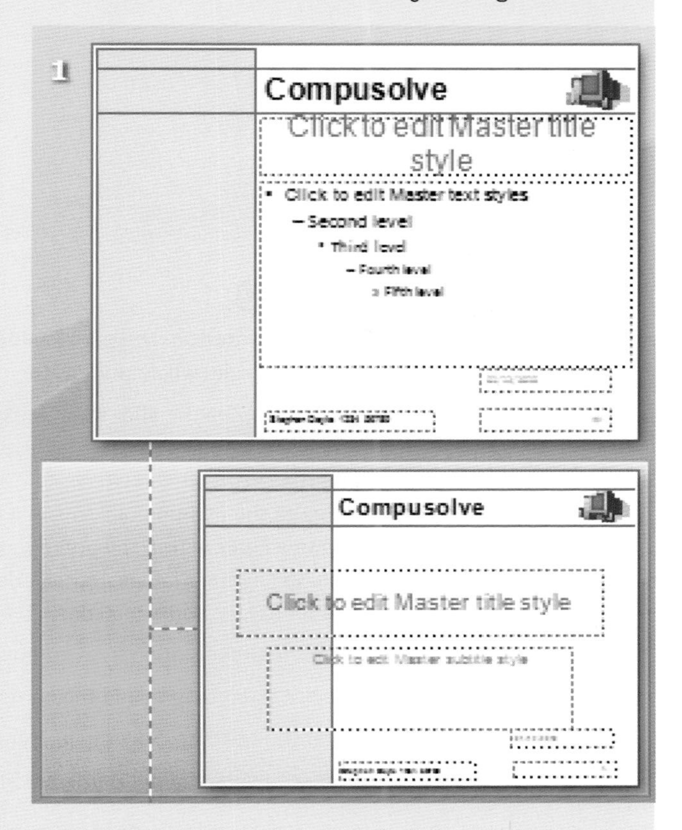

This slide appears like this:

⇨

You can see that some of the placeholder areas overlap the grey rectangle, so you need to move and resize them so that they look similar to that shown below:

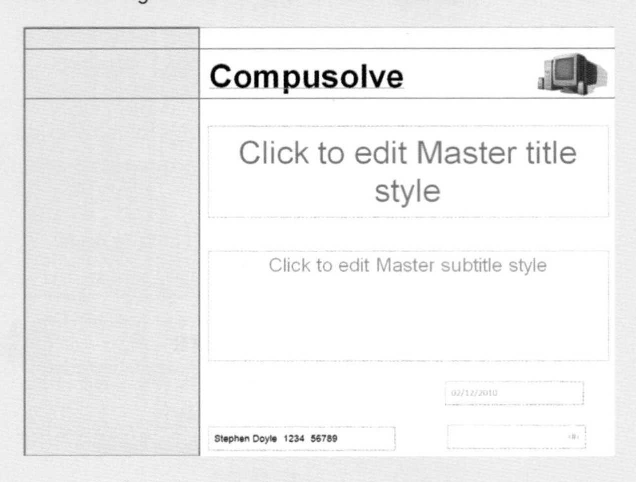

Highlight the text 'Click to edit Master subtitle style'.

Change the font if necessary to a sans serif font such as Arial and change the size to 32 points like this: Arial 32.

Centre the text by clicking on [centre icon] if the text is not already centred and change the font colour to black [A].

12 You have now completed the master slides which lay out the designs, but not the content for the slides in the presentation.

Save your work using the filename **Compusolve_v1**

Activity 19.3

Adding the content in the placeholder areas to create the slides

In this activity you will learn how to:

- include text on the slides including: headings, subheadings, and bulleted lists
- add text to a bulleted list in two columns
- insert a text box
- animate the addition of items in a bulleted list
- create a chart within the presentation package
- include segment labels, remove the legend, move and resize the chart
- add presenter notes to slides.

1 Load the presentation file **Compusolve_v1** if it is not already loaded.

2 You are now going to add the content for the first slide. Here is what you have to do:

Enter the heading: Solving your computer problems

Enter the subheading: Hardware and software problems solved

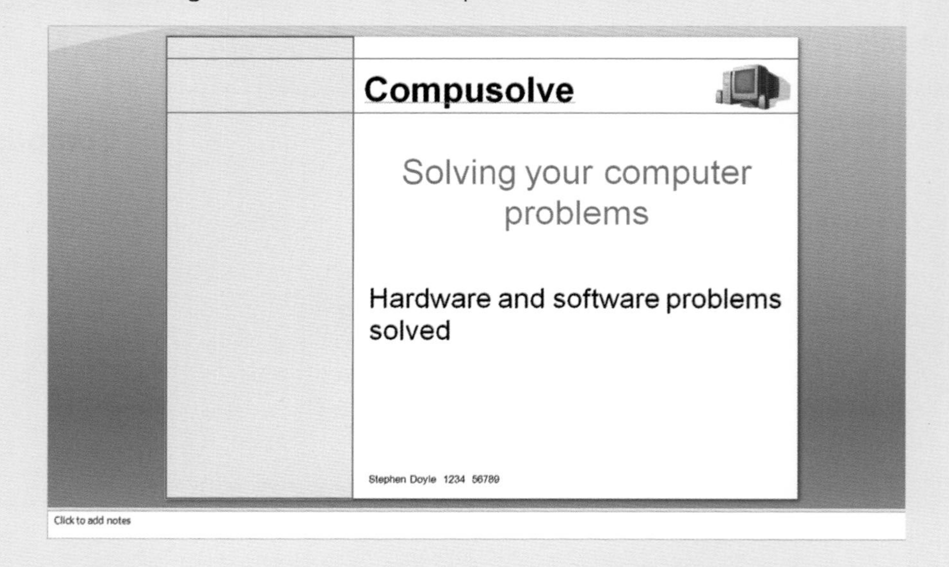

3 Notice the area below the slide containing the message 'Click to add notes'.

This is where you type in your presenter notes. These will appear on the presenter's computer screen but the audience will only see the slides.

In this area type in the following text:

> This brief presentation will introduce our business and outline the benefits of our unique service

4 You are now going to add the content to the second slide.

This slide is going to contain two bulleted lists – both in the white area of the slide with one list down the left-hand side and the other down the right-hand side.

You therefore need to pick a layout which includes this feature.

Click on  New Slide and then on Layout and the following slide layouts will be shown:

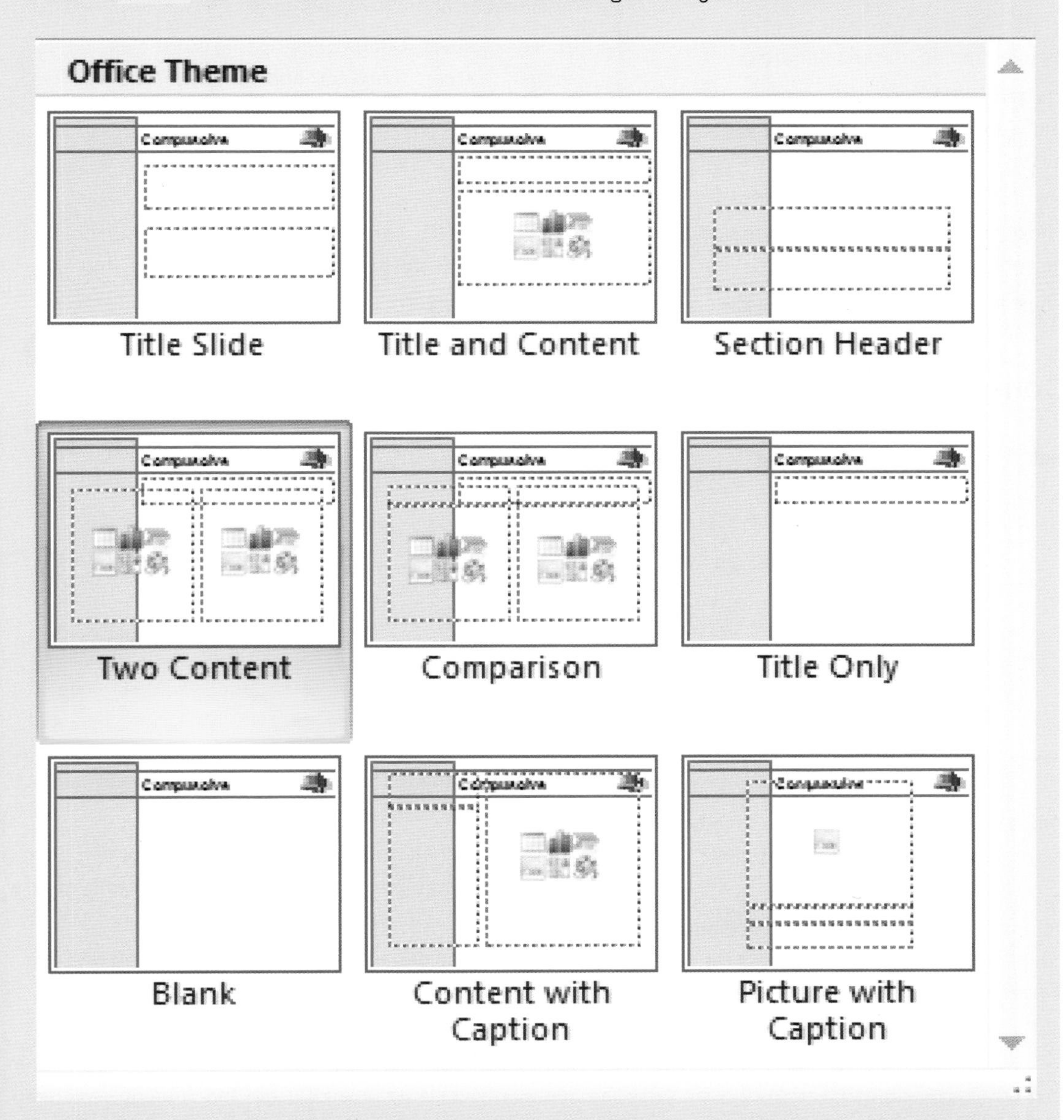

Choose the 'Two content' layout by clicking on it, as this layout includes two bulleted lists side by:side.

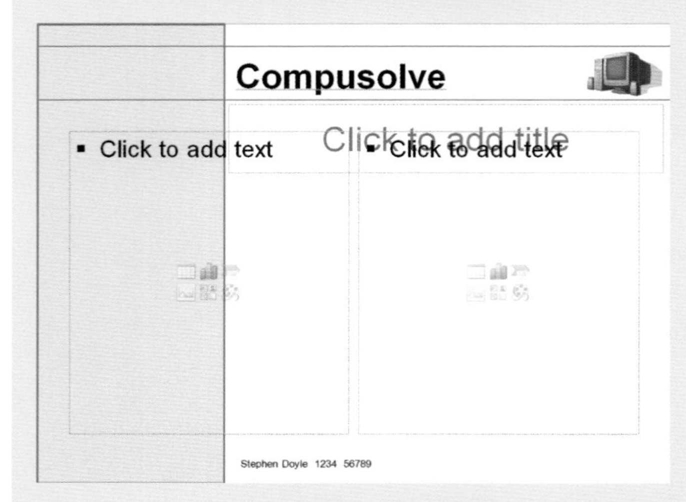

This slide contains overlapping placeholder areas.

Also, you have to put the subheading 'Helping businesses and individuals' on the slide. There is no heading for this slide.

Click on the outline of the placeholder area for the heading (i.e., the box containing the text 'Click to add title' and press the backspace key to delete it.

Your slide should now look like this:

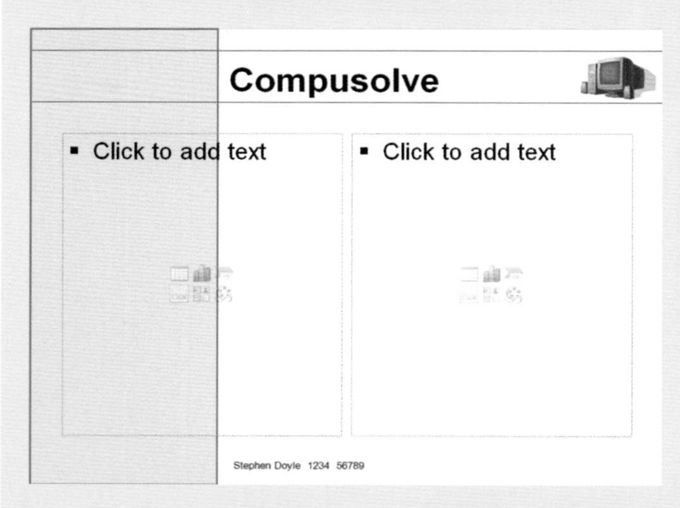

You now have to add a subtitle to the slide. This is to be positioned just above the two columns for the bullet points.

Before this is done you need to create some space and at the same time move the two placeholder areas for the bullets. When you have done this your slide should look like this:

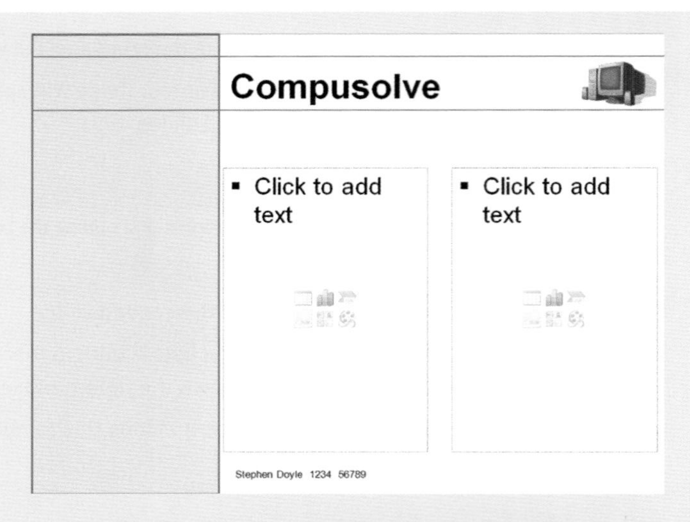

You now need to add a subtitle with the text 'Helping businesses and individuals' in it.

There are several ways of doing this but the easiest is simply to create a text box and check that the text has the formatting of a subtitle.

Click on Insert and then on Text Box and then create a text box in the position shown here:

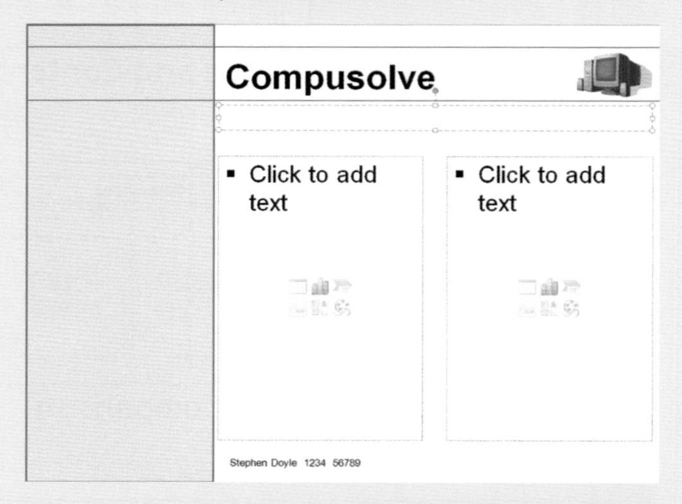

Now type in the following text for the subtitle: Helping businesses and individuals

Now format the text for the subheading to the text used for subheadings (i.e., subheading: sans serif font, centre aligned, and 32 point).

If the place holder areas for the bullets overlap with the subtitle, you will need to move the placeholder areas down slightly.

Now type in the bulleted text as shown here – check that the point size is 32 and if it is not this then you will need to change it.

Compusolve

Helping businesses and individuals

- Installing hardware
- Copying files
- Installing software
- Removing viruses
- Setting up networks
- Repairing hardware
- Setting up Internet access
- Recovering lost data

Stephen Doyle 1234 56789

5 You are now going to animate the bullets so that they appear one at a time.

Select the first column of bullet points like this:

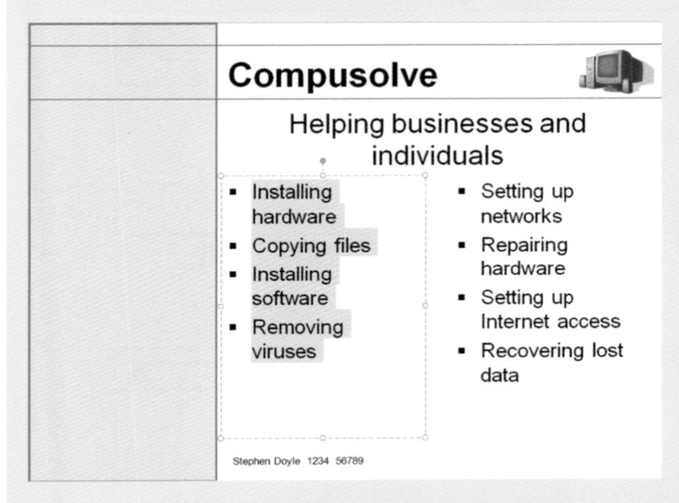

Click on Animate: No Animation and the following animation choices appear:

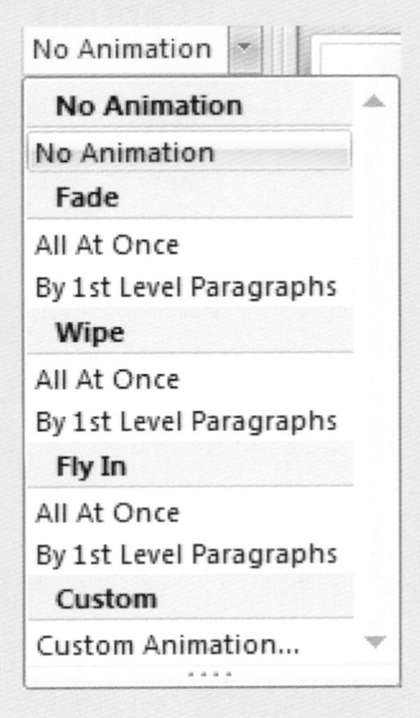

Move your cursor over each of the animations in turn to see how they animate the bullet points on the slide.

Select the animation shown here:

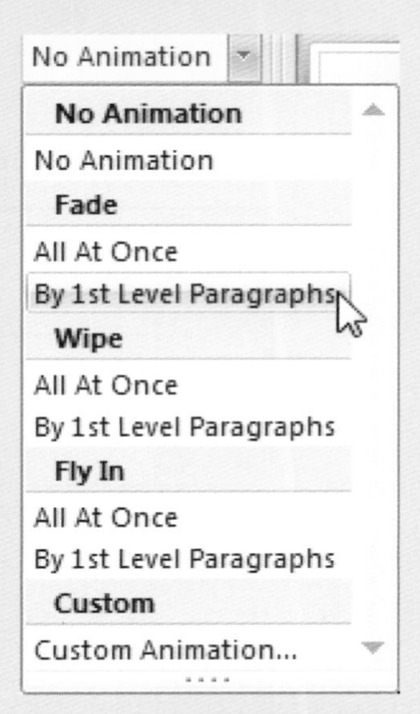

This will animate the bullet points by making each bullet appear gradually.

Animate the bullets in the second column in a similar way. For consistency make sure that you use the same type of animation for both columns of bullets.

Now add the following presenter notes:

Here are some of many services performed by our organization.

6 You are now going to enter the content for the third slide.

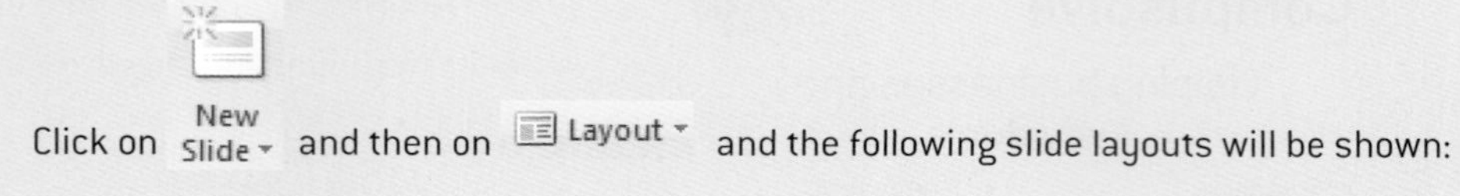

Click on New Slide and then on Layout and the following slide layouts will be shown:

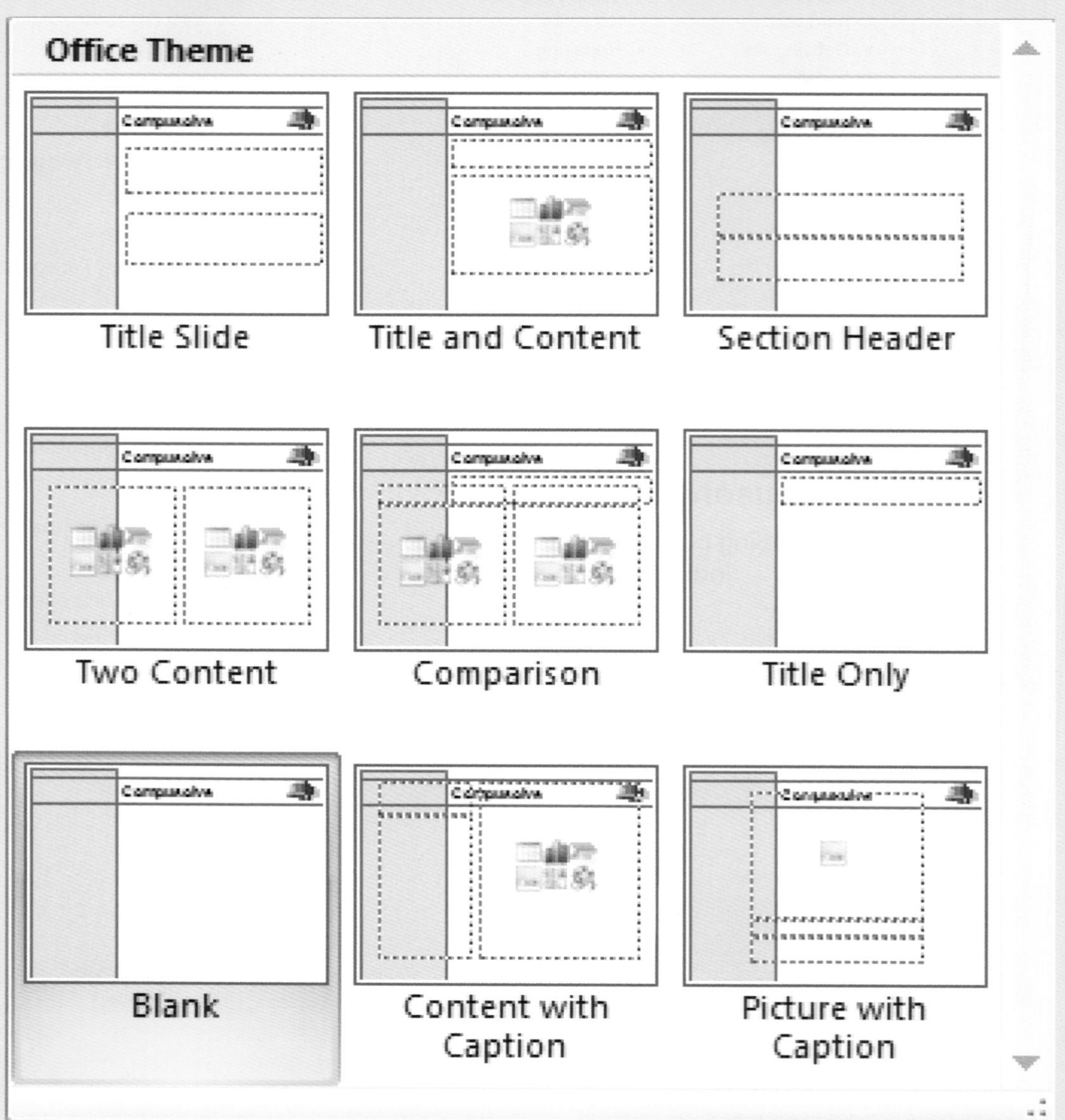

Click on Blank.

7 You are going to create a pie chart using the data in the following table on this slide:

Problem	% of calls
Virus infection	10
Hardware failure	28
Lost data	15
Lack of network	30
Other problems	17

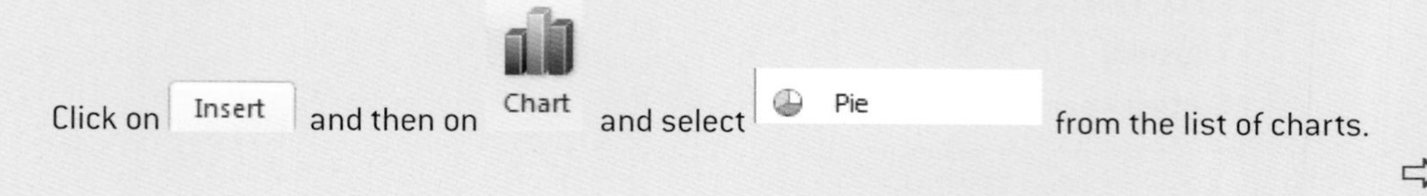

Click on Insert and then on Chart and select Pie from the list of charts.

Choose  and click on OK.

The following screen appears:

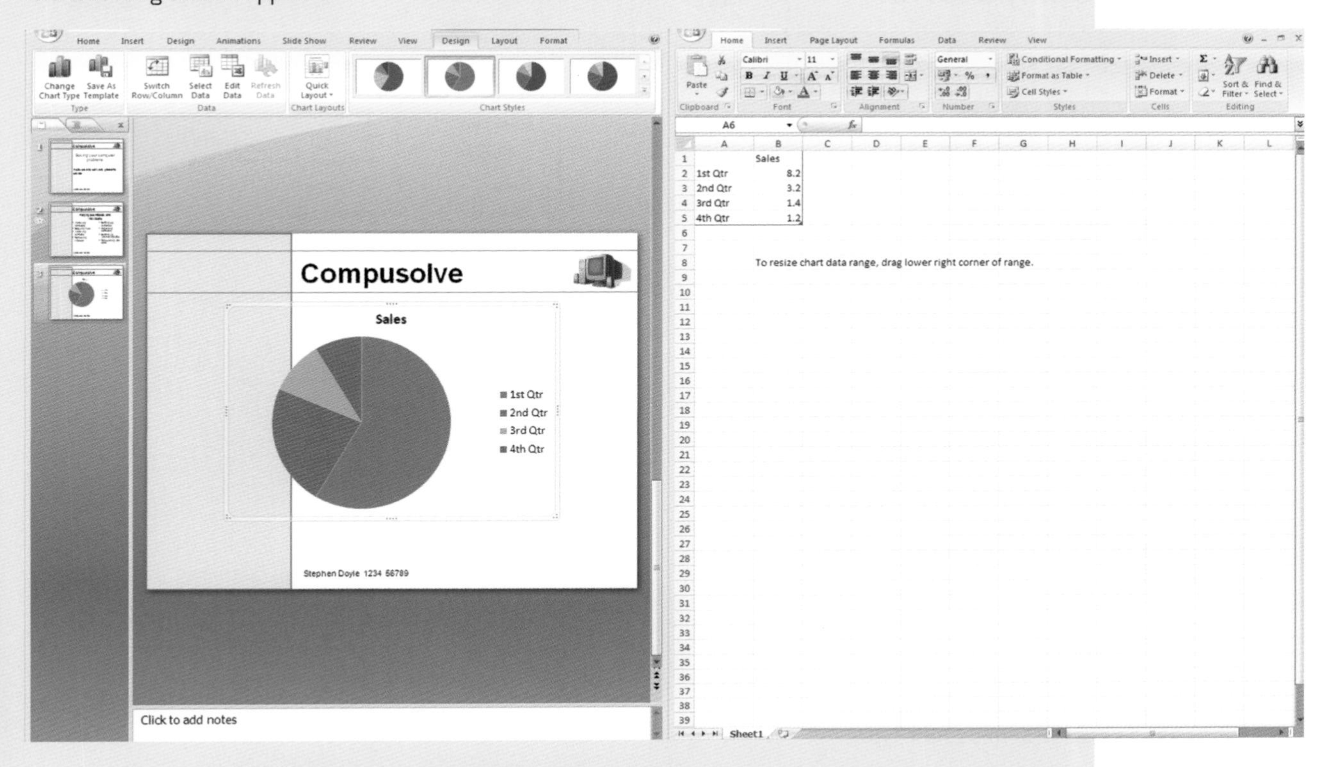

Notice that an Excel spreadsheet has appeared at the side of the slide. This spreadsheet is used to enter the data used to create the pie chart.

You can see that there is a table of data:

	A	B
1		Sales
2	1st Qtr	8.2
3	2nd Qtr	3.2
4	3rd Qtr	1.4
5	4th Qtr	1.2

You need to replace this data with your own data like this:

	A	B
1	**Problem**	**% of calls**
2	Virus infection	10
3	Hardware failure	28
4	Lost data	15
5	Lack of network	30
6	Other problems	17

⇨

You will then notice that the pie chart is produced automatically on the slide like this:

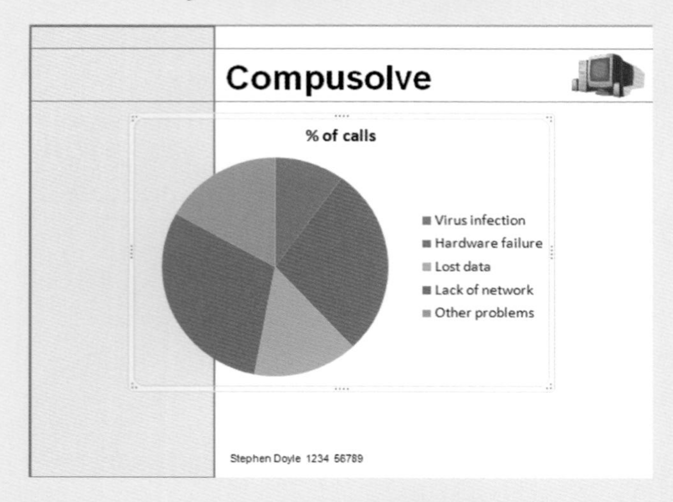

8 You are now required to put the percentages on the segments as well as the problem and then delete the legend.

Right-click on one of the sectors of the pie chart and this menu appears:

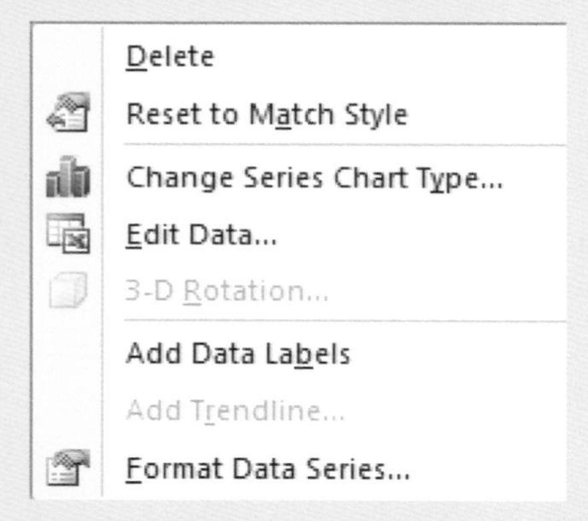

Click on Add Data Labels.

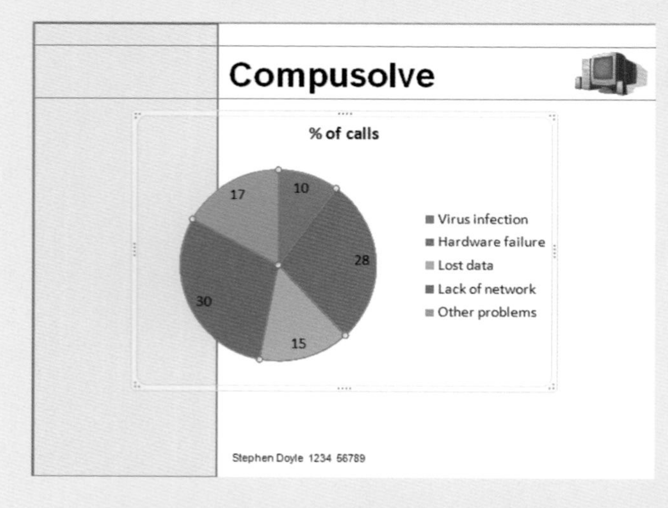

The numbers for the percentages are now added to the chart.

Right-click on a segment in the chart again and select Format Data Labels.

The format labels dialogue box now appears like this:

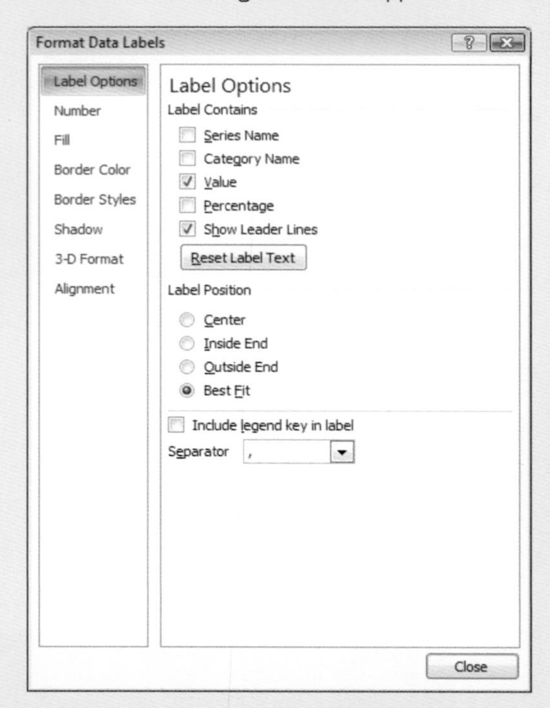

9 You are required to show the value and the category names.

Apply these to the dialogue box like this:

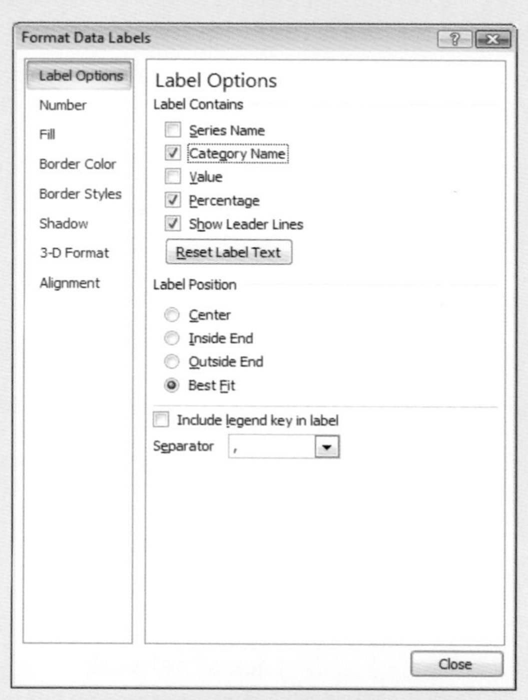

Click on Close to apply the settings.

The chart now appears like this:

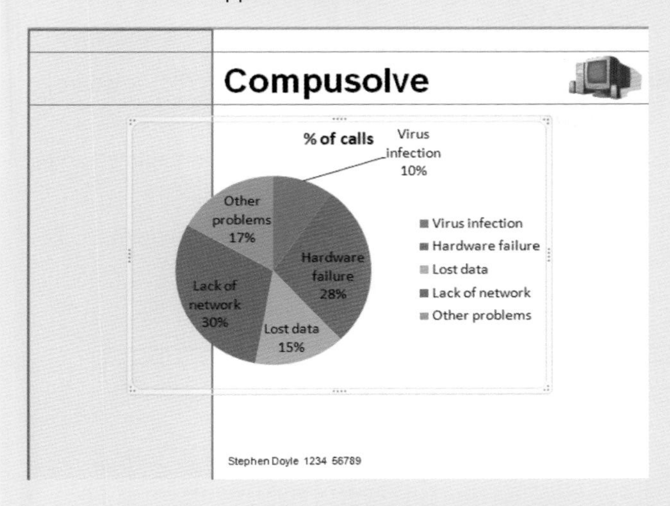

Notice that the text for 'Virus infection 10%' has not appeared in the segment. This is because the text is too wide to fit in the segment.

10 You have been asked to remove the legend from the chart.

To remove the legend, click on it so that the handles appear like this:

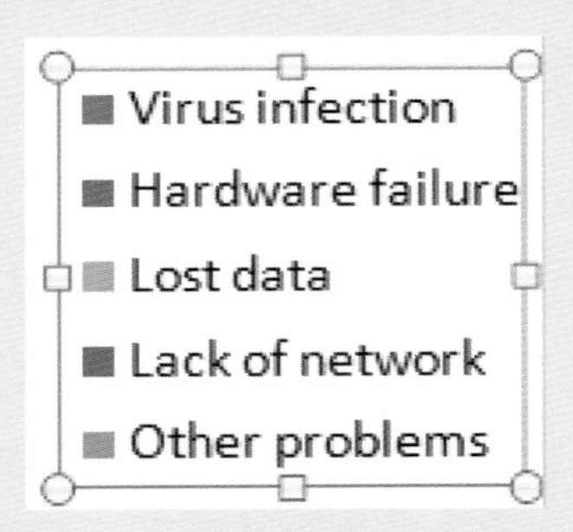

Press the backspace key and the legend is removed.

Position the cursor on the corner of the chart border like this:

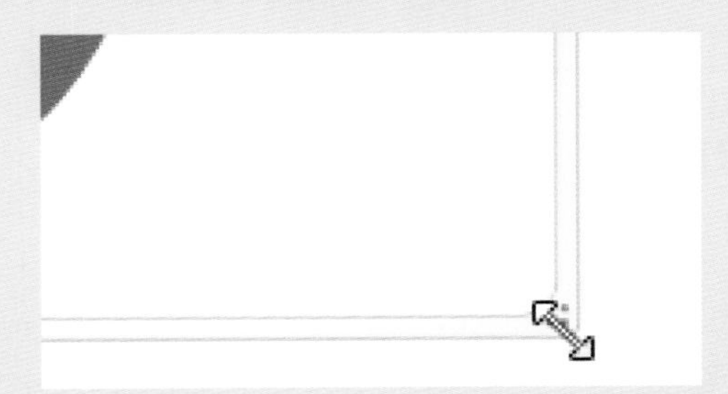

Now keep the right mouse button pressed down and drag the corner to make the pie chart larger. Also ensure that the pie chart is central in area.

Make the chart as large as possible to fit the white space like this:

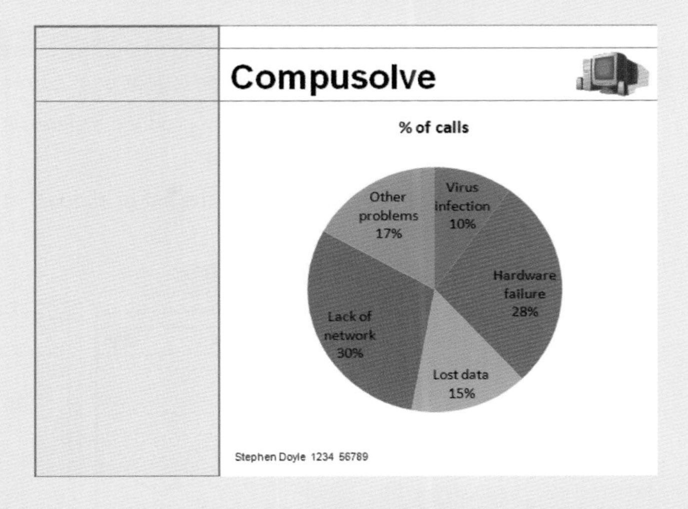

11 Now add the following presenter notes to complete this slide:

You can see from the chart that the majority of the calls for our services are about network problems, although calls about hardware failure are a close second.

12 Click on  New Slide to start adding content on the fourth slide and then click on Layout where you can choose the layout best suited to the slide you want to produce.

You are required to produce a slide with a subheading, a single bulleted list that is left aligned and a piece of clipart.

You can see from the list of slides that there is no one slide which meets all the requirements:

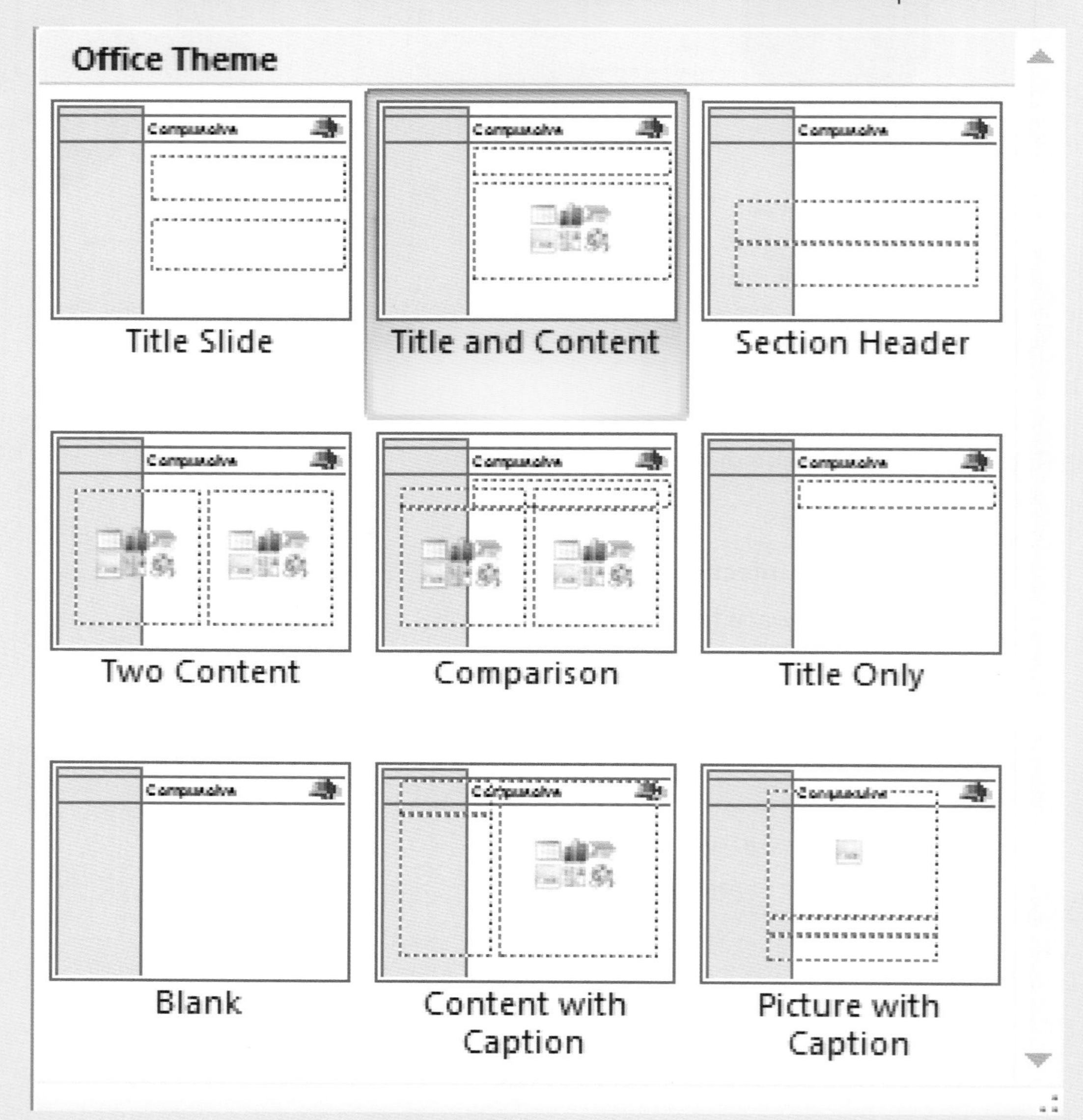

The most appropriate slide is the one called 'Title and Content' but this has a heading (called a title) rather than a subheading but it is easy to change it.

Click on the 'Title and Content' slide and the following slide appears:

Highlight the text 'Click to add title' and make the following changes to the text so that it is formatted as a subheading:

Change the font if necessary to a sans serif font such as Arial and change the size to 32 points like this:

Arial 32.

Centre the text by clicking on ≡ if the text is not already centred and change the font colour to black

.

Now enter the following text for the subheading:

In addition to dealing with computer problems we can:

Enter the following bulleted list on the left-hand side of the slide:

- Offer advice
- Provide training
- Provide backup services

Place a different clipart image of a person using a computer or a group of people using computers in the white section at the bottom of the bulleted list and in a similar position to that shown here:

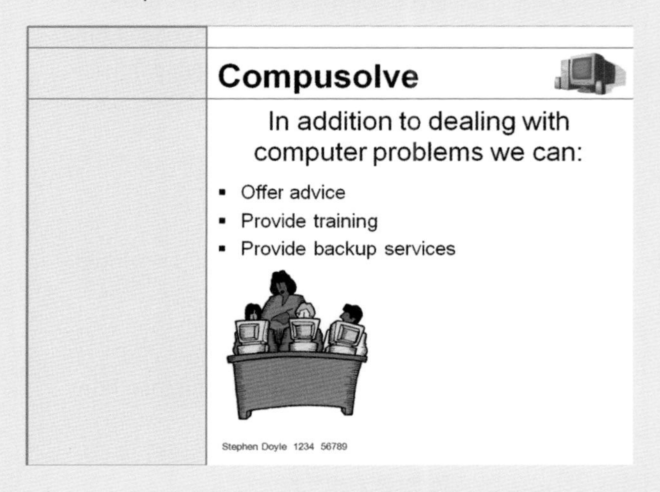

Now add these presenter notes to this slide:

Here are some of the other services that Compusolve can provide.

13 Save your presentation using the filename **Compusolve_v2**

Activity 19.4

Adding slide transitions, printing presenter notes, and providing screen shots for evidence

In this activity you will learn how to:

- automate the transitions between slides
- print a copy of the presentation slides including the presenter notes
- print a copy of the audience notes
- provide printed evidence of animation of bullet points
- provide printed evidence of animation of slide transitions.

1 Load the presentation software Microsoft PowerPoint and the file **Compusolve_v2** if it is not already loaded.

2 You now have to use a transitional effect between each slide in the presentation.

3 Click on Animations and then move your cursor onto each of the slide transitions (shown below) in turn to see their effect.

Choose one of them (you choose which one) and then click on Apply To All, which will apply the transition you have selected to all the slides in the presentation.

Notice the following settings which can be changed:

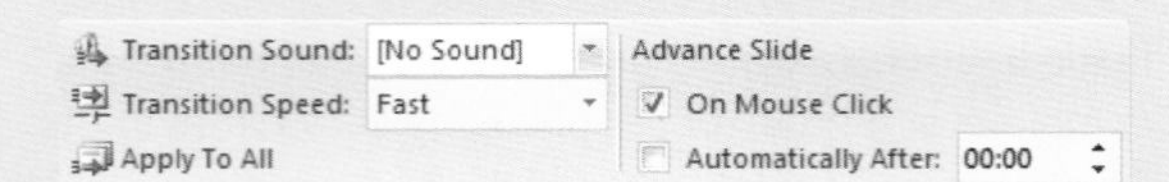

Notice that you can set the slide transitions to be performed automatically after a certain period of time if the presenter has not clicked on the mouse. You can also alter the speed of the transitions and also to make a sound during the transition.

4 Save the presentation using the filename:

Compusolve_final_version

5 A copy of the presentation slides needs to be printed which must show the presenter notes.

Click on [Office button] and then choose Print and then Print Preview like this:

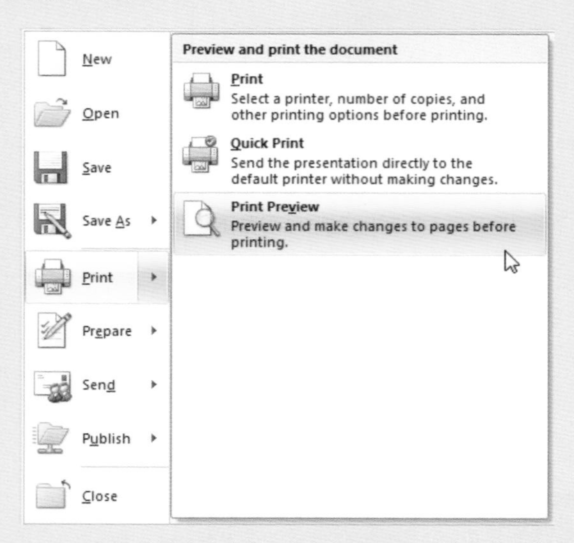

Now click on Print What: Slides and from the drop-down menu choose Notes Pages. You will now see each slide with the presenter notes on a separate page like this:

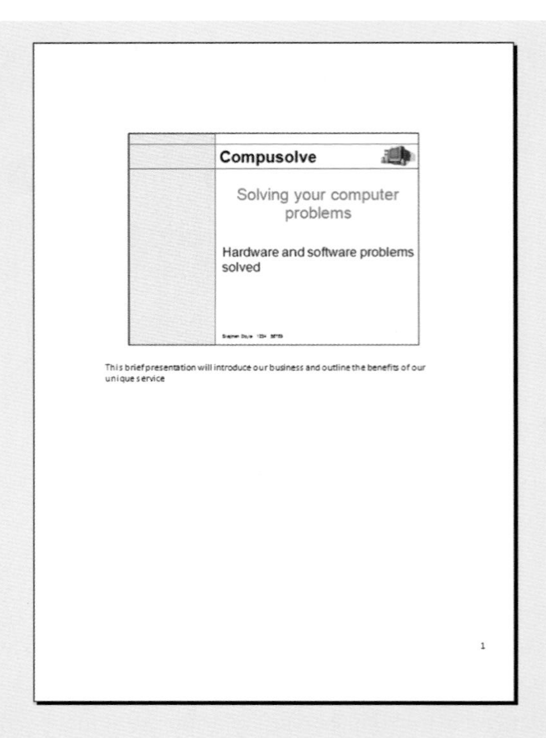

Click on Print to print all the pages.

The Print dialogue box appears similar to this:

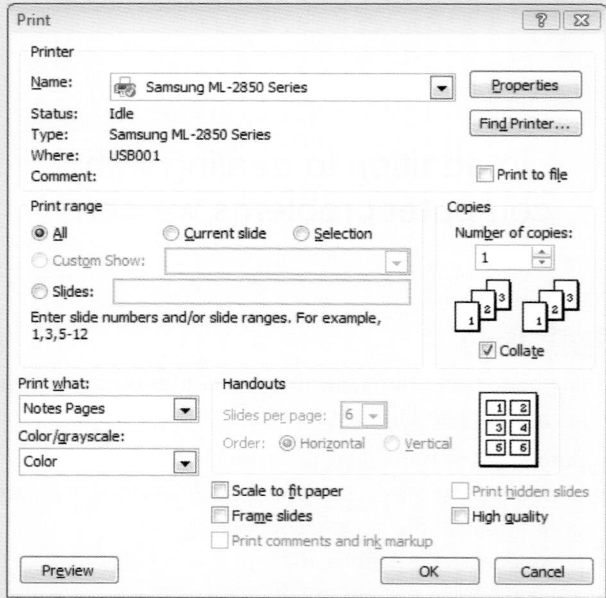

Notice all the settings which can be changed if needed here.

As all the settings are correct click on OK to print the slides.

These slides with presenter notes would usually be used by the presenter to refer to.

6 You now need to produce some printouts of the slides with space for the audience to make notes on each page. You have been asked to produce 3 slides on each page.

Print What:

Notes Pages

Click on the drop-down arrow in the [Notes Pages] section and choose Handouts (3 Slides Per Page).

Slides
Handouts (1 Slide Per Page)
Handouts (2 Slides Per Page)
Handouts (3 Slides Per Page)
Handouts (4 Slides Per Page)
Handouts (6 Slides Per Page)
Handouts (9 Slides Per Page)
Notes Pages
Outline View

You will now see how the slides and note area will appear when printed:

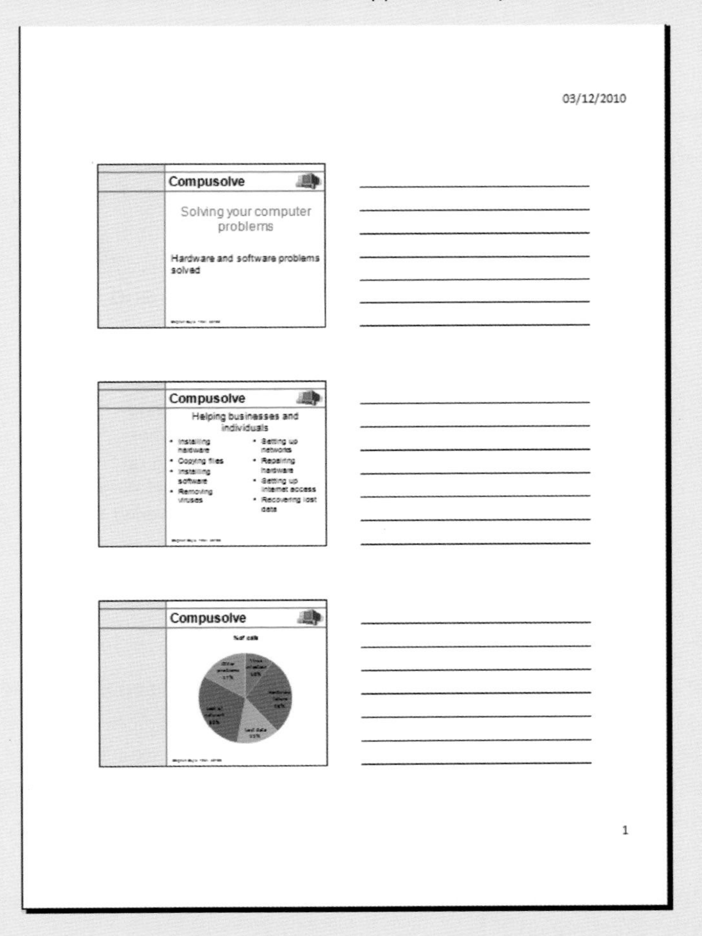

Print this document.

7 You now have to provide evidence that you have included the animations correctly.

Slide 2 contains the animations for the bulleted points to appear one at a time.

Click on this slide so that it is displayed like this:

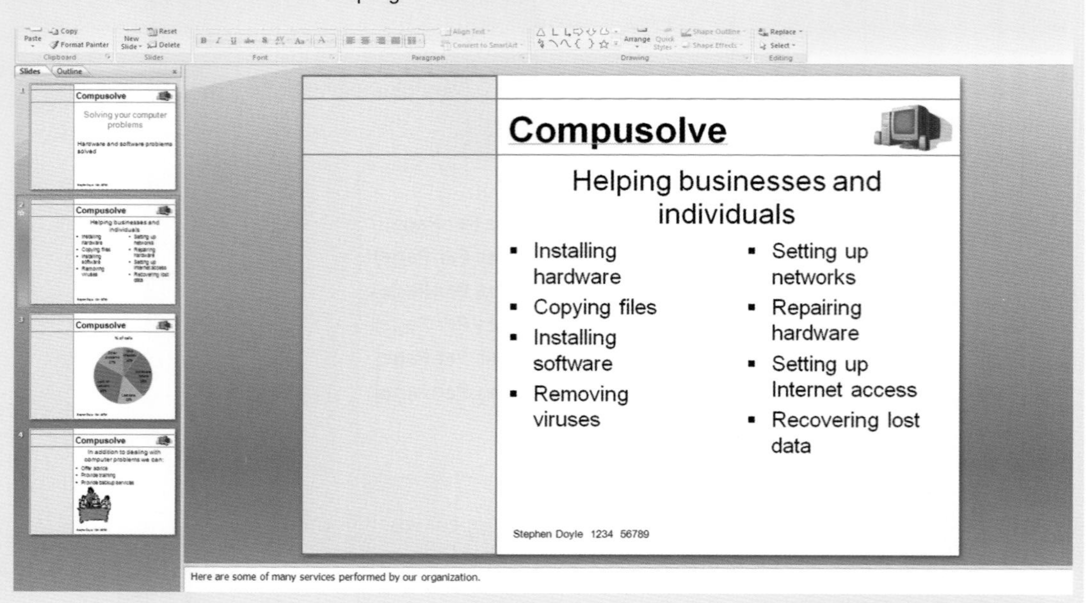

Click on Animations and then on Custom Animation .

The slide now appears with the bulleted points numbered like this:

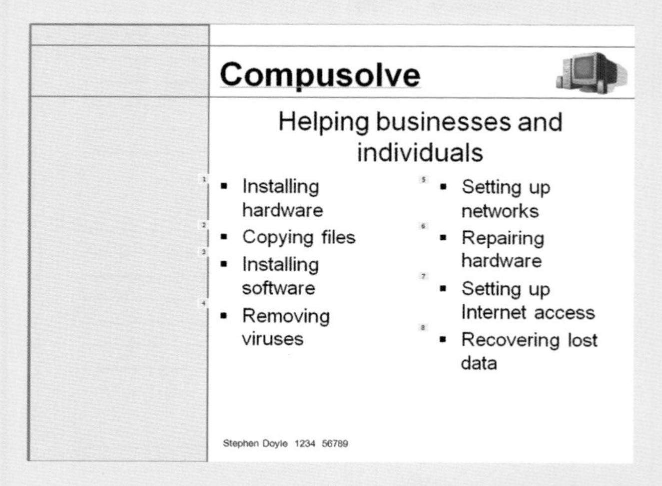

The numbering 1, 2, 3, etc., indicates that each bullet is animated separately and it is the evidence that you have to supply. You now have to take a screenshot of this screen by pressing Prt Scrn (i.e., Print Screen) key on your keyboard. The image of the screen is stored in the clipboard. You can then open a document using your word-processing software and then paste the image into a suitable position. You will need to make sure that you include your name, centre number, and candidate number on the document before printing it out.

8 You now have to produce evidence that the slide transitions have been applied according to the instructions in the examination.

Click on Normal if you are not looking at the normal slide view already.

Click on  Slide Sorter and you will see the slides listed in the presentation appear like this:

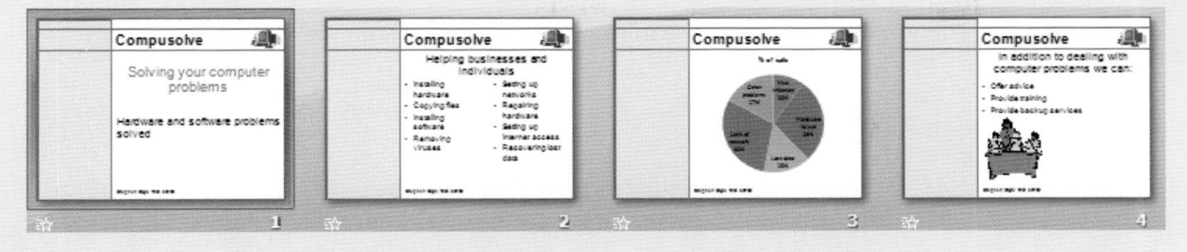

You can see the small symbol appear on each slide showing that a slide transition has been applied to each slide.

9 To obtain a printout of this evidence press the Prt Scrn (i.e., Print Screen) button on your keyboard to copy it, and open a new document in your word-processing software. Paste the screenshot image in a suitable place and make sure that you add your name, centre number, and candidate number to the document which you should then print.

Manipulating images for use on slides

Sometimes in the examination you will be given an image that has to be manipulated in some way such as:

- Resizing (i.e., made bigger or smaller in order to fit a certain space on the slide)
- Positioning (i.e., the image needs to be selected and dragged into the correct position)
- Cropping (i.e., only using part of the image)
- Copying (i.e., so that the same image can be used in different places)
- Changing the image contrast (i.e., adjusting the difference between the light and dark parts of the image)
- Changing the brightness (i.e., making the whole image lighter or darker).

You can do all of the above using PowerPoint rather than use specialist image editing software.

Activity 19.5

Manipulating images

In this activity you will learn how to:

- insert an image onto a slide
- crop an image
- resize an image
- flip an image
- adjust the contrast and brightness of an image

1 Load the presentation software Microsoft PowerPoint.

2 You need a blank slide on which to load an image.

Click on Layout and then from the list of layouts, select Blank.

3 Click on Insert and then Picture – you now need to follow instructions from your teacher who will tell you where the image file called **Caribbean** can be found.

Once located, double left-click on the image and it will be loaded onto the slide like this:

This image occupies the entire slide and there is also an unwanted object in the picture.

If the picture has not been selected (as shown by the small circles around the edges) then left-click on the image to select it.

The picture formatting toolbar will automatically appear like this:

4 Click on the Crop tool in this toolbar .

Notice the handles around the edge of the picture. Click on the top right handle and drag it to the left. Use these handles to crop the picture so that the part containing the bottle is not used:

Click on the white part of the slide to deselect the cropping tool.

Click on the image again so that the handles are shown.

Click on the bottom right of the image and drag the corner until the image has a height of about half of the original like this:

5 You are now going to flip the image.

Select the image if it is not already selected.

Click on Rotate .

From the menu select Flip Horizontal:

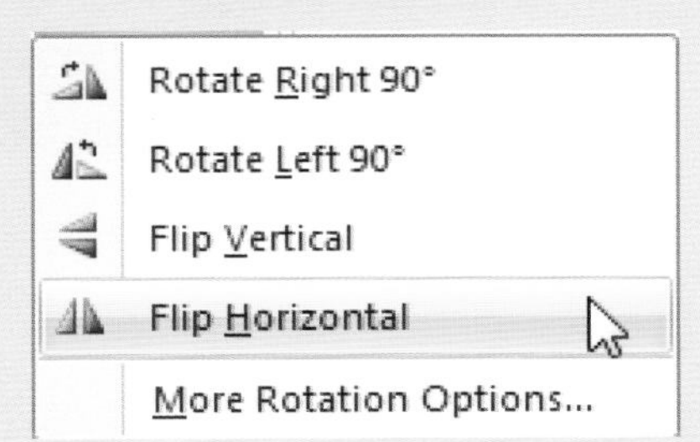

Notice the image has flipped with the boats in the image appearing on the right:

6 With the image selected move it to the top right-hand corner of the slide like this:

7 You now have to alter the contrast and the brightness of the image.

Ensure that the image is selected and the picture format tools are shown.

Click on Contrast

Increase the contrast to +20%:

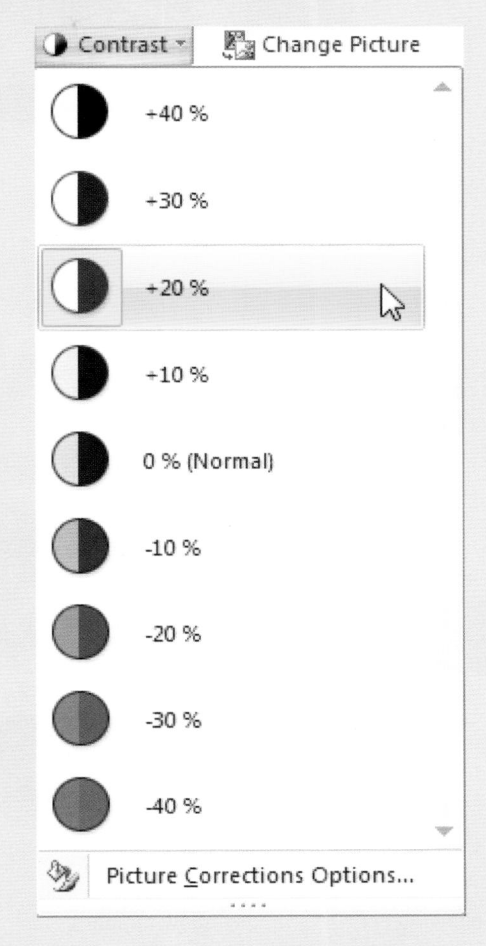

Click on Brightness and select +10%.

8 You now have to add a callout box containing text.

Click on Insert and then on  and from the list of shapes click on the Rectangular callout box shown here:

⇨

You will see the cursor change to a cross. Click on a position near to the image and drag the cursor creating the callout box like this:

Click on the yellow diamond and drag onto the image in a similar position to that shown here:

With the callout selected, right-click on it and choose Format Shape from the list.

The following Format Shape window appears:

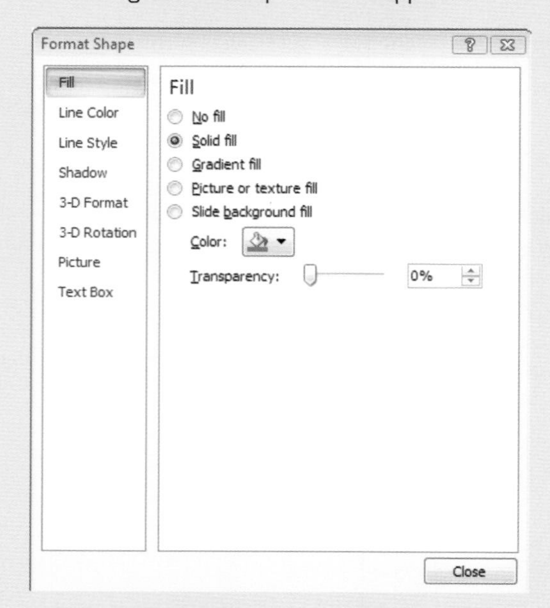

Notice Fill has been selected. Click on Color: and choose White.

Click on Line Color and click on Color: and choose Black.

Click on Close.

The callout box needs turning into a text box. Click on Text Box.

The cursor changes shape.

Click on Home and then click on the font colour and choose the colour red. Change font to Arial and size to 10pt.

Click inside the callout and type in the following text:

Natural coves are used by ships to shelter from storms.

The callout box now appears on the slide like this:

9 In this step you will be putting a copyright symbol followed by the name (i.e., © S Doyle) underneath the image.

First create a text box by clicking on Text Box and position the box as shown here:

Change the font size to 10 pt.

Click on Insert and then on Symbol and click on the copyright symbol.

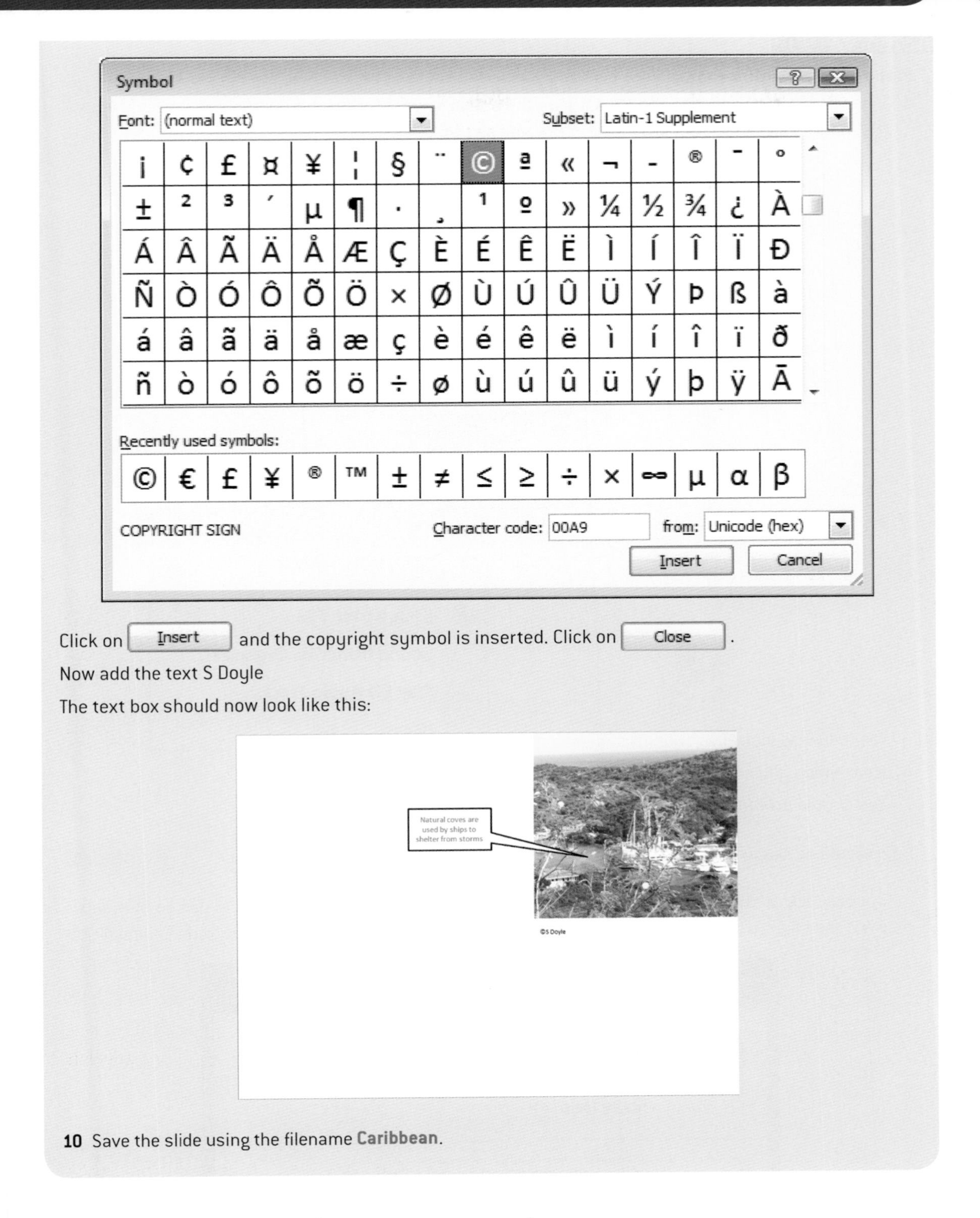

Click on Insert and the copyright symbol is inserted. Click on Close.

Now add the text S Doyle

The text box should now look like this:

10 Save the slide using the filename **Caribbean**.

Adding sound or music to a presentation

It is easy to add sound or music to a particular slide or even to the whole presentation.

Before you add sound you will need some sound files. In the examination, these sound files will be included and you will be given instructions where to find them.

Sound can make a presentation more fun but it can also detract from the message being given so it needs to be used with care.

Activity 19.6

Placing sound within a slide

1 Load PowerPoint and load the file **A life on the ocean wave**. Your teacher/lecturer will tell you where to find this file. Check the slide looks the same as this:

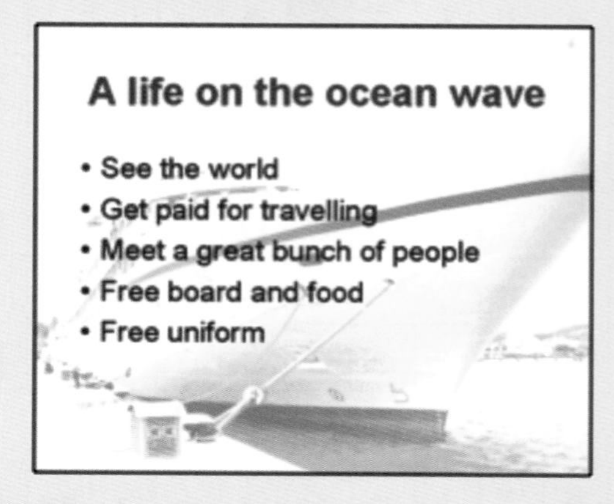

A life on the ocean wave
• See the world
• Get paid for travelling
• Meet a great bunch of people
• Free board and food
• Free uniform

2 To add some sound to this slide, click on Insert and then on the drop-down arrow on the following icon Sound. This menu appears

from which you should select Sound from File.

You should now find the following sound file.

Double-click on the file to insert it. The following menu appears where you have to decide how you want the slide to be played. Click on Automatically so that it plays as soon as the slide appears.

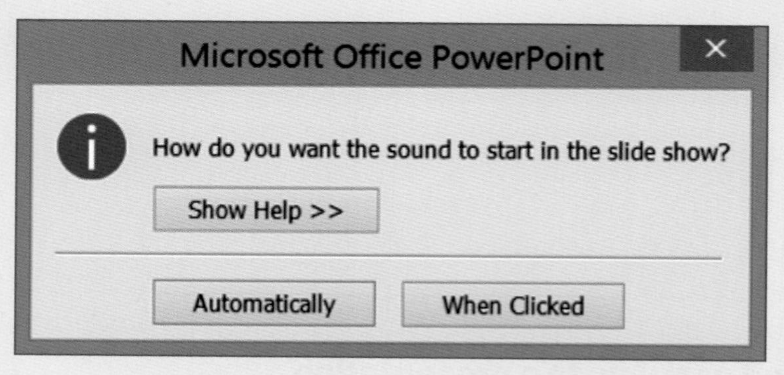

Notice the icon of a speaker appears. Here it is making the text hard to read so we will move it.

3 Move the cursor over the speaker icon where it changes shape to the following .

Click with the left mouse button and keeping the button pressed down drag the icon to the bottom right of the slide.

The slide will now look like this.

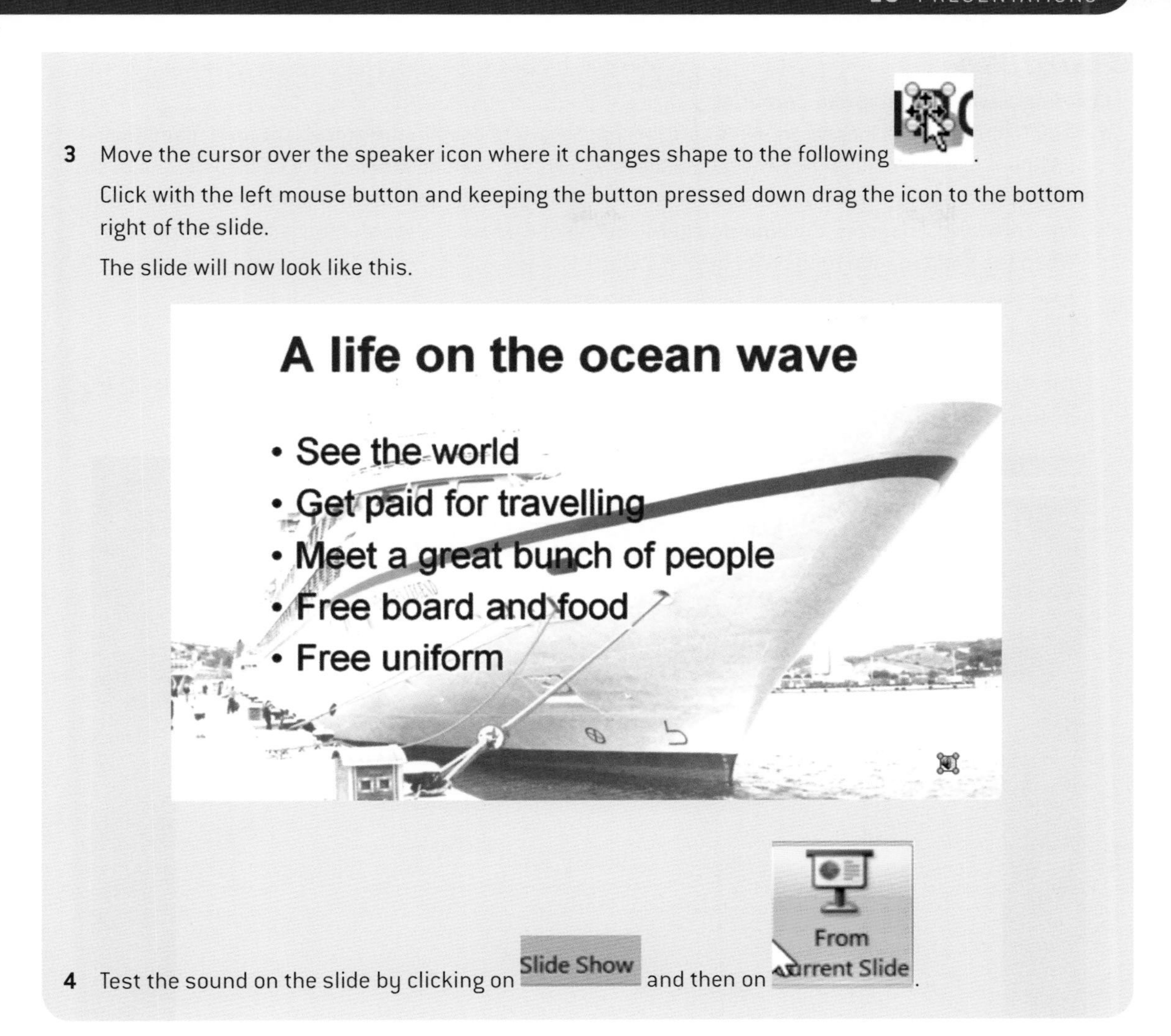

4 Test the sound on the slide by clicking on Slide Show and then on From Current Slide.

Adding movies (video clips and animated images) on the slide

Video clips are added in a similar way to sound and the steps are as follows:

1 On the slide where you want the video clip inserted click on Insert and then on the drop-down arrow on this

icon . Now select 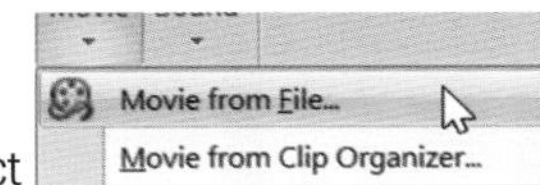.

2 You now have to find the movie file and double-click on the file name to insert it on the slide.

3 Test the video/animated image on the slide by clicking

on 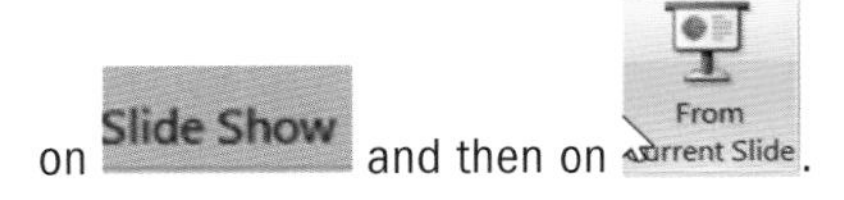.

Creating a controlled presentation

When a presentation is being run, there are some tools on the screen which appear very feint at the bottom left of the screen. When the cursor is moved over them they appear clearer and their functions are as follows:

View the previous slide.

Ink tool – allows the presenter to draw on the slide and highlight certain points.

Used to access the control menu which can be used to navigate the slides.

View the next slide.

All the above can control the running of the presentation.

Activity 19.7

Creating a looped on-screen carousel

1 Before creating the looped presentation, it is necessary to specify the amount of time that should be spent on each slide before the next slide is shown. This can be different for each slide. The best way to determine this is to rehearse the presentation and read the information on each slide slowly. The software can record the amount of time you spend on each slide and these timings are used when the presentation runs on its own in a loop.

Run the presentation software and load the file **What_are_graphics**

Click on Slide Show and then on Rehearse Timings.

The Rehearsal window appears at the top left. You should now read the content of each slide slowly and then click the slide to move onto the next slide until all the slides have been viewed.

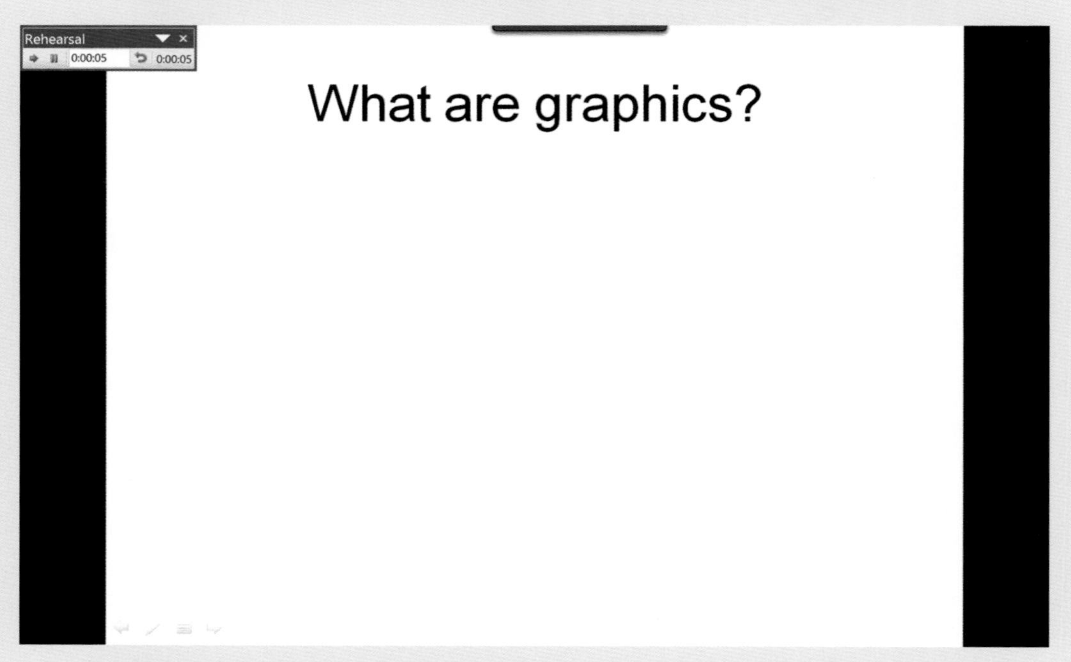

After the last slide the following window appears.

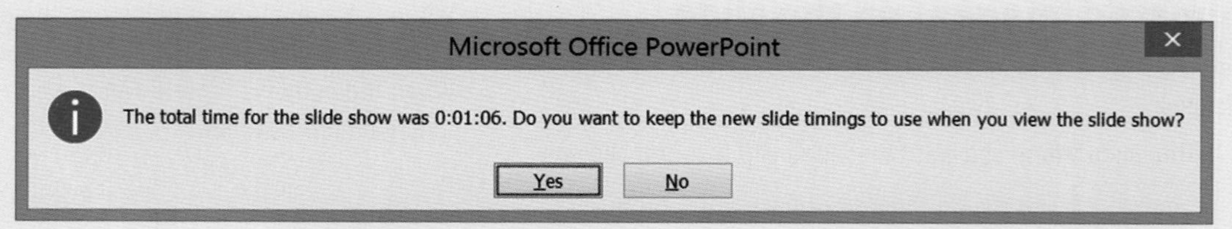

Click on Yes.

2 You will then be presented with a screen showing all the slides along with their timings in seconds.

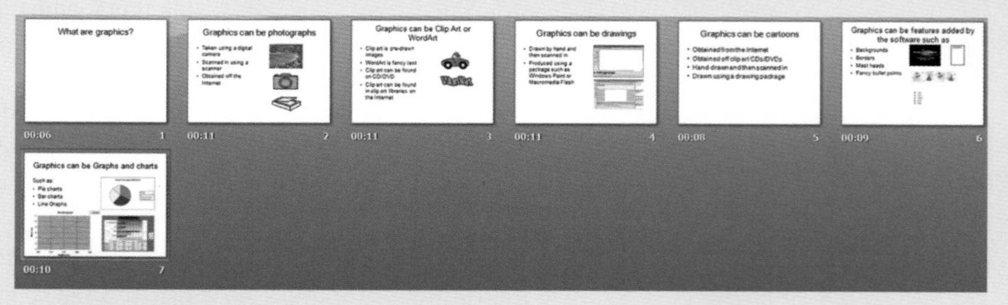

You have now completed the timings for all the slides.

3 The next step is to complete the loop so the presentation runs automatically according to the slide timings and repeats over and over until the Esc button is pressed.

Click Slide Show and then Set Up Slide Show when the following menu appears.

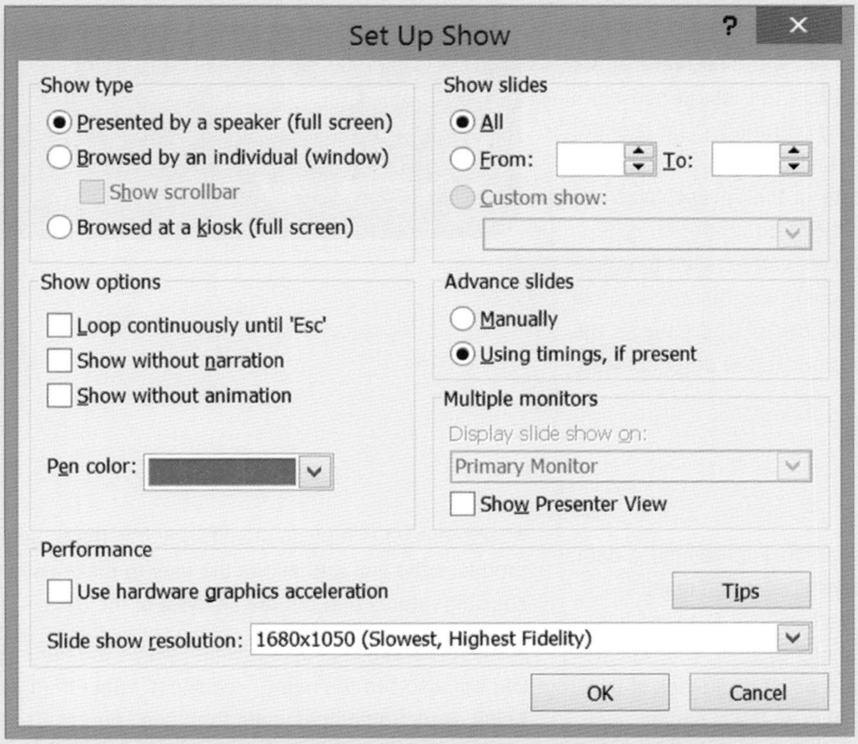

Take some time to look at all the options on this menu. As we want the presentation to play continuously without user intervention, select the following option.

Browsed at a kiosk (full screen)

Click on OK

4 To run the looped on-screen carousel presentation click on Slide Show and then From Current Slide.

The presentation now runs continuously and to stop it, press the Esc key.

20 Data analysis

In this chapter you will be learning about spreadsheets and how they are used to create data models. You will also learn about how such models need to be tested. Once the model has been created the results from the model can be obtained and presented in a suitable way. Spreadsheet software can also be used to create a simple database and the data in the spreadsheet can be extracted using search criteria or sorted into different orders. The tools provided in spreadsheet software can be used to adjust the display features in a spreadsheet to improve the appearance of the spreadsheet. Microsoft Excel has been used here as the spreadsheet software. There are other brands, and other software packages that do much the same.

The key concepts covered in this chapter are:

- Creating a data model
- Testing a data model
- Manipulating data
- Presenting data

What is a data model?

A data model is used to mimic a real situation. For example, a data model can be created using spreadsheet software that mimics the money coming into and going out of a business. Data models can be used to provide answers to questions such as 'what would happen if I did this?' For example, an economist could look at the effects that raising interest rates would have on the economy.

Creating a data model

In the examination you will be required to produce or adapt a computer model using the spreadsheet skills learned during your IGCSE course.

Entering the layout of the model

You will be following a set of instructions in the examination and it is important that you enter and position the items that form the model exactly as specified. There are a number of things you will need to know about when laying out a spreadsheet model and these are explained here.

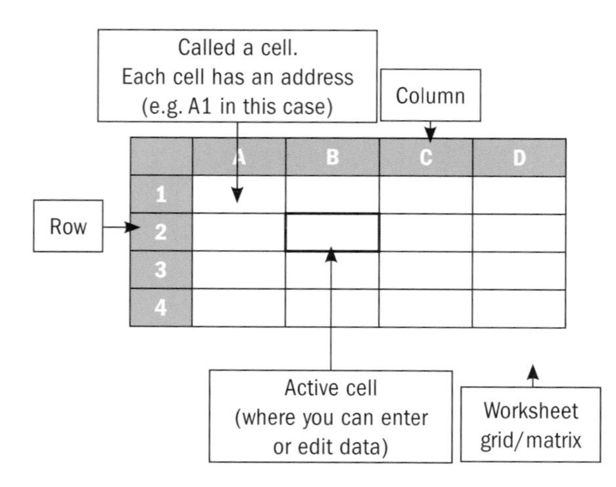

Sheets

A spreadsheet is a piece of software. It is the program that produces the grid and allows the user to make use of all the special features that enable analysis of the data.

A sheet is a single grid/matrix into which you enter the data and you will also see the name 'worksheet' rather than just 'sheet' being used. Sometimes more than one sheet is used to hold the data and a group of sheets form a workbook.

Sheet tab

A sheet tab appears at the bottom of the worksheet. **Sheet1** Sheet2 Shee

It has the name of the sheet on it, which can be changed by right-clicking on it and picking the option Rename from the menu that appears. By clicking a sheet tab, you can make that sheet the active sheet so that you can work on it.

Values

Values are the numbers that are entered into the cells of the spreadsheet.

Labels

Labels are the text next to cells that explain what it is that the cell contains. You should never have a value on a spreadsheet on its own as the user will be left wondering what it represents.

Formulae

Formulae are used to perform calculations on the cell contents. In Excel, in order to distinguish between text and formulae, a symbol (the equals sign) needs to be typed in first, like this =B3+B4.

Here are some calculations and what they do. Notice that you can use upper or lower case letters (i.e. capital or small letters):

= C3+C4 (adds the numbers in cells C3 and C4 together)
= A1*B4 (multiplies the numbers in cells A1 and B4 together)
= 3*G4 (multiplies the number in cell G4 by 3)
= C4/D1 (divides the number in cell C4 by the number in cell D1)
= 30/100*A2 (finds 30% of the number in cell A2)
= A2^3 (finds the cube of the number in cell A2)

What must be done with the numbers or contents of cells is determined by the operator. Here is a table of operators and what they do in a formula:

Operator	What it does
+	Add
-	Subtract
*	Multiply
/	Divide
^	Power

Functions
Functions are calculations that the spreadsheet has memorised. For example, the function =sum(b3:b10) adds up all the cells from b3 to b10 inclusive. Functions will be covered a bit later.

Copy and paste
Copy and paste can be used to move items around on the same spreadsheet or to copy data from a completely different document or file such as the table in a word-processed document or from a database.

Drag and drop
Drag and drop is a quick way of moving an item. You simply left-click the mouse button on the item and then, keeping the left mouse button down, drag the item to the new position. When the mouse button is released the item appears in the new position.

AutoFill
Suppose you want to type the days of the week or months of the year down a column or across a row. Excel is able to anticipate what you probably want to do by the first word alone. So, if you type Monday, the chances are that you want Tuesday in the next column or row, and so on. The main advantage in using AutoFill is that the data being entered is less likely to contain errors than if you type in the data yourself. There are many other ways you can use spreadsheet software to fill in data for you, so use the Help to find out more about AutoFill.

Manually verifying data entry
In the examination you will be asked to obey a set of instructions. It is extremely important that you obey the instructions exactly. So if you are asked to use a certain font, you must use this even if you feel a different one would be better.

Verifying data entry means checking that you have entered the data accurately from the examination paper or other source. You do this carefully by visually checking. A small mistake in a number in a spreadsheet can change the data in the whole spreadsheet so that it is all incorrect if there are formulae that use the incorrect number.

You will also need to check that you have obeyed all the instructions correctly. It is very easy to miss something out that you should have done.

Mathematical operations and formulae
Spreadsheets can perform all the usual mathematical operations such as:

- Add +
- Subtract -
- Multiply *
- Divide /
- Power ^ (e.g., square, cube, square root, cube root, etc.).

The order of mathematical operations in a formula
Spreadsheets perform calculations based on the formulae entered. Spreadsheets carry out the operations in calculations in the following order:

1. Brackets (called parentheses) are carried out first. Where there are brackets inside brackets (i.e. there are nested brackets), the calculations in the innermost brackets are carried out first. Then the calculations in the next brackets are carried out and so on.
2. Percentages
3. Indices (i.e. powers and roots)
4. Multiplication or division
5. Addition or subtraction

Suppose we have the following values contained in these cells in a spreadsheet.

a1 = 10
b7 = 4
g10 = 24
g11 = 3
h2 = 8

The table below shows how the spreadsheet would use each formula with the contents of the cells to calculate the result.

Formula	Calculation	Result
=a1+g10/g11	=10 + 24/3 = 10 + 8	18
=a1*b7+h2	=10 × 4 + 8 = 40 + 8	48
=a1*(b7+h2)	=10 × (4 + 8) = 10 × 12	120
=(a1*g11)+(g10/g11)	=(10 × 3) + (24 ÷ 3) = 30 + 8	38
=a1^2+g10/g11	$=10^2 + 24 \div 3 = 100 + 8$	108
=g10+10%	=24 + 0.1 × 24	26.4
=20%*h2	=0.2 × 8	1.6

Activity 20.1

Copying formulae relatively and showing the formulae for a spreadsheet

In this activity you will learn how to:

- enter the layout of a model
- use mathematical operations
- copy formulae
- display formulae.

1 Load your spreadsheet software and enter the following data exactly as it is shown here:

	A	B	C	D	E	F
1	Product	Sept	Oct	Nov	Dec	Total
2	Lawn rake	121	56	23	12	
3	Spade (Large)	243	233	298	288	
4	Spade (Medium)	292	272	211	190	
5	Spade (Small)	131	176	149	200	
6	Fork (Large)	208	322	178	129	
7	Fork (Small)	109	106	166	184	
8	Trowel	231	423	311	219	
9	Totals					

2 Enter the function =sum(b2:e2) in cell F2 like this.

	A	B	C	D	E	F
1	Product	Sept	Oct	Nov	Dec	Total
2	Lawn rake	121	56	23	12	=sum(b2:e2)
3	Spade (Large)	243	233	298	288	
4	Spade (Medium)	292	272	211	190	
5	Spade (Small)	131	176	149	200	
6	Fork (Large)	208	322	178	129	
7	Fork (Small)	109	106	166	184	
8	Trowel	231	423	311	219	
9	Totals					

3 Copy the function in cell F2 down the column to cell F8. You do this by moving the cursor to cell F2 containing the formula. Now click on the bottom right-hand corner of the cell and you should get a black cross shape. Hold the left mouse button down and move the mouse down the column until you reach cell F8. You will see a dotted rectangle around the area where the copied formula is to be inserted. Now take your finger off the button and all the results of the calculation will appear. This is called relative copying because the formula is changed slightly to take account of the altered positions of the numbers which are to be added together.

4 Enter the function =sum(b2:b8) into cell B9.

5 Copy the function in cell B9 relatively across the row until cell F9.

Check your spreadsheet looks the same as this:

	A	B	C	D	E	F
1	Product	Sept	Oct	Nov	Dec	Total
2	Lawn rake	121	56	23	12	212
3	Spade (Large)	243	233	298	288	1062
4	Spade (Medium)	292	272	211	190	965
5	Spade (Small)	131	176	149	200	656
6	Fork (Large)	208	322	178	129	837
7	Fork (Small)	109	106	166	184	565
8	Trowel	231	423	311	219	1184
9	Totals	1335	1588	1336	1222	5481

6 You are now going to display the formulae used rather than the values wherever there are calculations.

Click on  in the toolbar.

Look at the Formula Auditing section shown here:

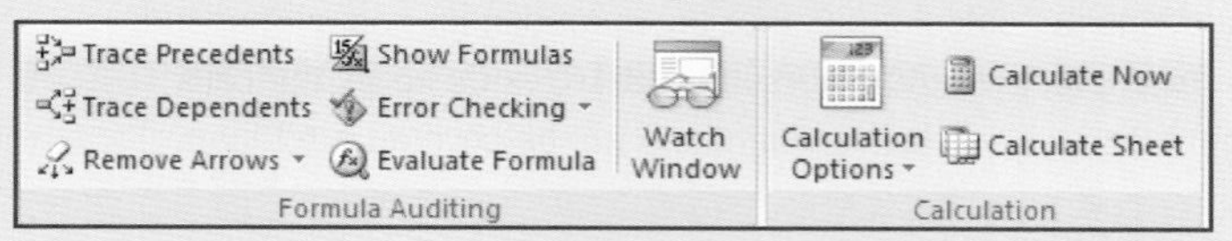

Click on Show Formulas.

You should now see the formulae being displayed like this. This is handy as the formulae can be checked.

	A	B	C	D	E	F
1	Product	Sept	Oct	Nov	Dec	Total
2	Lawn rake	121	56	23	12	=SUM(B2:E
3	Spade (Large)	243	233	298	288	=SUM(B3:E
4	Spade (Medium)	292	272	211	190	=SUM(B4:E
5	Spade (Small)	131	176	149	200	=SUM(B5:E
6	Fork (Large)	208	322	178	129	=SUM(B6:E
7	Fork (Small)	109	106	166	184	=SUM(B7:E
8	Trowel	231	423	311	219	=SUM(B8:E
9	Totals	=SUM(B2:B8)	=SUM(C2:C8)	=SUM(D2:D8)	=SUM(E2:E8)	=SUM(F2:F
10						

7 Click on Show Formulas again and the spreadsheet will return to showing the values.

8 Save this spreadsheet using the filename **Product_sales**.

Activity 20.2

Setting up a simple model

In this activity you will learn how to:

- enter data with 100% accuracy
- format cells.

A university student is living away from home for the first time and she wants to make sure that she budgets the limited amount of money she will have. She is going to create a spreadsheet model. You are going to follow her steps.

1 Load the spreadsheet software Excel.

2 Enter the details shown exactly as they appear below. Note: you will need to format the text in some of the cells. You will need to format cells D3 and D10 so that the text is wrapped in the cell. (This keeps the cell width the same by moving text so that it fits the width.) You can do this by clicking on the cell where you want the text to be wrapped and then on Home and then on Wrap Text

	A	B	C	D	E
1	**Weekly budget**				
2					
3	**Income**	**Amount**	**Weeks**	**Income per week**	
4	Student loan (per term)	$2250	12		
5	Weekly wage from part-time job	$105	1		
6					
7	**Total income per week**				
8					
9					
10	**Spending**	**Amount**	**Weeks**	**Spending per week**	
11	Monthly rent	$250	4		
12	Books (per term)	$210	12		
13	Food (per month)	$280	4		
14	Clothes (per term)	$240	12		
15	Entertainment (per term)	$50	1		
16	Travel (per term)	$145	12		
17					
18	**Total spending per week**				
19					
20	**Balance left/owed at end of week**				
21					

3 In cell D4 enter the following formula to work out the amount the student gets per week from her student loan.

=b4/c4

4 In cell D5 enter the following formula to work out the amount of money the student gets each week from her part-time job.

=b5/c5

5 In cell D7 enter the formula =d4+d5. This calculates the total weekly income.

6 Enter a formula that will calculate the weekly spending on rent and put the answer in cell D11.

7 Copy the formula you have entered in D11 relatively down the column as far as cell D16.

8 Put a formula in cell D18 to add up the spending from cells D11 to D16.

9 Enter a formula in cell D20 which will subtract the total spending from the total income.

10 Your completed spreadsheet will now look like that shown here.

	A	B	C	D	E
1	**Weekly budget**				
2					
3	**Income**	**Amount**	**Weeks**	**Income per week**	
4	Student loan (per term)	$2,250	12	$187.50	
5	Weekly wage from part-time job	$105	1	$105.00	
6					
7	**Total income per week**			$292.50	
8					
9					
10	**Spending**	**Amount**	**Weeks**	**Spending per week**	
11	Monthly rent	$250	4	$62.50	
12	Books (per term)	$210	12	$17.50	
13	Food (per month)	$280	4	$70.00	
14	Clothes (per term)	$240	12	$20.00	
15	Entertainment (per term)	$50	1	$50.00	
16	Travel (per term)	$145	12	$12.08	
17					
18	**Total spending per week**			$232.08	
19					
20	**Balance left/owed at end of week**			$60.42	
21					

11 Save your spreadsheet using the filename **Budget_model**.

Activity 20.3

Using the spreadsheet model to find the answers to 'what if' questions

In this activity you will learn how to:

- amend data in a spreadsheet
- print out formulae used.

1. Load the spreadsheet called **Budget_model** if it is not already loaded.
2. You are going to make some changes to the spreadsheet. It is important that you do **not** save any of these changes.
3. The monthly rent has been increased to £275 per month and the student's employers have reduced her hours for the part-time job, which means she now earns only £50 per week. Will she now be spending more money than she receives? Make these alterations to the spreadsheet to find out. How much does she have left at the end of the week?
4. Her grandparents decide they can give her £50 per week to help her. Add this amount to the spreadsheet in a suitable place and using a suitable label. Make any necessary changes to formulae. How much does she now have at the end of the month?
5. Save your spreadsheet using the filename **Revised_budget_model**.
6. Print a copy of this spreadsheet on a single page.
7. Print a copy of this spreadsheet showing all the formulae used.

Inserting rows and columns

To insert a column between two columns already containing data, take these steps.

1. Use the mouse to position the cursor on any cell in the column to the right of where you want the column to be inserted.
2. Right-click, and from the menu that appears, click Insert.

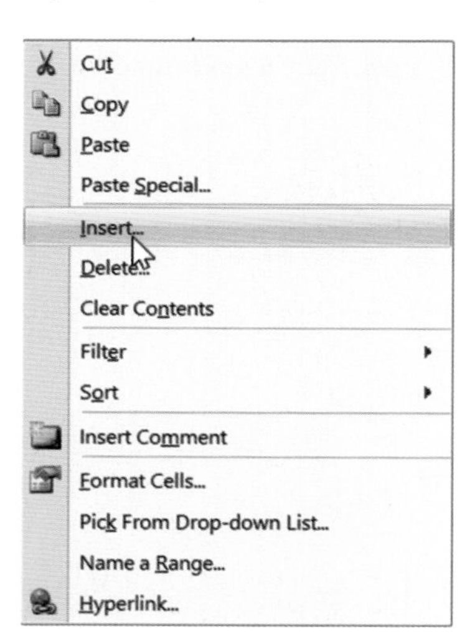

The following menu appears.

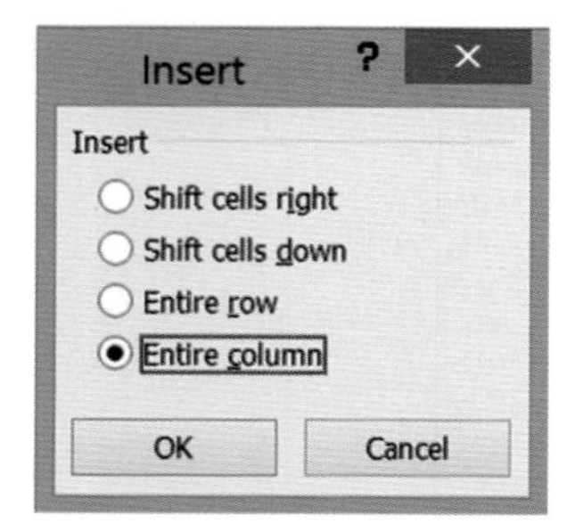

Select Entire column and click on OK.

3. Check that the column has been inserted in the correct place. Any formulae in these cells will be automatically adjusted.

To insert a row, take these steps.

1. Place the cursor on any of the cells in the row below where the new row needs to be placed and right-click the mouse button.
2. Click on Insert and select Entire row from the list and finally click on OK.
3. Check that the row has been inserted in the correct place.

Deleting a row or column

To delete a column containing data, take these steps.

1. Use the mouse to position the cursor on any cell in the column you want deleting, right-click and select Delete from the menu. The following menu appears where you select Entire column and then click on OK.

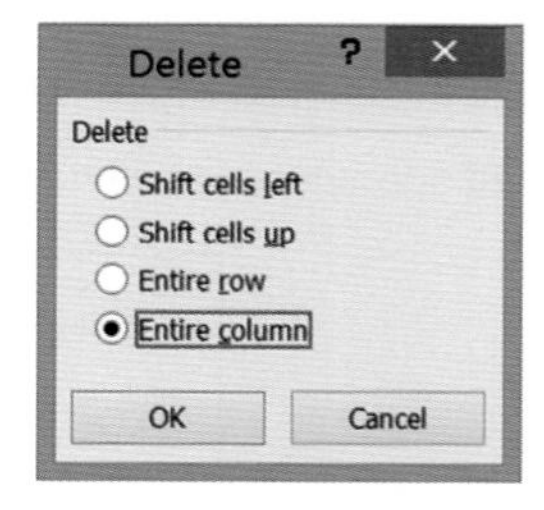

Select Entire column and click on OK.

2. Check that correct column has been deleted.

To delete a row, position the cursor on any cell in the row you want to delete and right-click and select Delete from the menu. The following menu appears where you select Entire row and then click on OK.

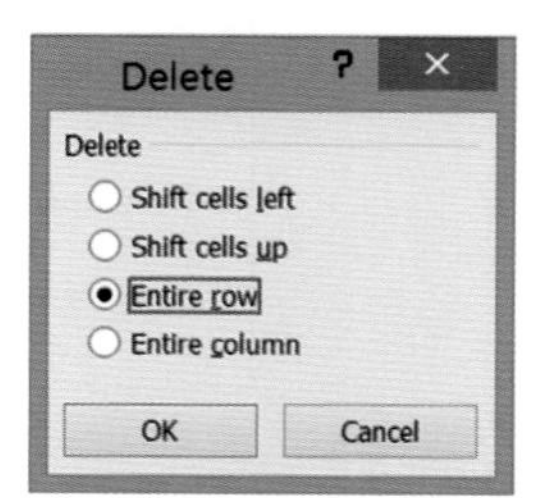

Deleting groups of cells

Select the group of cells you want to delete by left-clicking on one of the cells and dragging to select all the cells for deletion. Now right-click and select Delete and then choose from the options in the following menu.

Click on OK to confirm.

A note about deleting cells

You have to be quite careful about deleting cells because the remaining cells are shifted to fill the gap. This can cause unexpected results, so remember you can use the (undo button) or Ctrl+Z to undo an unwanted action.

If you have just a couple of cells to delete, then you are best using the backspace key (the key at the top with the arrow pointing to the left). Another way is to use the Del key when the cursor in positioned on the cell.

Absolute and relative cell referencing

There are two ways in which you can make a reference to another cell and it is important to know the difference if you want to copy or move cells. When the current cell is copied or moved to a new position, the cell to which the reference is made will also change position.

Look at the spreadsheet in Figure 1. In relative referencing (which is the normal method of referencing) cell B4 contains a relative reference to cell A1. This reference tells the spreadsheet that the cell to which it refers is 3 cells up and one cell to the left of cell B4. If cell B4 is copied to another position, say E5, then the reference will still be to the same number of cells up and to the left, so the reference will now be to cell D2.

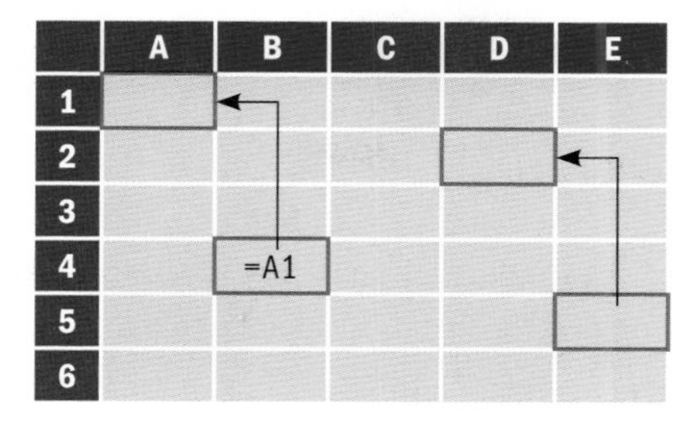

Figure 1

With absolute cell referencing, if cell B4 contains a reference to cell A1, then if the contents of B4 are copied to a new position the reference will not be adjusted and it will still refer to cell A1.

In most cases you want to use relative cell references and the spreadsheet will assume that ordinary cell references are relative cell references. Sometimes you want to refer to the same cell, even when the formula referring to the cell is copied to a new position. You therefore need to make sure that the formula contains an absolute cell reference. To do this, a dollar sign is placed in front of the column and row number.

In Figure 1, Cell A1 is a relative cell reference. To change it to an absolute cell reference we would add the dollar signs like this: $A$1 (Figure 2). An absolute reference always refers to the same cell whereas a relative reference, refers to a cell that is a certain number of rows and columns away.

Figure 2

Activity 20.4

Absolute and relative cell referencing

In this activity you will learn how to:

- identify when a relative or absolute cell reference is needed.

1 Load your spreadsheet software and enter the data as shown here:

	A	B	C	D
1	**Currency exchange**			
2				
3	Current exchange rate	£1	is equivalent to	$1.68
4				
5	Cost of:	Pounds	Dollars	
6	Airline ticket	£299		
7	Hotel	£654		
8	Transfers from airport	£134		
9				

2 The idea of this spreadsheet is to convert the cost of certain items in pounds into dollars.

In order to do this you would put a formula in cell C6 which multiplies the cost of the item in cell B6 by the conversion factor which is in cell D3.

Enter the formula in cell C6 as shown here:

	A	B	C	D
1	**Currency exchange**			
2				
3	Current exchange rate	£1	is equivalent to	$1.68
4				
5	Cost of:	Pounds	Dollars	
6	Airline ticket	£299	=D3*B6	
7	Hotel	£654		
8	Transfers from airport	£134		
9				

To save having to type in similar formulae for C7 and C8 you can simply copy the formula from C6 down the column. Try this and you will get the following result.

	A	B	C	D
1	**Currency exchange**			
2				
3	Current exchange rate	£1	is equivalent to	$1.68
4				
5	Cost of:	Pounds	Dollars	
6	Airline ticket	£299	$502.32	
7	Hotel	£654	$0.00	
8	Transfers from airport	£134	$0.00	
9				

You can see that copying the formula did not work.

3 Delete the contents of cells C7 and C8.

4 Now change the formula in cell C6 so that it includes an absolute cell reference to cell D3.

The formula to be entered is: =$D$3*B6.

The formula looks like this in the spreadsheet:

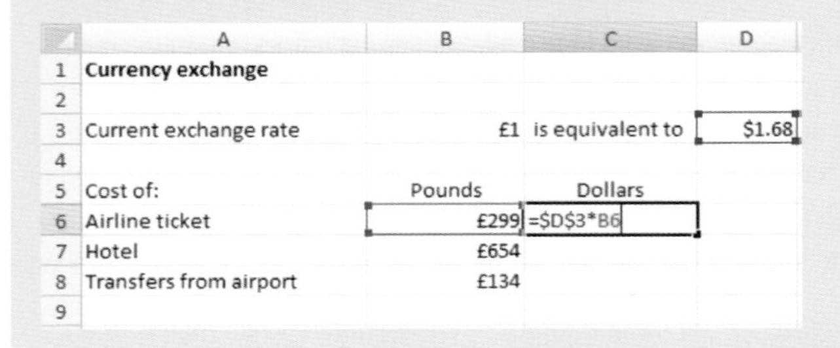

	A	B	C	D
1	**Currency exchange**			
2				
3	Current exchange rate	£1	is equivalent to	$1.68
4				
5	Cost of:	Pounds	Dollars	
6	Airline ticket	£299	=$D$3*B6	
7	Hotel	£654		
8	Transfers from airport	£134		
9				

5 As cell D3 has an absolute cell reference (i.e. it appears as $D$3 in the formula) the spreadsheet will keep referring to cell D3 even when the formula is moved or copied.

Copy the formula down the column as far as cell C8.

You will now see the formula copied to give the correct results as shown here:

	A	B	C	D
1	**Currency exchange**			
2				
3	Current exchange rate	£1	is equivalent to	$1.68
4				
5	Cost of:	Pounds	Dollars	
6	Airline ticket	£299	$502.32	
7	Hotel	£654	$1,098.72	
8	Transfers from airport	£134	$225.12	
9				

6 Save the spreadsheet using the filename **Relative_and_absolute_referencing**.

Named cells

Rather than refer to a cell by its cell address (e.g. B3) we can give it a name and then use this name in formulae or whenever else we need to refer to it. For example, if cell B3 contains an interest rate then we can use the name Interest_rate.

Activity 20.5

Naming a cell

In this activity you will learn how to:

- name a cell
- use named cells in formulae.

1 Load Excel and create a new document and input the following data exactly as appears here:

	A	B	C
1	A spreadsheet to work out interest received		
2			
3	Amount of money invested	$20,000	
4	Interest rate	5%	
5	Amount of interest received		
6			
7			

2 Click on cell B4 as this is the cell being given a name.

3 Click on Formulas in the toolbar and then choose Define Name. The following window opens:

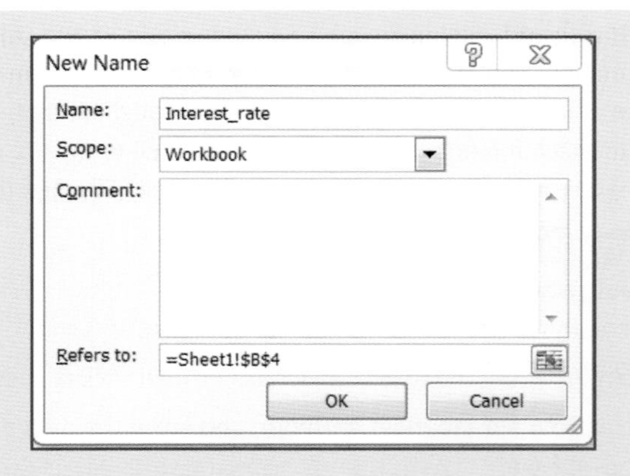

Notice that a name has been suggested by using the name in the adjacent cell. You can either keep the name suggested or change it. Here, we will keep the name so click on OK.

Important note: you cannot have blank spaces in a name for a cell. This means rather than use 'Interest rate' we use 'Interest_rate' as the name.

4 Using what you have learned from step 3. Give cell B3 the name:

Amount_of_money_invested

5 You will now create the formula for working out the interest in cell B5 using the cell names in the formula rather than the cell addresses.

Click on cell B5 and type in an equals sign (=) and click on Formulas (if the formula menus and icons are not shown) and then click on Use in Formula.

You will see that the names for the two cells are displayed. Click on Amount_of_money_invested.

You will see this is inserted into the formula.

Type in an asterisk (*) and then click on Use in Formula again and this time click on Interest_rate. These two named cells are used in the formula to calculate the interest received.

6 Your final spreadsheet should now look like this:

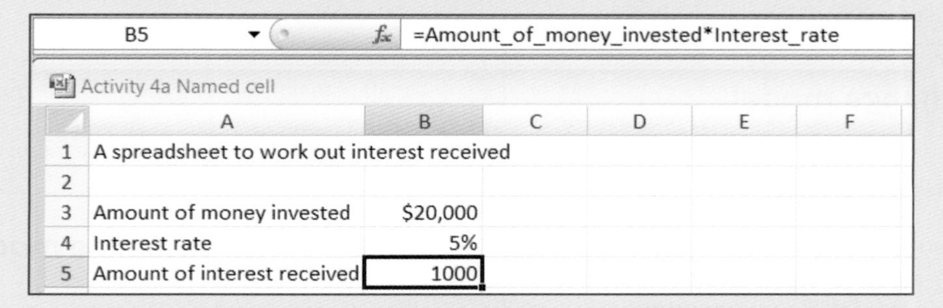

B5 =Amount_of_money_invested*Interest_rate

Activity 4a Named cell

	A	B	C	D	E	F
1	A spreadsheet to work out interest received					
2						
3	Amount of money invested	$20,000				
4	Interest rate	5%				
5	Amount of interest received	1000				

Naming cells makes it easier to understand how the formula works and makes it easier to test.

7 Cell B5 needs to be formatted to currency (dollars with no decimal places).

Do this and you have completed this simple spreadsheet.

Save this spreadsheet using the filename **Activity_5_Named_cell**.

CSV files

In this file format, which can be identified by the file extension .csv, only text and values are saved. If you save an Excel file in this file format all rows and characters are saved. The columns of data are saved separated by commas, and each row of data in the CSV file corresponds to a row in the spreadsheet. CSV files are used to enable data to be imported or exported between different applications without the need for re-typing if there is no direct way of doing this.

Here is a CSV file that has been created using a non-Microsoft spreadsheet package. It was saved in CSV file format so that it can be loaded into Microsoft Excel.

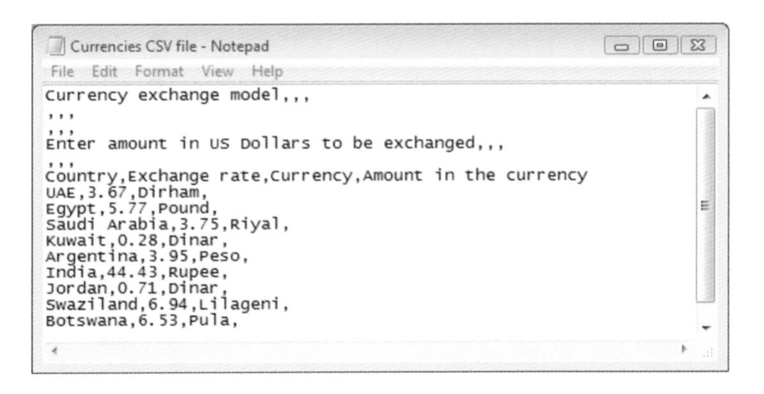

Currencies CSV file - Notepad

File Edit Format View Help

```
Currency exchange model,,,
,,,
,,,
Enter amount in US Dollars to be exchanged,,,
,,,
Country,Exchange rate,Currency,Amount in the currency
UAE,3.67,Dirham,
Egypt,5.77,Pound,
Saudi Arabia,3.75,Riyal,
Kuwait,0.28,Dinar,
Argentina,3.95,Peso,
India,44.43,Rupee,
Jordan,0.71,Dinar,
Swaziland,6.94,Lilageni,
Botswana,6.53,Pula,
```

Activity 20.6

Naming cell ranges

In this activity you will learn how to:

- import data in CSV file format
- name a cell range
- format cells
- sort data.

1 Load Excel and open the file in CSV file format called **Currencies_CSV_file**.

Check you have the following displayed on your screen:

	A	B	C	D	E	F
1	Currency exchange model					
2						
3						
4	Enter amount in US Dollars to be exchanged					
5						
6	Country	Exchange	Currency	Amount in the currency		
7	UAE	3.67	Dirham			
8	Egypt	5.77	Pound			
9	Saudi Aral	3.75	Riyal			
10	Kuwait	0.28	Dinar			
11	Argentina	3.95	Peso			
12	India	44.43	Rupee			
13	Jordan	0.71	Dinar			
14	Swaziland	6.94	Lilageni			
15	Botswana	6.53	Pula			

2 When this file was saved in csv format, the formatting was lost, so you need to tidy up this spreadsheet. Format the spreadsheet as follows:

Change the font size of the main heading (i.e. Currency exchange model) to 20 pt and make the text bold.

Make the text 'Enter amount in US Dollars to be exchanged' bold.

Widen the columns so that all the column headings and the text in the columns are shown properly.

Make all the column headings bold.

Check that your spreadsheet now looks like this:

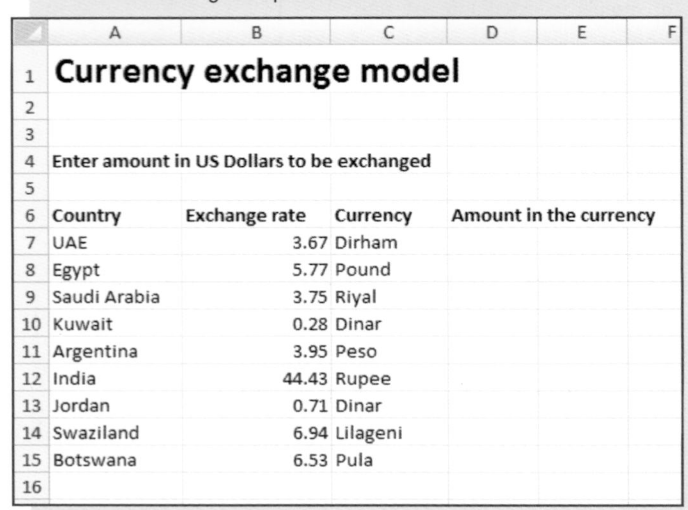

	A	B	C	D	E	F
1	**Currency exchange model**					
2						
3						
4	**Enter amount in US Dollars to be exchanged**					
5						
6	**Country**	**Exchange rate**	**Currency**	**Amount in the currency**		
7	UAE	3.67	Dirham			
8	Egypt	5.77	Pound			
9	Saudi Arabia	3.75	Riyal			
10	Kuwait	0.28	Dinar			
11	Argentina	3.95	Peso			
12	India	44.43	Rupee			
13	Jordan	0.71	Dinar			
14	Swaziland	6.94	Lilageni			
15	Botswana	6.53	Pula			
16						

3 Give cell D4 the following name: Amount_in_Dollars

Look back at the previous activity if you have forgotten how to name a cell.

4 You are now going to name a range of cells.

Click on cell B7 and drag down to cell B15 so that all the cells are highlighted.

Click on Formulas in the toolbar and then choose Define Name.

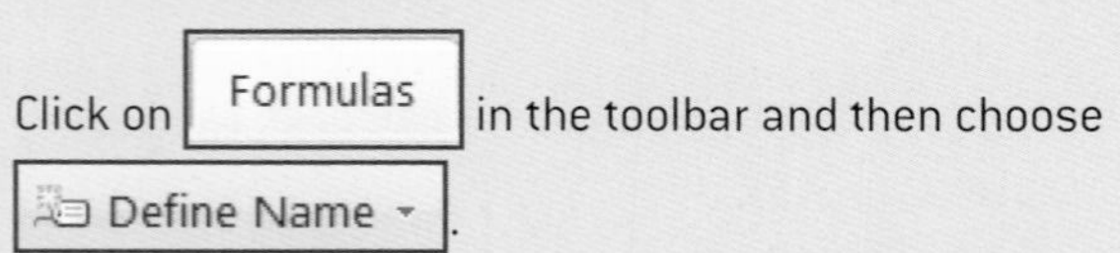

The following window opens:

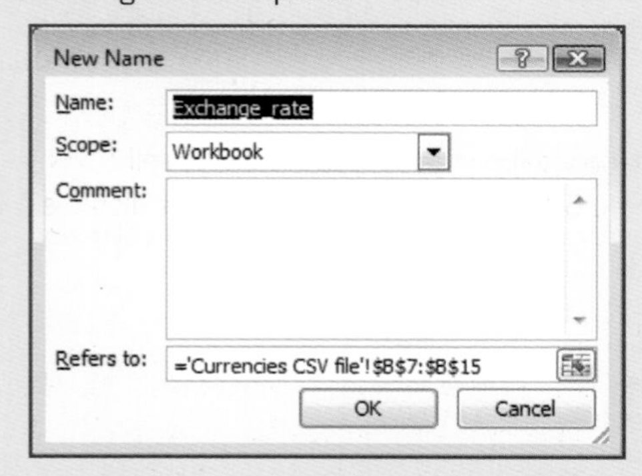

Notice that a name has been suggested and this name is suitable, so just click on OK to give the range of cells this name.

5 Now give the cells in the range D7 to D15 the name

Amount_when_converted

6 Once the cells have been named they can be used in the formulae, and doing this makes them easier to understand.

In cell D7 enter the following formula:

=Amount_in_Dollars*Exchange_rate

Notice that you do not use a relative reference for cell D4. The computer knows that this cell must have an absolute reference because the name refers to a single cell and not a range.

7 Enter the number 1 into cell D4. This is the amount in dollars to convert into the other currencies.

8 Now copy the formula in cell D4 down the column as far as cell D15.

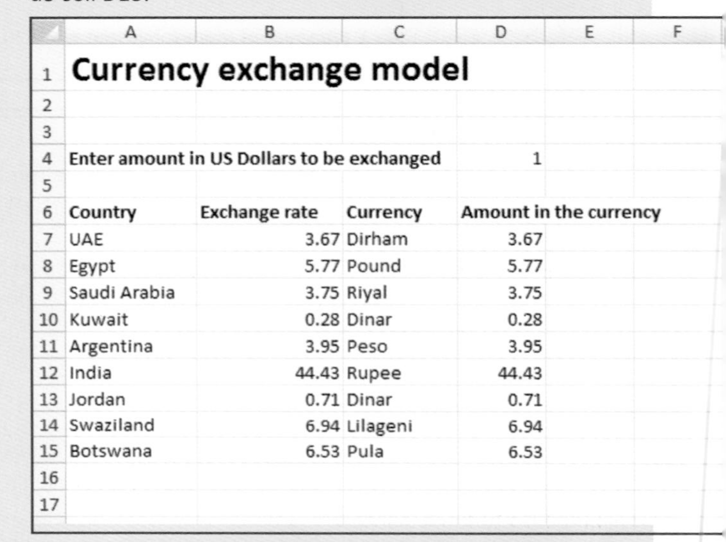

	A	B	C	D	E	F
1	**Currency exchange model**					
2						
3						
4	**Enter amount in US Dollars to be exchanged**			1		
5						
6	**Country**	**Exchange rate**	**Currency**	**Amount in the currency**		
7	UAE	3.67	Dirham	3.67		
8	Egypt	5.77	Pound	5.77		
9	Saudi Arabia	3.75	Riyal	3.75		
10	Kuwait	0.28	Dinar	0.28		
11	Argentina	3.95	Peso	3.95		
12	India	44.43	Rupee	44.43		
13	Jordan	0.71	Dinar	0.71		
14	Swaziland	6.94	Lilageni	6.94		
15	Botswana	6.53	Pula	6.53		
16						
17						

9 Change the amount in US Dollars to exchange in cell D4 to 250.

Notice that all the values now change in cells D7 to D15.

Check your spreadsheet looks like this:

	A	B	C	D	E	F
1	**Currency exchange model**					
2						
3						
4	**Enter amount in US Dollars to be exchanged**			250		
5						
6	**Country**	**Exchange rate**	**Currency**	**Amount in the currency**		
7	UAE	3.67	Dirham	917.5		
8	Egypt	5.77	Pound	1442.5		
9	Saudi Arabia	3.75	Riyal	937.5		
10	Kuwait	0.28	Dinar	70		
11	Argentina	3.95	Peso	987.5		
12	India	44.43	Rupee	11107.5		
13	Jordan	0.71	Dinar	177.5		
14	Swaziland	6.94	Lilageni	1735		
15	Botswana	6.53	Pula	1632.5		
16						
17						

⇨

10 It is easier to find data in a list if it is sorted into alphabetical order.

The countries need to be put into alphabetical order. To do this, highlight the data in cells A7 to D15.

Important note: If you just highlight the countries these will be put into order but the accompanying data will stay where it is thus jumbling the data up.

Once the data is highlighted click on Data in the toolbar and then on Sort.

Click on the ascending order (i.e. A to Z) button.

11 Save the file as an Excel workbook (i.e. not a csv file) using the filename **Currencies**.

Formulae and functions

A function is a calculation that the spreadsheet software has memorised. There are many of these functions, some of which are very specialised. A function must start with an equals sign (=) and it must have a range of cells to which it applies in brackets after it.

Sum: =SUM(E3:P3) displays the total of all the cells from cells E3 to P3 inclusive.

Average: For example, to find the average of the numbers in a range of cells from A3 to A10 you would use: =AVERAGE(A3:A10)

Maximum: =MAX(D3:J3) displays the largest number in all the cells from D3 to J3 inclusive.

Minimum: =MIN(D3:J3) displays the smallest number in all the cells from D3 to J3 inclusive.

Mode: =MODE(A3:A15) displays the mode (i.e. the most frequent number) in the cells from A3 to A15 inclusive.

Median: =MEDIAN(B2:W2) displays the median of the cells from cells B2 to W2 inclusive.

COUNT: Suppose we want to count the number of numeric entries in the range C3 to C30. We can use =COUNT(C3:C30). Any blank lines or text entries in the range will not be counted.

COUNTA: This counts the text, numbers, and blank lines. For example it could be used to calculate the number of people in a list of names like this =COUNTA(C3:C30).

ROUND

The ROUND function rounds a number correct to a number of decimal places that you specify. ROUND is used in the following way:

ROUND(number, number of digits)
where number is the number you want rounded off and number of digits is the number of decimal places.

Here are some examples:

=ROUND(3.56678,2) will return the number 3.57
=ROUND(5.43,1) will return the number 5.4

INTEGER

An integer is a positive or negative whole number or zero. Excel uses the INT function to give only the whole number part of a number. It is important to note that it simply chops off the decimal part of a number to leave just the integer part; it does not round up.

=INT(12.99) will return the integer 12
=INT(0.31022) will return the integer 0

IF

The IF function is called a logical function because it makes the decision to do one of two things based on the value it is testing. The IF function is very useful because you can use it to test a condition, and then choose between two actions based on whether the condition is true or false.

The IF function makes use of something called relational operators. You may have come across these in your mathematics lessons but it is worth going through what they mean.

Relational operators

Symbol	Meaning	Examples
=	equals	5 + 5 = 10
>	greater than	5*3 > 2*3
<	less than	-6 < -1 or 100 < 200
<>	not equal to	'Red' <> 'White' or 20/4 <> 6*4
<=	less than or equal to	'Adam' <= 'Eve'
>=	greater than or equal to	400 >= 200

Here are some examples of the use of a single IF function:

=IF(B3>=50,"Pass","Fail")

This function tests to see if the number in cell B3 is greater than or equal to 50. If the answer is true, Pass is displayed and if the answer is false, Fail is displayed.

=IF(A2>=500,A2*0.5,A2)

This tests to see if the number in cell A2 is greater than or equal to 500. If true, the number in cell A2 will be multiplied by 0.5 and the answer displayed. If false, the number in cell A2 will be displayed.

Nested functions

A nested function is a function which is inside another function.

Suppose we want to create a spreadsheet which will tell a teacher whether a student has passed, failed or needs to resit a test. To do this we first need a list of students and their marks.

A formula will need to be created that will do the following:

- If a student scores 40 marks or less then the message 'Fail' appears.
- If a student scores between (and including) 41 to 50 marks then the message 'Resit' appears.
- If a student scores 51 or more marks then the message 'Pass' appears.

Assuming that a mark to be tested is in cell B2, the following formula making use of nested functions will do this:

```
=IF(B2>=51,"Pass",IF(B2>=41,"Resit","Fail"))
```

This formula works like this: the mark in cell B2 is tested to see if it is equal to or more than 51 marks, and if it is the message 'Pass' appears in the cell where the formula is placed. If this condition is not true then the formula moves to the next IF statement which tests the mark to see if it is greater or equal to 41 in which case a 'Resit' message appears. If this is false, the message 'Fail' appears.

Activity 20.7

Nested formulae

In this activity you will learn how to:

- use nested formulae
- copy formulae relatively
- demonstrate that the model works using test data.

1 Load Excel and open the file called **Nested_functions**.

2 Check you have the following on your screen:

	A	B	C
1	**Name**	**Mark**	**Action**
2	Rachel Liu	27	
3	Rujav Singh	40	
4	Rcardo Vega	78	
5	Steven Gibbs	54	
6	Marietta Fortuni	89	
7	Vasilios Spanos	79	
8	Mustafa Karwad	25	
9	Amy Hughes	32	
10	Lesley Wong	45	
11	Amor Nanas	10	
12	Mohamed Bugalia	47	
13	Paul Wells	65	
14	Chelsea Dickinson	51	
15	Josuha Mathews	97	

3 Enter the following formula which contains nested functions into cell C2.

```
=IF(B2>=51,"Pass",IF(B2>=41,"Resit","Fail"))
```

	A	B	C	D	E	F
1	**Name**	**Mark**	**Action**			
2	Rachel Liu	27	=IF(B2>=51,"Pass",IF(B2>=41,"Resit","Fail"))			
3	Rujav Singh	40				

4 Copy this formula relatively down the column from cell C2 to C15.

5 Check that your spreadsheet looks like this:

	A	B	C
1	**Name**	**Mark**	**Action**
2	Rachel Liu	27	Fail
3	Rujav Singh	40	Fail
4	Rcardo Vega	78	Pass
5	Steven Gibbs	54	Pass
6	Marietta Fortuni	89	Pass
7	Vasilios Spanos	79	Pass
8	Mustafa Karwad	25	Fail
9	Amy Hughes	32	Fail
10	Lesley Wong	45	Resit
11	Amor Nanas	10	Fail
12	Mohamed Bugalia	47	Resit
13	Paul Wells	65	Pass
14	Chelsea Dickinson	51	Pass
15	Josuha Mathews	97	Pass

6 Save your spreadsheet using the filename **Nested_formulae**.

7 It is quite hard to construct nested formulae correctly. This means that testing is extremely important. You now have to produce a series of tests that will test that the formula being used is producing the correct results.

You need to check the border values for each message and you need to ensure that you use data which will test these. You now have to create a set of marks that will test the spreadsheet.

Enter these marks by replacing the marks in the previous spreadsheet and check that each message being produced is correct.

Save your spreadsheet using the filename **Checking_nested_formulae**.

COUNTIF

This is used to count the number of cells with data in them that meet a certain condition.

For example, it could be used to find the total number of males in a group of students by counting the number of entries of 'Male'.

Activity 20.8

Using the COUNTIF function

In this activity you will learn how to:

- use COUNTA functions
- use COUNTIF functions.

1 Load the spreadsheet file **Checking_nested_formulae** saved in Activity 14.7.

2 In cell A18 enter the text 'Number of students'.

3 You have been asked to create a formula to display the number of students.

Enter the following formula into cell C18

=COUNTA($C$2:$C$15)

Notice the absolute cell reference in this formula. This is done so that if the formula is copied it will always refer to cells C2 to C15.

4 In cell A19 enter the text 'Number of students who passed' and in cell C19 enter the formula =COUNTIF($C$2:$C$15,"Pass")

5 Enter the text for the labels as shown here:

18	Number of students
19	Number of students who passed
20	Number of students who failed
21	Number of students to resit

6 Add the formulae in cells C20 and C21 to count the number of students who failed and who need to resit. Remember that it is best to use absolute cell references in these formulae.

7 Check that your spreadsheet looks like this:

	A	B	C
1	Name	Mark	Action
2	Rachel Liu	27	Fail
3	Rujav Singh	40	Fail
4	Rcardo Vega	78	Pass
5	Steven Gibbs	54	Pass
6	Marietta Fortuni	89	Pass
7	Vasilios Spanos	79	Pass
8	Mustafa Karwad	25	Fail
9	Amy Hughes	32	Fail
10	Lesley Wong	45	Resit
11	Amor Nanas	10	Fail
12	Mohamed Bugalia	47	Resit
13	Paul Wells	65	Pass
14	Chelsea Dickinson	51	Pass
15	Josuha Mathews	97	Pass
16			
17			
18	Number of students		14
19	Number of students who passed		7
20	Number of students who failed		5
21	Number of students to resit		2

8 Save your spreadsheet using the filename **Spreadsheet_using_countif**.

Lookup functions

Lookup functions search for an item of data in a table and then extract the rest of the information relating to that item. The table of items of data is stored either in the same spreadsheet or a different one. The table can also be a completely different file.

For example, if each different item in a shop has a unique product number, a table of product details can be produced with information such as product number, product description, and product price. If a particular product number is entered, the lookup function will retrieve the other details automatically from the table.

This all sounds a bit complicated but hopefully by completing activities 20.9 and 20.10.

Activity 20.9

Using VLOOKUP

In this activity you will learn how to:

- use the VLOOKUP function
- format cells.

In this activity you will be using a function called VLOOKUP. The idea of this spreadsheet is to type in a product code, and the details corresponding to that code will be looked up by looking for a match for the product code in a vertical column.

1 Load Excel and create a new document.

2 Look carefully at the following spreadsheet and enter the details exactly as they appear here:

	A	B	C
1	**A spreadsheet using the VLOOKUP function**		
2			
3			
4	**Product number**		
5	**Description**		
6	**Price**		
7			
8			
9			
10			
11	**Product Number**	**Product description**	**Product price**
12			
13	12	Red pen	£0.25
14	13	Black pen	£0.25
15	14	Blue pen	£0.25
16	15	Ruler	£0.75
17	16	Correction fluid	£1.99
18	17	Note pad	£1.25
19	18	Stapler	£7.99
20	19	Staples	£1.20
21	20	Paper clips	£1.55
22	21	String	£1.65
23			

The idea is that the user will type a product number into cell B4, and the software will look for a match in the vertical column of the table. It will then get the product description and product price and enter them automatically into cells B5 and B6.

3 Enter the product number 15 into cell B4.

The VLOOKUP function needs to be placed in cells B5 and B6. This will use the product number in cell B4 to obtain the rest of the details about the product.

In cell B5 enter the formula =VLOOKUP(B4,A13:C22,2)

This tells the computer to search for the data vertically in the table from cells A13 to C22 until a match with the product number entered in B4 is found. The '2' at the end means that once a match is found to use the data in the second column.

In cell B6 enter the formula =VLOOKUP(B4,A13:C22,3). Notice the last number in this formula is a 3 because the data to be used is now in the third column.

Format cell B6 to currency with 2 decimal places.

Check that your spreadsheet looks like this:

	A	B	C
1	**A spreadsheet using the VLOOKUP function**		
2			
3			
4	**Product number**	15	
5	**Description**	Ruler	
6	**Price**	£0.75	
7			
8			
9			
10			
11	**Product Number**	**Product description**	**Product price**
12			
13	12	Red pen	£0.25
14	13	Black pen	£0.25
15	14	Blue pen	£0.25
16	15	Ruler	£0.75
17	16	Correction fluid	£1.99
18	17	Note pad	£1.25
19	18	Stapler	£7.99
20	19	Staples	£1.20
21	20	Paper clips	£1.55
22	21	String	£1.65

4 Test your spreadsheet thoroughly by entering different product numbers in cell B4. As part of the testing you would need to test to see what happens when product numbers that do not exist in the list are entered.

5 Save the spreadsheet using the filename **VLOOKUP**.

Activity 20.10

Using HLOOKUP

In this activity you will learn how to:

- use the HLOOKUP function.

In this exercise you will be using a function called HLOOKUP. This is used when you want the software to look at the data in the table in the horizontal direction to obtain the match.

⇨

The idea of this spreadsheet is for the user to type in pupil names in a class with their forms, and the name of the form teacher automatically appears. The code for the form is looked up horizontally in the table until a match is found.

1 Load Excel, create a new document and type in the following spreadsheet exactly as it appears here.

	A	B	C	D	E	F	G	H
1	**A spreadsheet using the HLOOKUP function**							
2								
3	**Student Name**	**Form**	**Form teacher**					
4	A Leong	11F						
5	L Rae	11P						
6	F Lam	11R						
7	N Wilkes	11T						
8	S Rousos	11S						
9	F Li	11G						
10	H Patel	11M						
11								
12	**Form**	11F	11G	11H	11J	11K	11L	11M
13	**Form teacher**	Mrs Mullen	Mr Patel	Mr Li	Mr Vega	Mrs Hughes	Dr Singh	Mrs Fortuni

Important note: the data that is used to find the list (i.e. in cells B12 to H12) must be arranged in ascending order for the HLOOKUP function to work.

2 Enter the following formula in cell C4:

=HLOOKUP(B4,$B$12:$H$13,2)

This formula works by looking at the contents of cell B4 and then comparing it with the contents in the horizontal set of data in cells B12 to H13 until it finds a match. Once the match is found the computer places the data in the second row (i.e. the form teacher) below the matched data into cell C4.

3 Copy the formula in cell C4 down the column as far as cell C10.

Your spreadsheet now contains all the form teachers for the students like this:

	A	B	C	D	E	F	G	H
1	**A spreadsheet using the HLOOKUP function**							
2								
3	**Student Name**	**Form**	**Form teacher**					
4	A Leong	11F	Mrs Mullen					
5	L Rae	11M	Mrs Fortuni					
6	F Lam	11L	Dr Singh					
7	N Wilkes	11L	Dr Singh					
8	S Rousos	11K	Mrs Hughes					
9	F Li	11G	Mr Patel					
10	H Patel	11M	Mrs Fortuni					
11								
12	**Form**	11F	11G	11H	11J	11K	11L	11M
13	**Form teacher**	Mrs Mullen	Mr Patel	Mr Li	Mr Vega	Mrs Hughes	Dr Singh	Mrs Fortuni

4 Save the spreadsheet using the filename **HLOOKUP**.

Testing the data model

Models should always be thoroughly tested to make sure that they are producing the correct results. It is easy to make a mistake when constructing a formula. Here are some things to watch out for:

- Mistakes in the formula (e.g. mistakes in working out a percentage).
- Incorrect cell references being used.
- Absolute cell references being used in a formula instead of relative cell references and vice versa.
- The wrong output being produced when data is filtered, sorted, etc.
- Not formatting the data in a cell containing a formula.

Always check any calculations performed on data using a calculator and check that all the instructions have been obeyed exactly.

When testing spreadsheets, you should have a testing strategy and these were covered in Chapter 7. The testing strategy tests the validation checks using normal, abnormal and extreme data. A table is completed containing the test number, the test data, the purpose of the test, the expected result and the actual result.

Manipulating data

Spreadsheet software such as Excel can be used to produce a flat-file database. This means that all the data is stored in a single file or spreadsheet. In order to produce a database using Excel you need to ensure that the field names appear at the top of each column and that the rows of data are directly below these. You must not leave a blank line between the field names and the data. When creating a simple flat-file database using Excel, the field names are at the top of the column and the data representing a record is a row of the spreadsheet.

Activity 20.11

Searching to select subsets of data

In this activity you will learn how to:

- perform searches on a set of data
- use Boolean operators in searches.

1 Load Excel and open the file **Database_of_employees**.

Check that the file you are using is the same as this one:

	A	B	C	D	E	F	G	H	I	J	K
1	**Forename**	**Surname**	**Sex**	**DOB**	**No of IGCSEs**	**Includes GCSE Maths**	**Includes GCSE English**	**Position**	**Salary (US $)**	**Full or part time**	**Driving licence held**
2	Yasmin	Singh	F	12/03/1992	3	Y	Y	Web designer	43000	F	Y
3	Mohamed	Bugalia	M	01/09/1987	10	Y	Y	Programmer	48000	F	Y
4	Viveta	Karunakaram	F	09/10/1978	5	Y	Y	Programmer	16500	P	Y
5	Amor	Nanas	F	08/07/1987	6	N	Y	Network manager	67000	F	Y
6	Yuvraj	Singh	M	28/02/1990	11	Y	Y	Web designer	47000	F	Y
7	Sally	Sadik	F	12/03/1967	8	Y	N	Programmer	38000	F	Y
8	Mustafa	Karwad	M	01/02/1984	4	N	Y	Systems analyst	54000	F	Y
9	Alex	Gomaz	M	30/09/1993	0	N	N	Artist	41000	F	Y
10	Bianca	Schastok	F	03/11/1980	0	N	N	Systems analyst	56500	F	Y
11	Vyoma	Pathak	F	14/12/1956	1	N	Y	Technician	41000	F	N
12	Nakul	Borade	M	22/06/1960	5	Y	Y	Web designer	45000	F	Y
13	Rachel	Liu	F	13/12/1961	7	Y	Y	Programmer	38900	F	N
14	Sho Ling	Wong	F	17/09/1978	9	Y	Y	Animator	28000	P	N
15	Chloe	Burns	F	10/02/1972	5	N	Y	Security analyst	52000	F	Y
16	Rachel	Hughes	F	16/09/1991	0	N	N	Network administrator	34200	P	N
17	Grace	Hughes	F	25/12/1965	4	N	Y	Director	78000	F	Y
18	Marzena	Jankowski	F	31/12/1955	2	N	Y	Admin clerk	32000	P	Y
19	Bishen	Singh	M	16/01/1958	0	N	N	Assistant network manager	39500	F	N
20	Raol	Ncube	M	12/01/1974	5	Y	Y	Director	87000	F	Y
21	James	Murphy	M	30/06/1964	4	N	Y	Admin clerk	26000	P	N
22	Rajan	Uppal	M	22/08/1977	5	Y	Y	Web designer	14000	P	Y
23	Hamid	Zadeh	M	03/01/1992	10	Y	Y	Trainee analyst	23000	F	Y
24	Kevin	Fortuni	M	30/09/1990	7	Y	N	Trainee analyst	26000	F	Y
25	Maria	Fortuni	F	16/06/1989	11	Y	Y	Technician	27000	F	N
26	Rupinder	Singh	M	29/05/1990	6	Y	N	Finance clerk	24900	P	N
27	Emily	Wilson	F	27/12/1989	4	Y	N	Finance clerk	37000	F	Y
28	Osama	Diad	M	03/11/1993	2	N	N	Trainee network engineer	23000	P	N
29	Ahmed	Fathy	M	23/12/1990	7	Y	Y	Network engineer	41800	F	Y
30	Hassan	Sheata	M	09/11/1989	6	Y	N	Network engineer	43000	F	Y
31	Abdullah	Nordin	M	02/01/1969	2	N	N	Web designer	42000	F	N
32	Fay	Hoy	F	09/10/1988	8	Y	Y	Receptionist	21300	P	N
33	Robert	Marley	M	17/05/1965	0	N	N	Marketing administrator	21000	P	N
34	Hassouneh	Al Sheikh	M	01/01/1990	5	Y	N	Marketing administrator	27600	F	Y
35	Samantha	Jackson	F	09/12/1970	7	Y	Y	Programmer	30500	F	Y
36	Mia	Hamm	M	06/11/1978	2	N	Y	Finance clerk	12000	P	N

Notice the column names which act as the field names and the rows of data which are the records.

⇨

2 You are now going to search through this set of data and find a subset of it that satisfies certain criteria.

You need to select the data you want to use. In this case you are going to use all the data so click on at the top left of the spreadsheet as this is a quick way of selecting the entire set of data.

Check that you have the Home tab Home selected and if not, click on it.

Click on Sort & Filter in the toolbar and then select Filter. You will notice drop-down arrows appear on each field name.

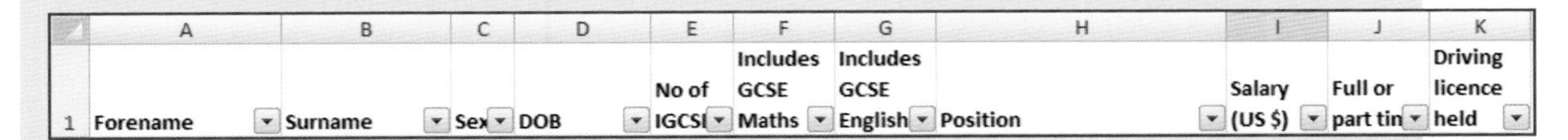

3 Click on the drop-down arrow for Surname and then select Text Filters and finally Equals:

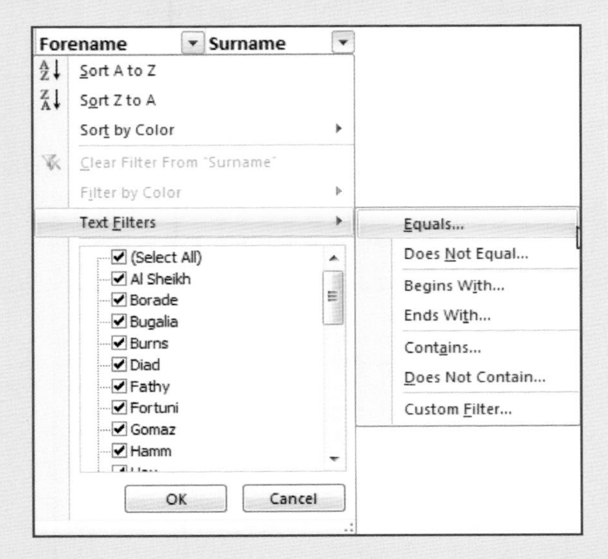

The following window appears where you view only those records that correspond to a certain Surname:

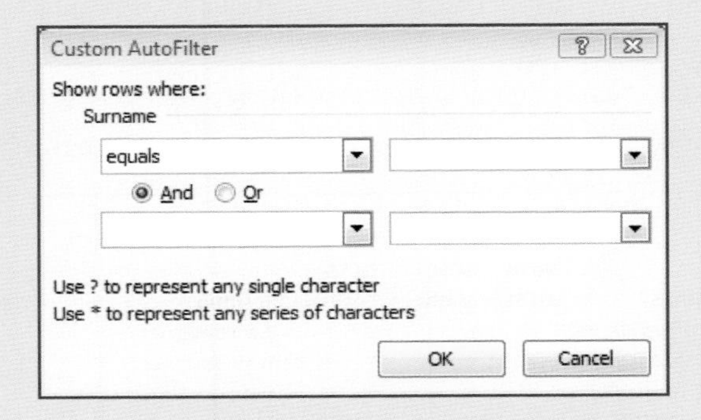

Enter the surname Singh into the box like this and then click on OK:

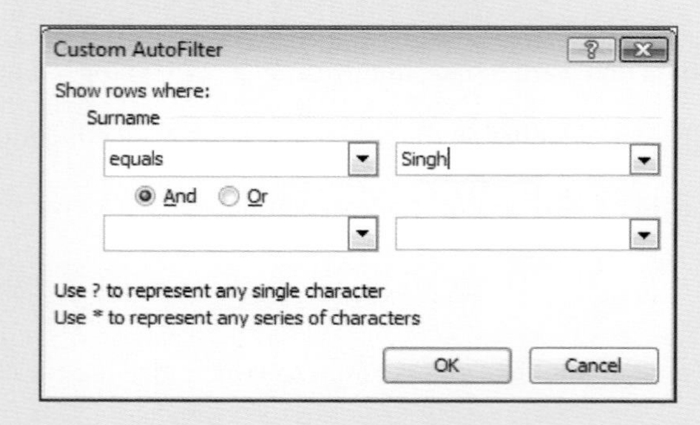

Notice that only the records having the Surname equal to Singh are shown like this:

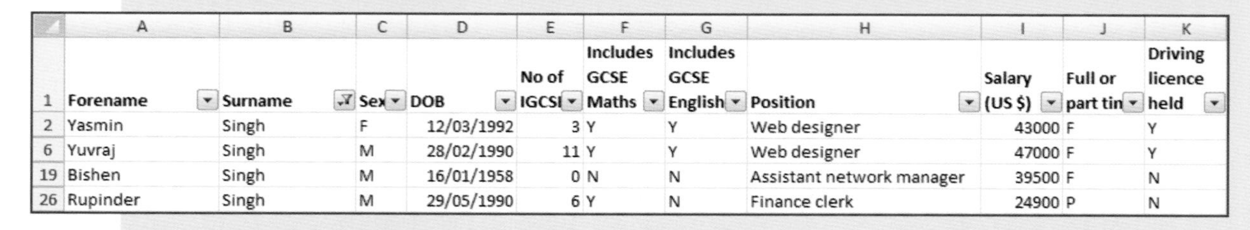

	A	B	C	D	E	F	G	H	I	J	K
1	Forename	Surname	Sex	DOB	No of IGCSE	Includes GCSE Maths	Includes GCSE English	Position	Salary (US $)	Full or part time	Driving licence held
2	Yasmin	Singh	F	12/03/1992	3	Y	Y	Web designer	43000	F	Y
6	Yuvraj	Singh	M	28/02/1990	11	Y	Y	Web designer	47000	F	Y
19	Bishen	Singh	M	16/01/1958	0	N	N	Assistant network manager	39500	F	N
26	Rupinder	Singh	M	29/05/1990	6	Y	N	Finance clerk	24900	P	N

4 You are now going to perform a search using multiple criteria.

You need to get back to the complete set of data and to do this you click on the undo button .

In this filter you will be displaying only employees having the surname Singh or the surname Hughes.

Click on the drop-down arrow for Surname and then select Text Filters and finally Equals.

Now enter the data as shown:

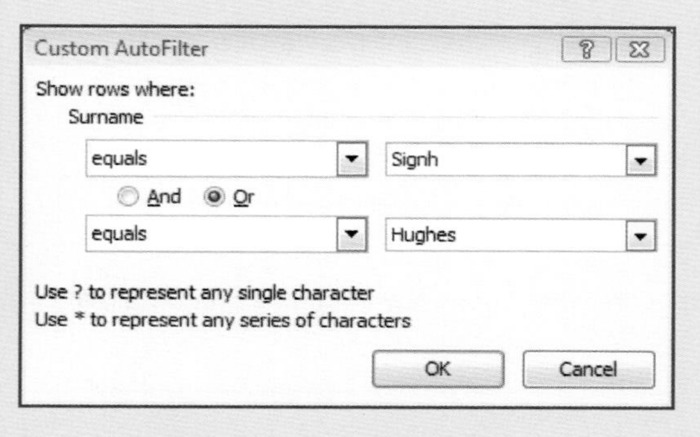

This selects all those records having the surname of either Singh or Hughes.

Click on OK and the data which meets the criteria will be displayed like this:

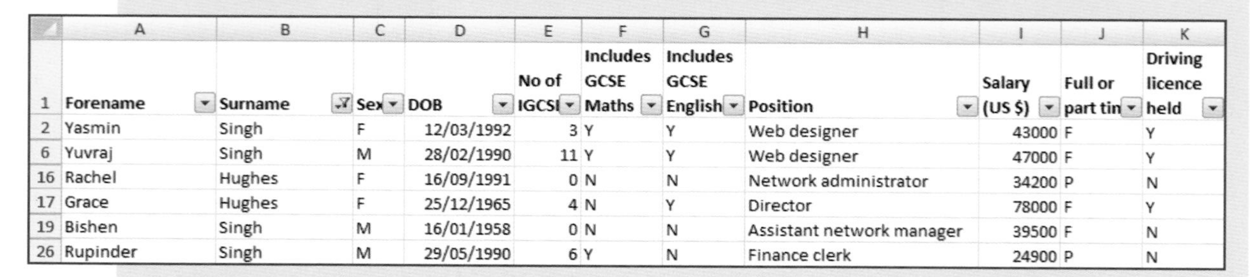

	A	B	C	D	E	F	G	H	I	J	K
1	Forename	Surname	Sex	DOB	No of IGCSE	Includes GCSE Maths	Includes GCSE English	Position	Salary (US $)	Full or part time	Driving licence held
2	Yasmin	Singh	F	12/03/1992	3	Y	Y	Web designer	43000	F	Y
6	Yuvraj	Singh	M	28/02/1990	11	Y	Y	Web designer	47000	F	Y
16	Rachel	Hughes	F	16/09/1991	0	N	N	Network administrator	34200	P	N
17	Grace	Hughes	F	25/12/1965	4	N	Y	Director	78000	F	Y
19	Bishen	Singh	M	16/01/1958	0	N	N	Assistant network manager	39500	F	N
26	Rupinder	Singh	M	29/05/1990	6	Y	N	Finance clerk	24900	P	N

5 Go back to the original set of data and ensure that all the data is selected.

Check you have selected all 36 records.

Click on the drop-down arrow for Salary, select Number Filters, and then select Greater Than Or Equal To, like this:

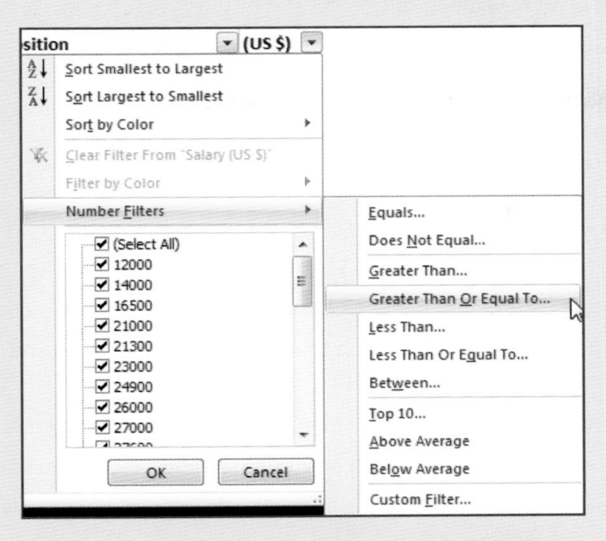

We want to display all the details for those employees earning $40000 or greater so enter the details as shown into the window:

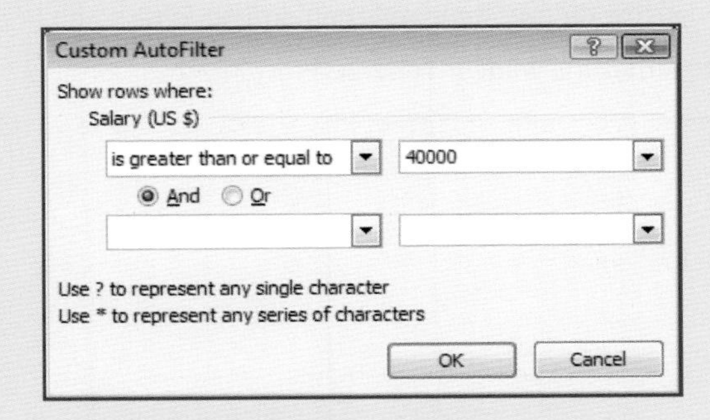

Click on OK and the filtered data will appear like this:

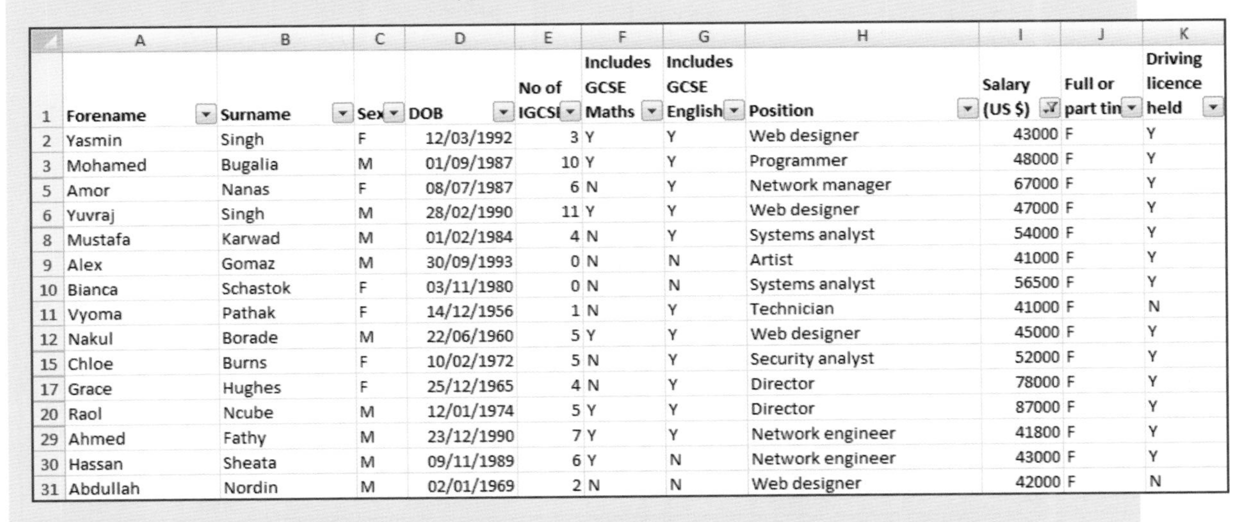

	A	B	C	D	E	F	G	H	I	J	K
1	Forename	Surname	Sex	DOB	No of IGCSE	Includes GCSE Maths	Includes GCSE English	Position	Salary (US $)	Full or part tim	Driving licence held
2	Yasmin	Singh	F	12/03/1992	3	Y	Y	Web designer	43000	F	Y
3	Mohamed	Bugalia	M	01/09/1987	10	Y	Y	Programmer	48000	F	Y
5	Amor	Nanas	F	08/07/1987	6	N	Y	Network manager	67000	F	Y
6	Yuvraj	Singh	M	28/02/1990	11	Y	Y	Web designer	47000	F	Y
8	Mustafa	Karwad	M	01/02/1984	4	N	Y	Systems analyst	54000	F	Y
9	Alex	Gomaz	M	30/09/1993	0	N	N	Artist	41000	F	Y
10	Bianca	Schastok	F	03/11/1980	0	N	N	Systems analyst	56500	F	Y
11	Vyoma	Pathak	F	14/12/1956	1	N	Y	Technician	41000	F	N
12	Nakul	Borade	M	22/06/1960	5	Y	Y	Web designer	45000	F	Y
15	Chloe	Burns	F	10/02/1972	5	N	Y	Security analyst	52000	F	Y
17	Grace	Hughes	F	25/12/1965	4	N	Y	Director	78000	F	Y
20	Raol	Ncube	M	12/01/1974	5	Y	Y	Director	87000	F	Y
29	Ahmed	Fathy	M	23/12/1990	7	Y	Y	Network engineer	41800	F	Y
30	Hassan	Sheata	M	09/11/1989	6	Y	N	Network engineer	43000	F	Y
31	Abdullah	Nordin	M	02/01/1969	2	N	N	Web designer	42000	F	N

6 Print out a copy of the filtered data, putting the data onto a single page and using landscape orientation.

7 Close the spreadsheet without saving.

Activity 20.12

Searching using wildcards

In this activity you will learn how to:

- use wildcards in searches.

1 Load Excel and open the file **Database_of_employees**.

Check that the dataset has 36 records.

2 Select the entire set of data.

3 Click on 

in the toolbar and then select Filter.

4 Suppose you want to find the subset of data that contains only those surnames having three characters. To do this you use wildcards.

Wildcards are characters that are used as substitutes for other characters in a search.

For example, if you want all surnames with exactly three characters, you use the wildcards ???

If you wanted all the surnames starting with the letter A you could use the wildcard like this A*

The difference between the wildcards ? and * are that with ? only a single character is shown but with * any number of characters are shown.

Click on the drop-down arrow for surname and then on Text Filters and finally on Equals.

In the box that appears, enter the wildcard characters ???:

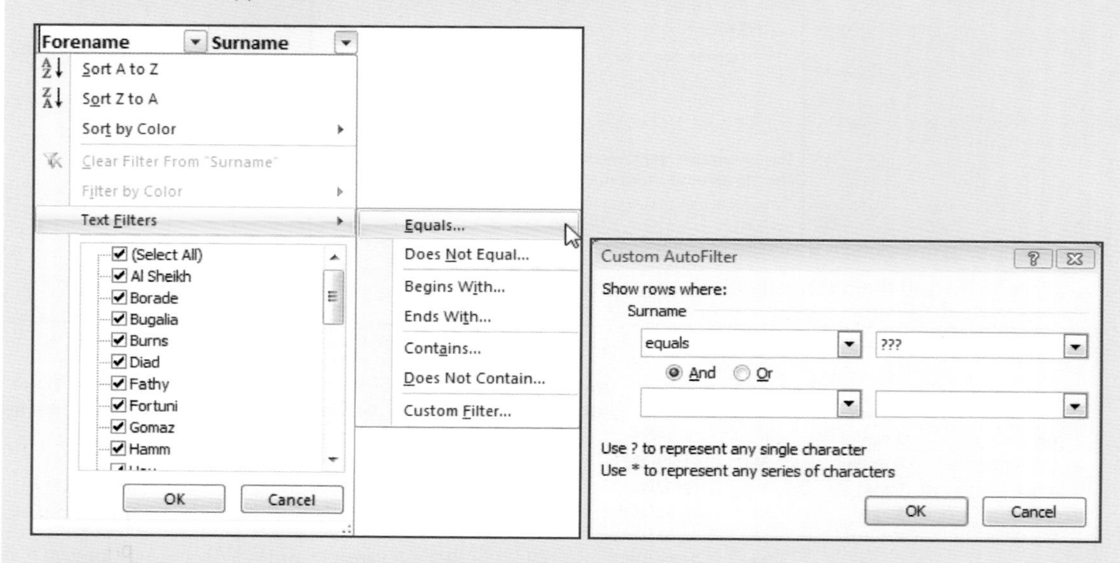

This will pick out the surnames having only three characters.

Click on OK to display the data:

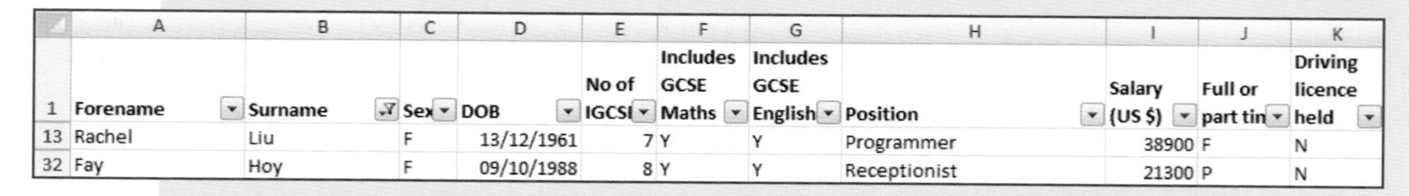

	A	B	C	D	E	F	G	H	I	J	K
1	Forename	Surname	Sex	DOB	No of IGCSEs	Includes GCSE Maths	Includes GCSE English	Position	Salary (US $)	Full or part time	Driving licence held
13	Rachel	Liu	F	13/12/1961	7	Y	Y	Programmer	38900	F	N
32	Fay	Hoy	F	09/10/1988	8	Y	Y	Receptionist	21300	P	N

5 To remove the filter and get back to the original data you can use Undo but here you will do this another way.

Click on **Surname**. Notice the filter symbol showing that data in this column has been filtered.

Click on Clear Filter From Surname.

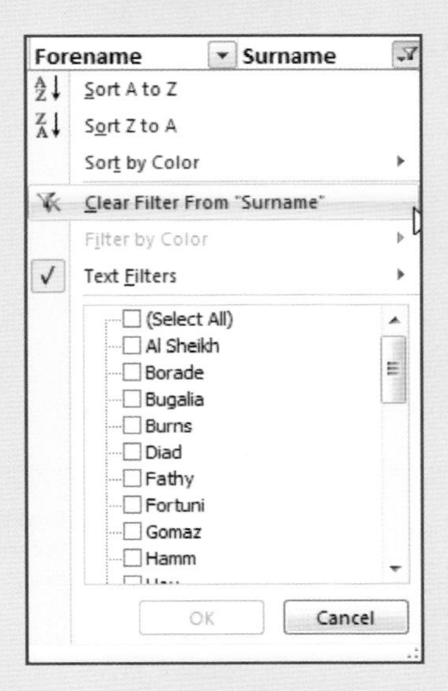

Check that the filter has been removed and you have 36 records on the screen.

6 Close without saving.

More about filters

There are lots of ways of filtering data.

This filter shows all the data for employees with surnames starting with the letter 'D'

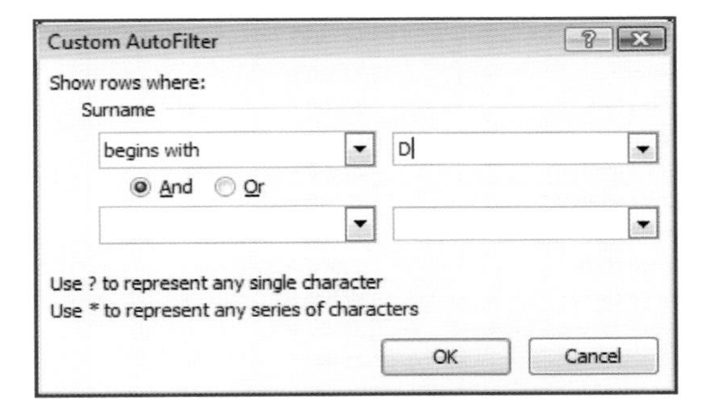

This shows all the details for employees who are either Web designers or Programmers.

A word about wildcards

To explain how wildcards can be used we will look at how they apply to the field Surname.

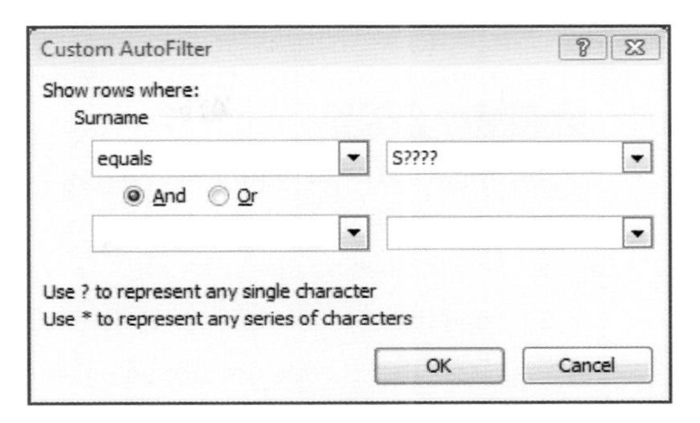

This will search for all surnames starting with the letter 'S' followed by any 4 characters. Thus, Smith would be included in the search, but not Simpson.

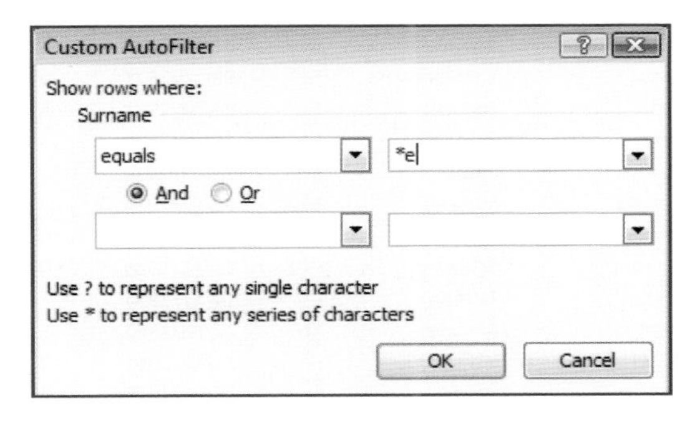

This will search for all surnames which have any combinations of characters providing they end with an 'e'.

Activity 20.13

Searching for subsets of data in the employee database

In this activity you will learn how to:

- perform searches using a single criterion and using multiple criteria
- perform searches using wildcards.

In this activity you need to load Excel and the file called **Database_of_employees** and then perform the following searches by filtering the data. For each search you are asked for, you should produce a printout in landscape orientation and fitted to a single page.

Here are the searches:

1 A list of details for employees who are female.
2 A list of details for employees who earn less than $30 000.
3 A list of all the part-time employees.
4 A list of all the employee details with surnames starting with the letter H.
5 A list of all the employee details with surnames that end with the letter y.
6 A list of all employees who earn less than $50 000 but more than $35 000.
7 A list of all the employees who are either Network engineers or Technicians.

Sorting data

When sorting data you have to select all the data and not simply the field you are performing the sort on. If you do not do this the whole set of data will become jumbled up.

Activity 20.14

Sorting a set of data into ascending or descending order

In this activity you will learn how to:

- sort data using one criterion.

1 Load the spreadsheet software Excel and open the file **Database_of_employees**.

Check that the file you are using is the same as this one:

	A	B	C	D	E	F	G	H	I	J	K
1	**Forename**	**Surname**	**Sex**	**DOB**	**No of IGCSEs**	**Includes GCSE Maths**	**Includes GCSE English**	**Position**	**Salary (US $)**	**Full or part time**	**Driving licence held**
2	Yasmin	Singh	F	12/03/1992	3	Y	Y	Web designer	43000	F	Y
3	Mohamed	Bugalia	M	01/09/1987	10	Y	Y	Programmer	48000	F	Y
4	Viveta	Karunakaram	F	09/10/1978	5	Y	Y	Programmer	16500	P	Y
5	Amor	Nanas	F	08/07/1987	6	N	Y	Network manager	67000	F	Y
6	Yuvraj	Singh	M	28/02/1990	11	Y	Y	Web designer	47000	F	Y
7	Sally	Sadik	F	12/03/1967	8	Y	N	Programmer	38000	F	Y
8	Mustafa	Karwad	M	01/02/1984	4	N	Y	Systems analyst	54000	F	Y
9	Alex	Gomaz	M	30/09/1993	0	N	N	Artist	41000	F	Y
10	Bianca	Schastok	F	03/11/1980	0	N	N	Systems analyst	56500	F	Y
11	Vyoma	Pathak	F	14/12/1956	1	N	Y	Technician	41000	F	N
12	Nakul	Borade	M	22/06/1960	5	Y	Y	Web designer	45000	F	Y
13	Rachel	Liu	F	13/12/1961	7	Y	Y	Programmer	38900	F	N
14	Sho Ling	Wong	F	17/09/1978	9	Y	Y	Animator	28000	P	N
15	Chloe	Burns	F	10/02/1972	5	N	Y	Security analyst	52000	F	Y
16	Rachel	Hughes	F	16/09/1991	0	N	N	Network administrator	34200	P	N
17	Grace	Hughes	F	25/12/1965	4	N	Y	Director	78000	F	Y
18	Marzena	Jankowski	F	31/12/1955	2	N	Y	Admin clerk	32000	P	Y
19	Bishen	Singh	M	16/01/1958	0	N	N	Assistant network manager	39500	F	N
20	Raol	Ncube	M	12/01/1974	5	Y	Y	Director	87000	F	Y
21	James	Murphy	M	30/06/1964	4	N	Y	Admin clerk	26000	P	N
22	Rajan	Uppal	M	22/08/1977	5	Y	Y	Web designer	14000	P	Y
23	Hamid	Zadeh	M	03/01/1992	10	Y	Y	Trainee analyst	23000	F	Y
24	Kevin	Fortuni	M	30/09/1990	7	Y	N	Trainee analyst	26000	F	Y
25	Maria	Fortuni	F	16/06/1989	11	Y	Y	Technician	27000	F	N
26	Rupinder	Singh	M	29/05/1990	6	Y	N	Finance clerk	24900	P	N
27	Emily	Wilson	F	27/12/1989	4	Y	N	Finance clerk	37000	F	Y
28	Osama	Diad	M	03/11/1993	2	N	N	Trainee network engineer	23000	P	N
29	Ahmed	Fathy	M	23/12/1990	7	Y	Y	Network engineer	41800	F	Y
30	Hassan	Sheata	M	09/11/1989	6	Y	N	Network engineer	43000	F	Y
31	Abdullah	Nordin	M	02/01/1969	2	N	N	Web designer	42000	F	N
32	Fay	Hoy	F	09/10/1988	8	Y	Y	Receptionist	21300	P	N
33	Robert	Marley	M	17/05/1965	0	N	N	Marketing administrator	21000	P	N
34	Hassouneh	Al Sheikh	M	01/01/1990	5	Y	N	Marketing administrator	27600	F	Y
35	Samantha	Jackson	F	09/12/1970	7	Y	Y	Programmer	30500	F	Y
36	Mia	Hamm	M	06/11/1978	2	N	Y	Finance clerk	12000	P	N

2 Click on [ ] at the top left of the spreadsheet to select the entire set of data.

3 Click on 

and select Custom sort from the drop-down menu and this Sort box appears:

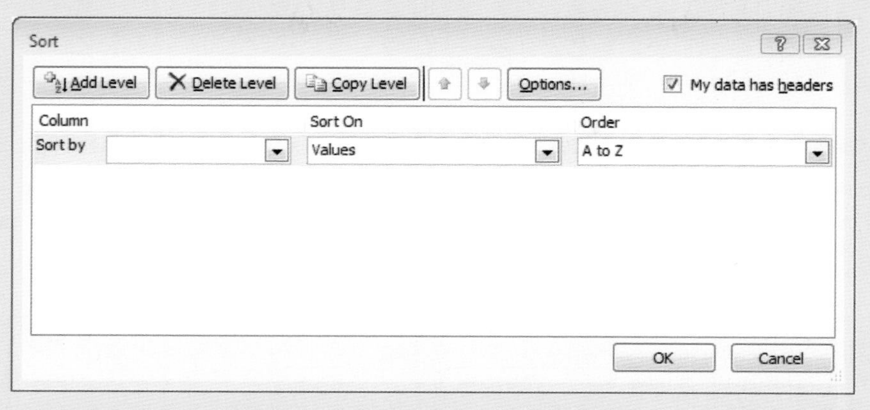

4 Click on the drop-down arrow for Sort by and select the field Surname from the list of fields. Notice that we will be sorting the data in ascending order (A to Z) according to surname:

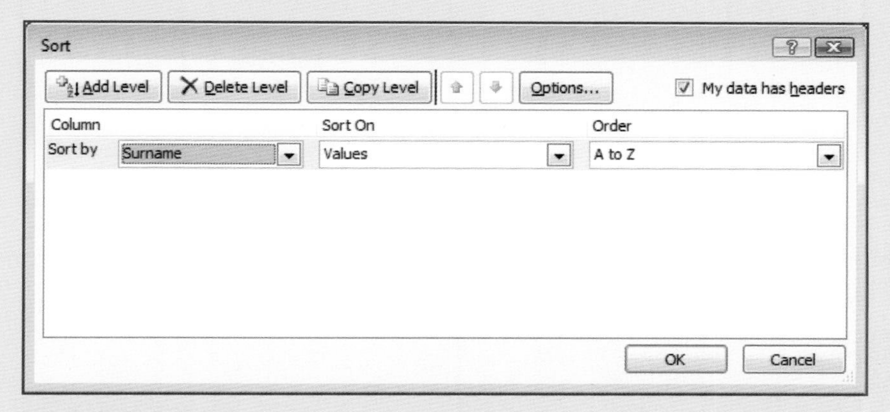

Click on OK to complete the sort.

5 Check that the entire set of data is still selected and sort the data into descending order of Surname. You do this by changing the order of the sort.

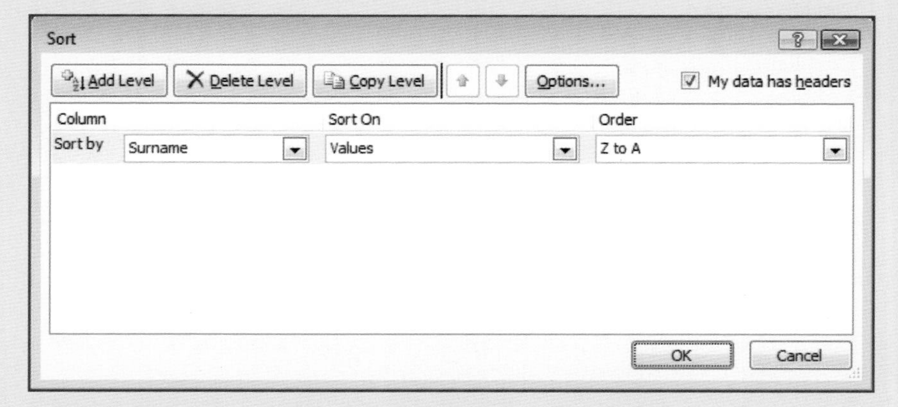

6 Close the file without saving as this will keep the file in its original order.

Activity 20.15

Sorting data using more than one criterion

In this activity you will learn how to:

- sort data using more than one criterion.

1 Load Excel and open the file **Database_of_employees**.

2 Select the entire set of data and click on Sort & Filter and select Custom sort from the drop-down menu.

You are now going to sort the data into alphabetical order according to position and also for those people in the same position, sort these into numerical order of Salary with the greatest salary first.

Enter the details as shown:

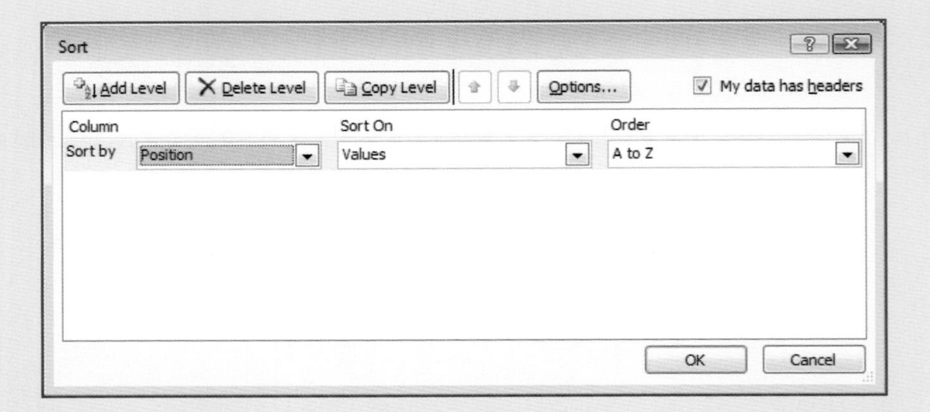

Click on Add Level and in the Then by box select the field Salary and finally in the right-hand box select Largest to smallest.

Check your settings are now the same as this:

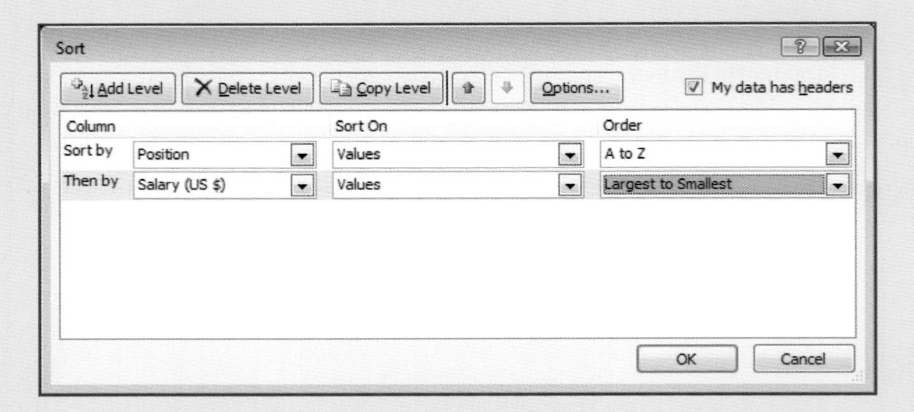

Click on OK to perform the sort.

Check your results are the same as this:

	A	B	C	D	E	F	G	H	I	J	K
1	Forename	Surname	Sex	DOB	No of IGCSEs	Includes GCSE Maths	Includes GCSE English	Position	Salary (US $)	Full or part time	Driving licence held
2	Marzena	Jankowski	F	31/12/1955	2	N	Y	Admin clerk	32000	P	Y
3	James	Murphy	M	30/06/1964	4	N	Y	Admin clerk	26000	P	N
4	Sho Ling	Wong	F	17/09/1978	9	Y	Y	Animator	28000	P	N
5	Alex	Gomaz	M	30/09/1993	0	N	N	Artist	41000	F	Y
6	Bishen	Singh	M	16/01/1958	0	N	N	Assistant network manager	39500	F	N
7	Raol	Ncube	M	12/01/1974	5	Y	Y	Director	87000	F	Y
8	Grace	Hughes	F	25/12/1965	4	N	Y	Director	78000	F	Y
9	Emily	Wilson	F	27/12/1989	4	Y	N	Finance clerk	37000	F	Y
10	Rupinder	Singh	M	29/05/1990	6	Y	N	Finance clerk	24900	P	N
11	Mia	Hamm	M	06/11/1978	2	N	Y	Finance clerk	12000	P	N
12	Hassouneh	Al Sheikh	M	01/01/1990	5	Y	N	Marketing administrator	27600	F	Y
13	Robert	Marley	M	17/05/1965	0	N	N	Marketing administrator	21000	P	N
14	Rachel	Hughes	F	16/09/1991	0	N	N	Network administrator	34200	P	N
15	Hassan	Sheata	M	09/11/1989	6	Y	N	Network engineer	43000	F	Y
16	Ahmed	Fathy	M	23/12/1990	7	Y	Y	Network engineer	41800	F	Y
17	Amor	Nanas	F	08/07/1987	6	N	Y	Network manager	67000	F	Y
18	Mohamed	Bugalia	M	01/09/1987	10	Y	Y	Programmer	48000	F	Y
19	Rachel	Liu	F	13/12/1961	7	Y	Y	Programmer	38900	F	N
20	Sally	Sadik	F	12/03/1967	8	Y	N	Programmer	38000	F	Y
21	Samantha	Jackson	F	09/12/1970	7	Y	Y	Programmer	30500	F	Y
22	Viveta	Karunakaram	F	09/10/1978	5	Y	Y	Programmer	16500	P	Y
23	Fay	Hoy	F	09/10/1988	8	Y	Y	Receptionist	21300	P	N
24	Chloe	Burns	F	10/02/1972	5	N	Y	Security analyst	52000	F	Y
25	Bianca	Schastok	F	03/11/1980	0	N	N	Systems analyst	56500	F	Y
26	Mustafa	Karwad	M	01/02/1984	4	N	Y	Systems analyst	54000	F	Y
27	Vyoma	Pathak	F	14/12/1956	1	N	Y	Technician	41000	F	N
28	Maria	Fortuni	F	16/06/1989	11	Y	Y	Technician	27000	F	N
29	Kevin	Fortuni	M	30/09/1990	7	Y	N	Trainee analyst	26000	F	Y
30	Hamid	Zadeh	M	03/01/1992	10	Y	Y	Trainee analyst	23000	F	Y
31	Osama	Diad	M	03/11/1993	2	N	N	Trainee network engineer	23000	P	N
32	Yuvraj	Singh	M	28/02/1990	11	Y	Y	Web designer	47000	F	Y
33	Nakul	Borade	M	22/06/1960	5	Y	Y	Web designer	45000	F	Y
34	Yasmin	Singh	F	12/03/1992	3	Y	Y	Web designer	43000	F	Y
35	Abdullah	Nordin	M	02/01/1969	2	N	N	Web designer	42000	F	N
36	Rajan	Uppal	M	22/08/1977	5	Y	Y	Web designer	14000	P	Y

3 You now have to follow a series of similar steps to produce the following sort on two criteria:

Sort into numerical order according to salary with the largest first and then into alphabetical order according to surname starting with the letter A first.

4 Produce a printout on a single sheet of paper in landscape orientation.

5 Close the spreadsheet file without saving.

Activity 20.16

Further sorting into two criteria

In this activity you will learn how to:

- sort data
- print output according to instructions.

1 Load the spreadsheet **Database_of_employees**.

2 Sort the entire set of data into Position order according to alphabetical order (A to Z) and then sort into date of birth order with the oldest employees first.

3 Save and print a copy of the sorted data on a single page in landscape orientation.

4 Close the spreadsheet without saving.

Headers and footers

A header is an area between the very top of the page and the top margin. A footer is the area between the very bottom of the page and the bottom margin. Once you tell the spreadsheet you want to use headers and footers then you can insert text into one or both of these areas.

Here are some types of information commonly put into the headers and footers:

- Page numbers
- Today's date
- The title
- A company logo (it can be a graphic image)
- The author's name
- The filename of the file that is used to hold the document.

Activity 20.17

Creating a header and footer for the spreadsheet for the database of employees

In this activity you will learn how to:

- add headers and footers to a spreadsheet
- insert information such as date, page numbers, etc., into headers and footers.

1 Load Excel and open the file **Database_of_employees**.

2 Click on Insert and then on Header & Footer.

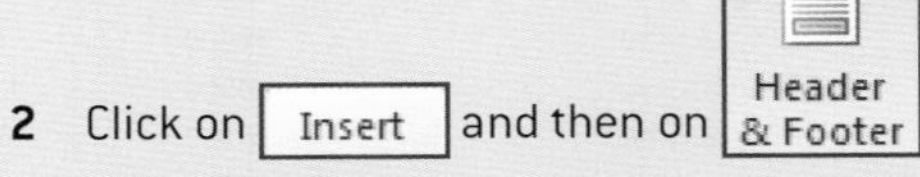

3 You will see an area at the top of the spreadsheet document like this:

Click to add header

Click on this area and a text box appears into which you can enter the details for the header. Enter the text shown here:

Database of employees last updated by S Doyle

With the cursor positioned on the line below this text line, click on Current Date.

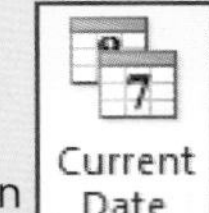

The details in the header will now look the same as this:

Database of employees last updated by S Doyle
&[Date]

4 You are now going to add the footer.

Click on Footer in the toolbar and a menu appears from which you need to click on Page 1 of ?

This will appear in the footer like this:

Page 1 of 1

5 You are now going to see how the header and footer appear if the data was to be printed.

Click Print Preview.

The spreadsheet containing the header and footer is now shown:

Database of employees last updated by S Doyle
21/09/2011

Forename	Surname	Sex	DOB	No of IGCSEs	Includes GCSE Maths	Includes GCSE English	Position	Salary (US $)	Full or part time	Driving licence held
Yasmin	Singh	F	12/03/1992	3	Y	Y	Web designer	43000	F	Y
Mohamed	Bugalia	M	01/09/1987	10	Y	Y	Programmer	48000	F	Y
Viveta	Karunakaram	F	09/10/1978	5	Y	Y	Programmer	16500	P	Y
Amor	Nanas	F	08/07/1987	6	N	Y	Network manager	67000	F	Y
Yuvraj	Singh	M	28/02/1990	11	Y	Y	Web designer	47000	F	Y
Sally	Sadik	F	12/03/1967	8	Y	N	Programmer	38000	F	Y
Mustafa	Karwad	M	01/02/1984	4	N	Y	Systems analyst	54000	F	Y
Alex	Gomaz	M	30/09/1993	0	N	N	Artist	41000	F	Y
Bianca	Schastok	F	03/11/1980	0	N	N	Systems analyst	56500	F	Y
Vyoma	Pathak	F	14/12/1956	1	N	Y	Technician	41000	F	N
Nakul	Borade	M	22/06/1960	5	Y	Y	Web designer	45000	F	Y
Rachel	Liu	F	13/12/1961	7	Y	Y	Programmer	38900	F	N
Sho Ling	Wong	F	17/09/1978	9	Y	Y	Animator	28000	P	N
Chloe	Burns	F	10/02/1972	5	N	Y	Security analyst	52000	F	Y
Rachel	Hughes	F	16/09/1991	0	N	N	Network administrator	34200	P	N
Grace	Hughes	F	25/12/1965	4	N	Y	Director	78000	F	Y
Marzena	Jankowski	F	31/12/1955	2	N	Y	Admin clerk	32000	P	Y
Bishen	Singh	M	16/01/1958	0	N	N	Assistant network manager	39500	F	N
Raol	Ncube	M	12/01/1974	5	Y	Y	Director	87000	F	Y
James	Murphy	M	30/06/1964	4	N	Y	Admin clerk	26000	P	N
Rajan	Uppal	M	22/08/1977	5	Y	Y	Web designer	14000	P	Y
Hamid	Zadeh	M	03/01/1992	10	Y	Y	Trainee analyst	23000	F	Y
Kevin	Fortuni	M	30/09/1990	7	Y	N	Trainee analyst	26000	F	Y
Maria	Fortuni	F	16/06/1989	11	Y	Y	Technician	27000	F	N
Rupinder	Singh	M	29/05/1990	6	Y	N	Finance clerk	24900	P	N
Emily	Wilson	F	27/12/1989	4	Y	N	Finance clerk	37000	F	Y
Osama	Diad	M	03/11/1993	2	N	N	Trainee network engineer	23000	P	N
Ahmed	Fathy	M	23/12/1990	7	Y	Y	Network engineer	41800	F	Y
Hassan	Sheata	M	09/11/1989	6	Y	N	Network engineer	43000	F	Y
Abdullah	Nordin	M	02/01/1969	2	N	N	Web designer	42000	F	N
Fay	Hoy	F	09/10/1988	8	Y	Y	Receptionist	21300	P	N
Robert	Marley	M	17/05/1965	0	N	N	Marketing administrator	21000	P	N
Hassouneh	Al Sheikh	M	01/01/1990	5	Y	N	Marketing administrator	27600	F	Y
Samantha	Jackson	F	09/12/1970	7	Y	Y	Programmer	30500	F	Y
Mia	Hamm	M	06/11/1978	2	N	Y	Finance clerk	12000	P	N

Page 1 of 1

6 Save the spreadsheet using the filename **Database_of_employees_with_header_and_footer**.

Presenting data

In this section you will learn about altering the appearance of the spreadsheet by making use of different fonts and font sizes, the use of styles such as bold, underline, etc., the use of colour, aligning data in cells and more. You will be given precise instructions in the examination as to how to present the data in the spreadsheet.

Adjusting the display features in a spreadsheet

The ways of altering the appearance of text in a spreadsheet can be accessed using the Home tab. In order to alter the data in the spreadsheet you click on the cell or highlight the data to have its appearance changed if it runs over more than one cell and then click on one or more of the icons in the font section.

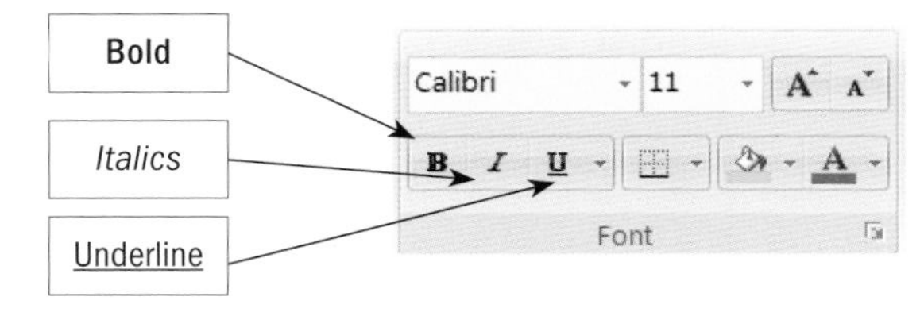

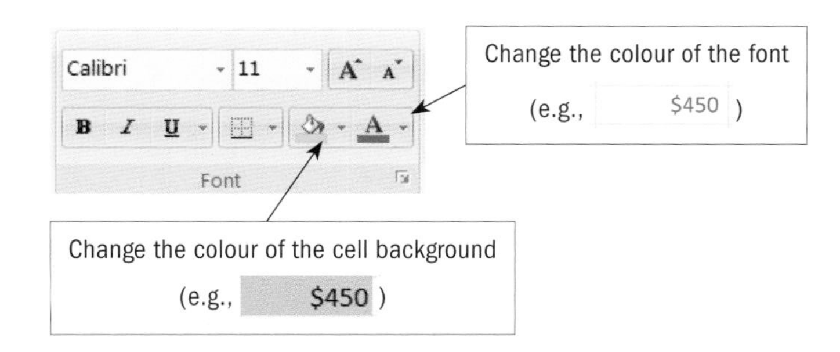

Always choose background and font colours carefully. Make sure there is enough contrast between them. If you print the spreadsheet out in black and white, there may not be enough contrast to be able to read the data.

Choosing a font (changing the shape of letters and numbers)

If you don't like the font that the computer has automatically chosen (called the default font) then you can change it. To do this you select the text you want to change and then click on the font box.

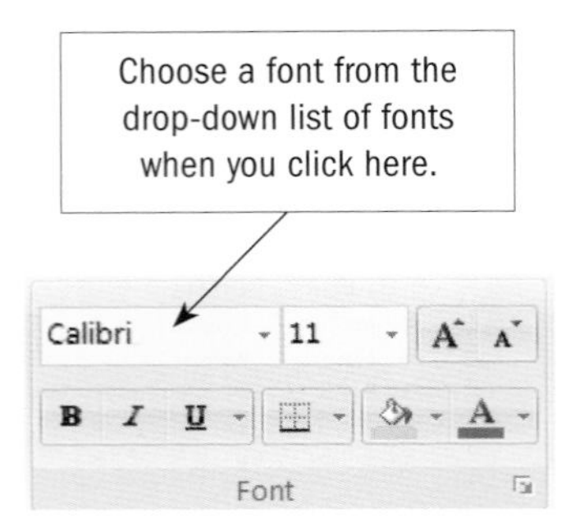

Altering the size of the font

There are a few ways to alter the size of a font.

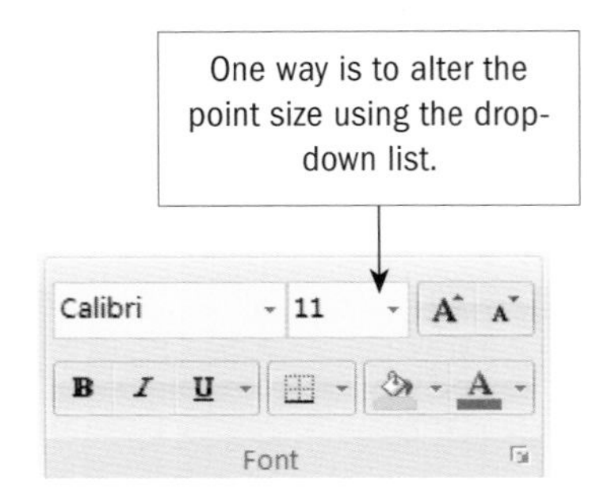

Another way is to use these two icons:

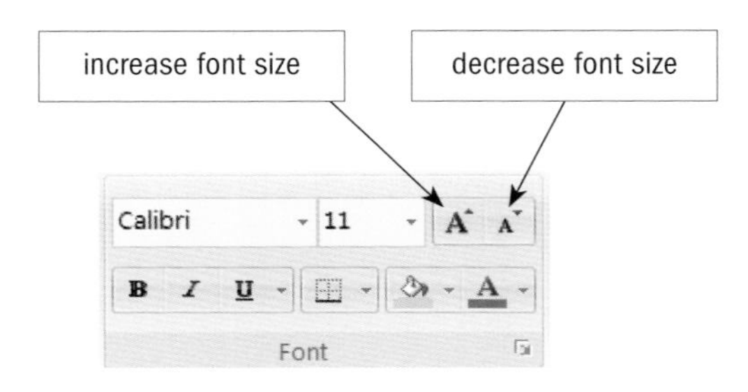

Using the Format Cells menu

You can format cells using the Format Cells menu by right-clicking and then selecting Format Cells. The selected cells in the spreadsheet can be formatted using this menu:

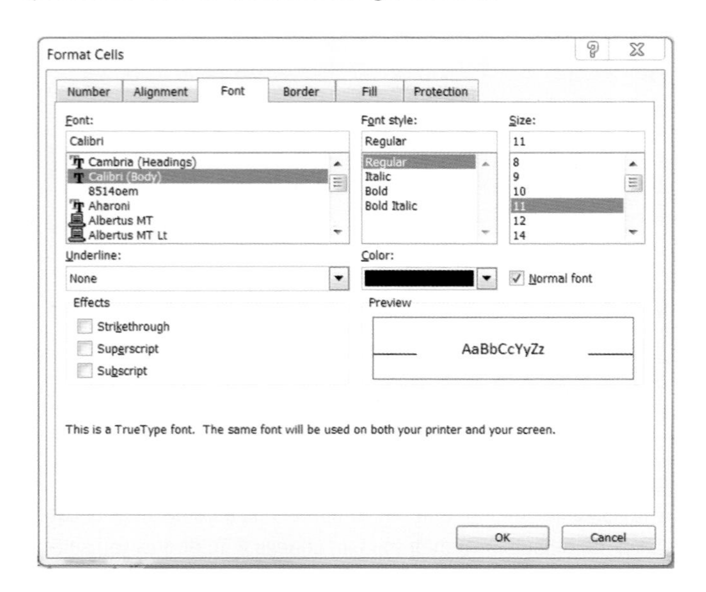

Adding colour/shading to enhance the spreadsheet

To add colour or apply shading to cells, you first select the cells by left-clicking on them and dragging until the cells are highlighted. Then right-click on the selected cells and the following menu appears, from which you select Format Cells:

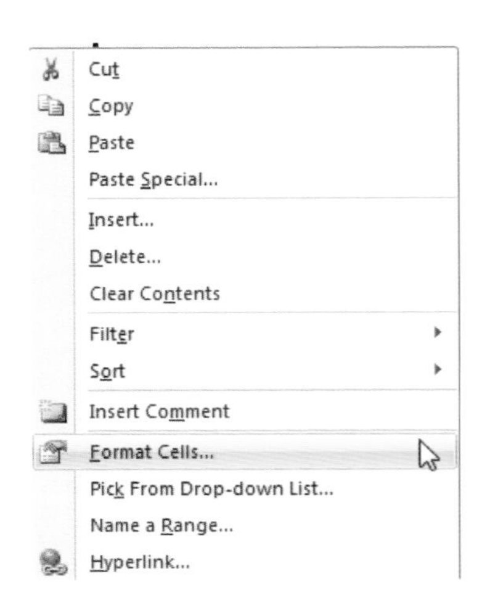

The Format Cells window appears:

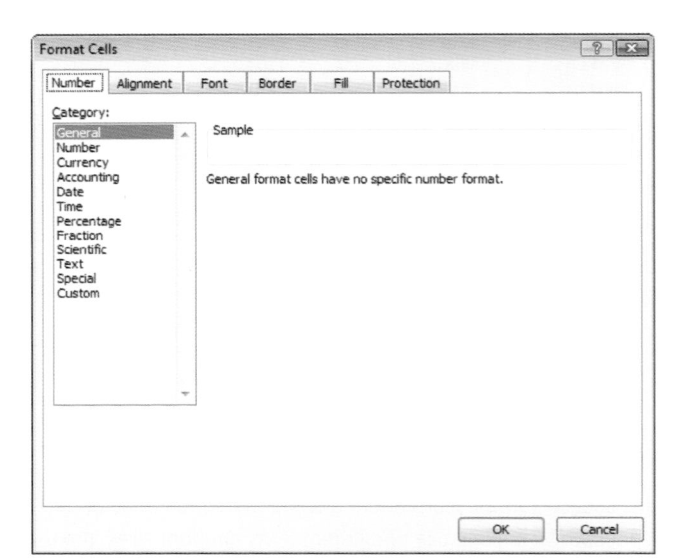

Select the Fill tab and the following window appears where you can select colours and patterns:

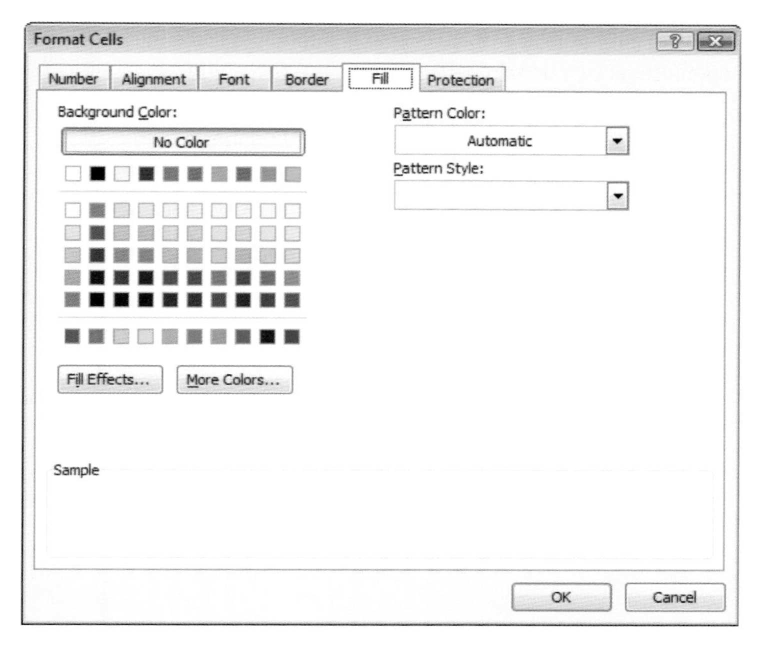

Aligning data and labels
Data in cells is normally positioned according to the following:

- Numbers are aligned to the right in a cell.
- Text is aligned to the left in a cell.

To align cells, click on the Home tab and then look at the alignment section of the screen.

To align the data or labels in a cell, select it and click one of these icons:

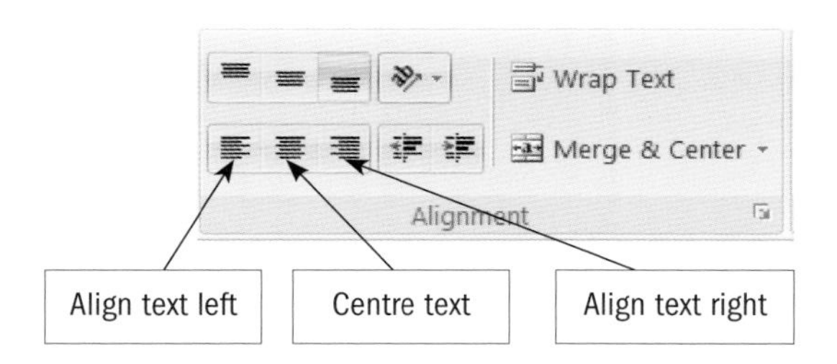

Adjusting column width and row height
To adjust the width of a column, position the cursor on the line between the column letters like this:

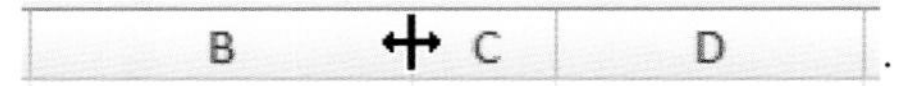

Keeping the left mouse button pressed down, drag the cursor to the left or right to change the width.

To adjust the row height, move the mouse onto the line joining the rows like this:

Keeping the left mouse button pressed down, drag the cursor up or down to change the row height.

If you want to adjust the column widths automatically so they accommodate the width of the field names and the data in the fields, there is the following quick way: Click on the following symbol at the top left of the spreadsheet . This will select the entire spreadsheet. When this is done, you will see all the spreadsheet cells change to light blue. Position the cursor between two of the columns (any will do) and double-click the left mouse button. You will see the columns automatically change so they fit all the contents.

Hiding rows or columns in a spreadsheet
You may be asked in the examination to show a spreadsheet with rows and columns of data hidden.

1 First select the row(s) or column(s) you want to hide.

2 Click on Home then click on Format from Cells group shown here

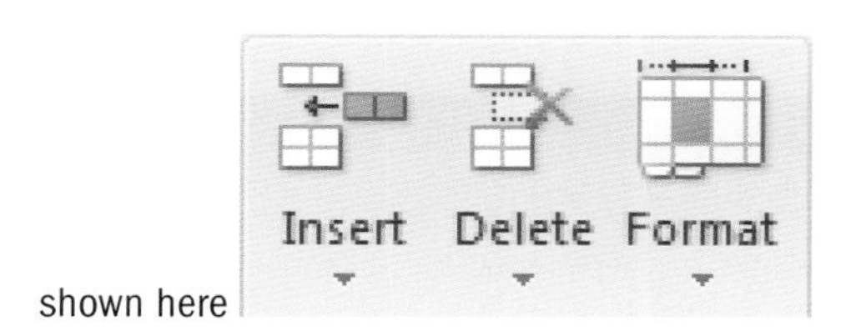

The following menu appears from which you should select Hide & Unhide.

In the menu on the left you can now decide whether you want to Hide columns or rows.

- Notice that the menu also allows you to adjust the row height and column width.

3 To unhide, you do not need to select any cells first. Simply go to Format Cells and then click on either Unhide Rows or Unhide Columns.

Setting the page layout so that it prints on a specified number of pages

It is always preferable to print out a spreadsheet on a single page, but in large spreadsheets this is not possible as the text could end up being too small to read.

You can specify the number of pages on which the spreadsheet is printed in the following way.

Click on Page Layout in the main menu. Look for the Scale to Fit section of the ribbon:

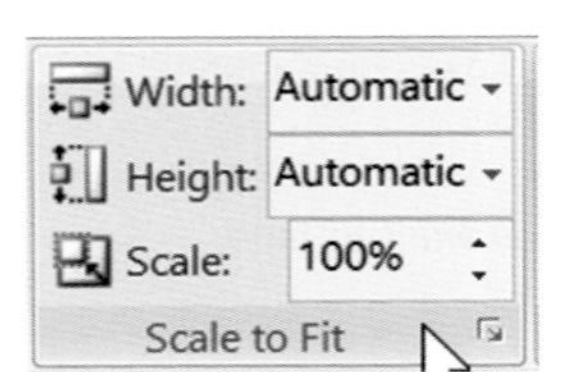

Notice that you can alter the width to 1, 2, 3, etc., pages and the height will be automatically adjusted. Alternatively you can alter the height to 1, 2, 3, etc., pages and the width will be automatically adjusted. There is also the option to scale the spreadsheet to a certain percentage of its original size. When adjusting the size, always check that you can read all the text and remember to preview the spreadsheet before printing.

Website authoring

In this chapter you will be learning how to create a structured website with stylesheets, tables and hyperlinks. You will be using a special code for this called HTML. There are other ways to create web pages using special web-authoring software such as Adobe Dreamweaver or Microsoft FrontPage. In this book you will be using just a text editor to create the HTML code and then use web browser software to view the web page you have produced.

The key concepts covered in this chapter are:

- Web development layers
- Creating web pages
- Using stylesheets
- Testing and publishing a website

Web development layers

When large websites are developed it is important to note that many people will be involved in the development of the site. For example, writers, illustrators, photographers, cartoonists, etc., may work supplying the content for the website. Designers put the content together in a way that is attractive and eye-catching to users. Technical staff will create the code/instructions that will make the various elements of the website behave in a certain way.

There are three layers in web development:

- Content layer
- Presentation layer
- Behaviour layer.

Content layer – when you view a web page the content layer is what you see. The content layer consists of text and images and also the pointers that allow you to navigate around the website.

Presentation layer – is how the document will look to the reader. The presentation layer defines how each page should look rather than what is on each page. The presentation layer is used to format an individual element on a web page, a whole web page or all the web pages that make up a website. Cascading style sheets (css) are used to define how the web pages should look. Rather than make a change to the format each time a particular element occurs, you can change it once in the style sheet and the changes are made automatically to the element wherever it occurs.

Behaviour layer – this is the layer of the web page that does something when the user clicks on something. There are different methods used to create this layer and in this chapter you will create sets of instructions in a code called HTML (Hypertext Markup Language).

HTML (HyperText Markup Language)

HTML is short for HyperText Markup Language and is the special code that is used for making web pages. HTML consists of special markers called tags that tell the computer what to do with the text, images or links that are entered. It could tell the computer to present the text in a certain way, or size and align an image on the page. For example, the tags could tell the computer that the text being entered is intended to be a heading or to make a certain block of text bold.

HTML is a text file, just like MS Word, except that it contains special markers called tags. HTML can be entered into a text editor, which enables tags to be entered as well as content to which the tags apply. You will see how to use a simple text editor for entering and editing HTML later.

Important note: HTML is not a programming language as such. It just tells the computer how to display text and pictures on web pages.

KEY WORDS

HyperText Markup Language (HTML) used to create documents on the world wide web. You use it to specify the structure and layout of a web document.
Tags special markers used in HTML to tell the computer what to do with the text. A tag is needed at the start and end of the block of text to which the tag applies.

Browser and editor software

Browser software is responsible for requesting text and graphics on web pages stored on servers and then assembling them for display. In this chapter, web browser software will be used to view the web pages you create. Editor software is used for creating and amending text. In this chapter, you will be using a text editor for the preparation of HTML web pages.

Here are the steps involved in creating and viewing web pages using HTML:

1. Load the text editor software.
2. Type in the HTML code.
3. Save the HTML file.
4. Load the browser software.
5. View the web page and see if there are any mistakes in the code.
6. Go back to the HTML code using the editor software.
7. Edit the HTML code using the editor.
8. Save the HTML file.
9. View the web page using the browser.
10. If there are any more corrections needed go to step 6.
11. Save the final HTML code.

Setting up a relevant folder structure

Before you start work on this chapter it is important that that you create a folder structure to hold all the files you will be making in the activities.

To create a folder structure for your work follow these steps:

1. Open a new Explorer window, either by selecting Computer from the Start menu or by using the keyboard shortcut: Windows key + E.
2. You will be presented with a screen showing the hard disk drives and devices with removable storage:

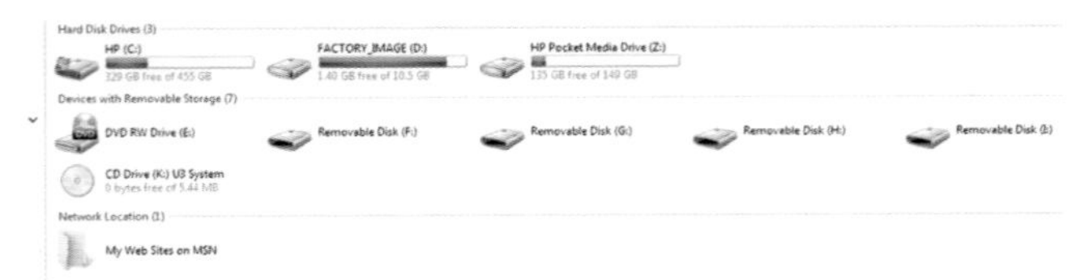

Click on the device where you want the HTML files to be stored. In most cases this will be one of the hard drives or a flash drive. Your teacher will tell you on which drive/device the files are to be stored.

3. Double-click on the name of the device and you will be presented with a list of folders currently on that device.
4. Right-click on the white area around the file list and the following menu appears:

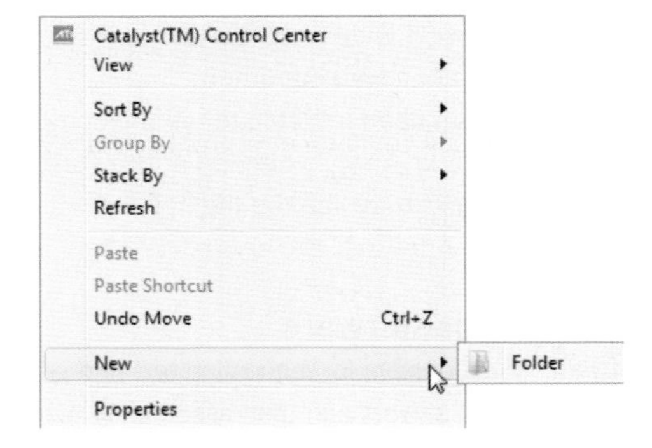

Click on New followed by Folder.

5. A box appears like this:

Delete the text inside the box and replace it with the name HTML like this:

6. The folder now appears in the list of folders like this:

IGCSE
Review and marketing
HTML
Ro

Click on the folder HTML.

7. The empty folder appears like this:

8. Click on Organize and then select New Folder from the menu.

The sub-folder appears like this as a folder within the folder called HTML. Note that it is called a sub-folder because it is a folder within a folder and lower down in the hierarchy.

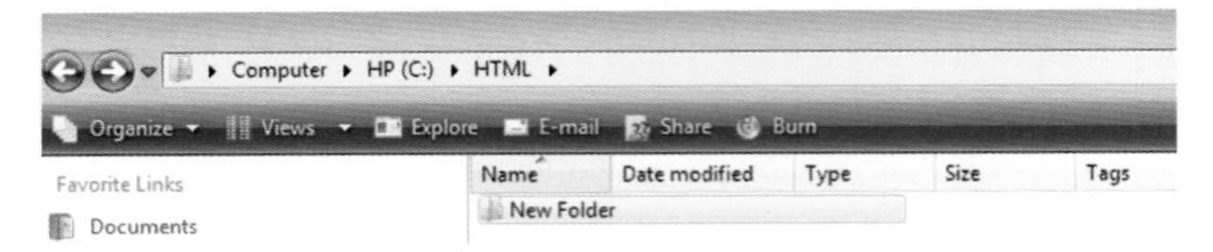

Click once on the New Folder.

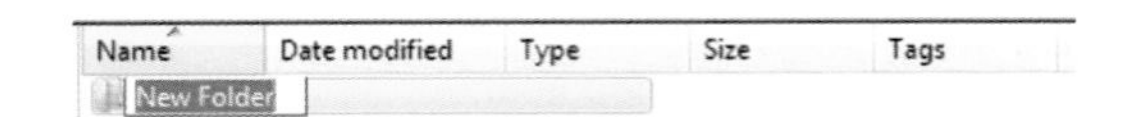

You can now type in the name for this sub-folder.

In this case we will call this sub-folder Activity 21.1. You can now add sub-folders for any of the other Activities as they are needed.

Saving all the web files in the same folder or sub-folder

When a web page is created, all the files used in the web page should be stored in the same folder or sub-folder as the HTML file. This means that if a web page contains an image then the image file should be stored in the same folder or sub-folder as the HTML file.

The structure of an HTML document

HTML documents need to be structured in a certain way. Here are the basics of the structure:

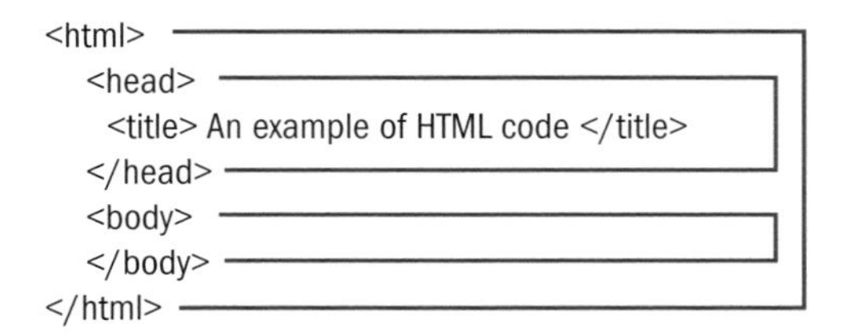

Look at this section of code carefully. You will see the way the tags are nested within each other (i.e. shown by the blue lines) and also how indentation is used to make it easier to see the various parts of the code.

The tags are the words enclosed between the < and > signs. They are HTML instructions and they tell the computer how to display or format the text.

Here is what each tag in this section of code does:

- <html> and </html> tells the computer that we are creating an HTML document. The <html> tag tells the computer that the document is starting and the </html> tells the computer that the document has ended.
- <head> and </head> is a section where you provide information about your document.
- <title> and </title> tells the computer that you want to put the text between the tags as a title. As the title is part of the 'head' section it needs to be inserted after the opening head tag <head> and before the closing tag </head>. The nesting of tags in this way is very important as not nesting the starting and closing tags is a frequent cause of problems when writing HTML. The text used as the title does not appear on the web page but instead is shown in the browser's window. If a user of the web page adds the page to their favourites, this title text is used as the name of the page.
- <body> and </body> is a section where you insert the content of your document and visual information about how it is to appear on the web page.

More code and what it does

- <h1> and </h1> indicates that you want the text to be a heading of type h1. There are other headings h2, h3, etc., to choose from. h1 headings are larger in size compared to h2 headings and so on. The text you want on the page is placed between these two tags like this: <h1>This text is size h1</h1>
- <p> and </p> mean start and end a new paragraph. A break between the paragraphs is inserted that is approximately two lines in length. The text for the paragraph is placed between the two tags.

More tags and what they do

Here are some other tags and what they do:

- <b> and </b> means start bold and stop bold style text.
- <u> and </u> means start underline and stop underline style text.
- <i> and </i> means start italics and stop italics style text.
-
 means insert a line break. There is no closing tag for this. This creates a smaller gap than the <p> tag.
- <hr> means insert a horizontal line and there is no closing tag.

Aligning text using HTML

Text can be aligned on the page to the right, left or centre. The text has to be in a separate paragraph in order for it to have specific alignment attributes attached to it. Here is an example of HTML using different alignments:

```
<html>
    <body>
        <p align="left">This is an example of left aligned text.</p>
        <p align="center">This text has been centred.</p>
        <p align="right">This text has been right aligned.</p>
    </body>
</html>
```

Creating an external stylesheet

Before the introduction of stylesheets, if you had a website containing 50 pages and wanted to change the text for all the h1 headings from black to blue, you had to look for all the h1 tags on all the web pages and alter them manually. This was very time consuming.

Luckily we now have stylesheets and you would need to make the change only once in the stylesheet and all the text for the h1 tags would change from black to blue automatically.

It is important to note that stylesheets are not HTML documents and do not use tags, and you will never see < and > used in stylesheets. Also the files used to save stylesheets do not have the file extension .htm but instead use the .css extension

The use of stylesheets has the following advantages:

- They save time – you change things in only one document (i.e. the stylesheet).
- They help give a website a consistent look (e.g. heading sizes, fonts, font sizes, etc., will be consistent across all pages).

There are a few rules you need to obey when creating a stylesheet using CSS (Cascading Style Sheets).

CSS codes have the following structure:

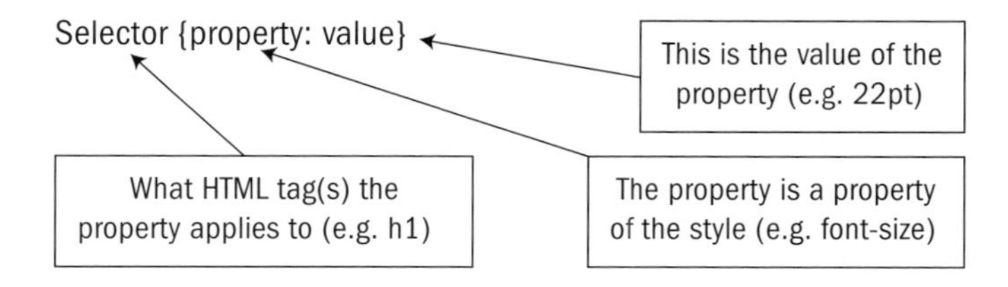
Selector {property: value}
This is the value of the property (e.g. 22pt)
What HTML tag(s) the property applies to (e.g. h1)
The property is a property of the style (e.g. font-size)

So to change the size of the font used for the h1 heading we would use the following CSS stylesheet:

```
h1 {font-size: 22pt}
```

To understand stylesheets you need to look at a few examples:

Here two different styles are applied to the h1 heading:

```
h1 {text-align: center, color: blue}
```

Notice the American spelling of both 'center' and 'color' which you must use.

This stylesheet sets the heading h1 to the colour blue and also aligns the text centrally. Notice the curly brackets at the start and end of the stylesheet and also notice that there is a comma separating the properties.

If you had a website and used a stylesheet and wanted all occurrences of the h1 heading to be changed from blue to red, you would only need to change the stylesheet slightly to this:

```
h1 {text-align: center, color: red}
```

Activity 21.1

Creating a simple web page using HTML

In this activity you will create a very simple web page using HTML and you will learn how to:

- use the editor to create HTML code
- use simple tags to create a web page
- alter the sizes of headings using h1, h2 and h3
- save an HTML document.

1 Notepad is an editor and we can use it to put the HTML together. If you don't have Notepad as a shortcut, then from the Start menu, type 'notepad' in the search field to locate and then open it.

Type the following text into the editor exactly as it is shown here:

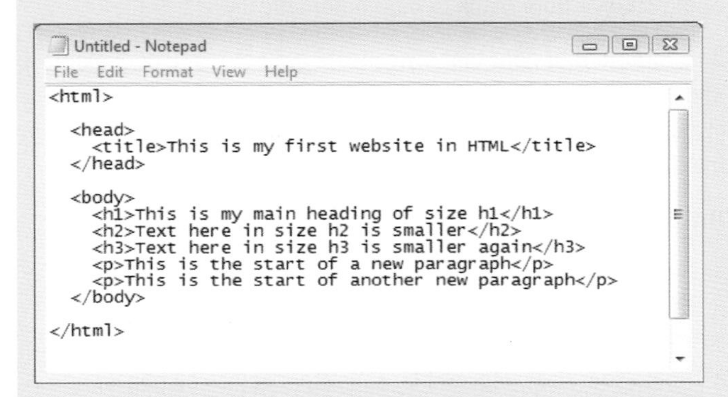


```
<html>

  <head>
    <title>This is my first website in HTML</title>
  </head>

  <body>
    <h1>This is my main heading of size h1</h1>
    <h2>Text here in size h2 is smaller</h2>
    <h3>Text here in size h3 is smaller again</h3>
    <p>This is the start of a new paragraph</p>
    <p>This is the start of another new paragraph</p>
  </body>

</html>
```

2 You need to save this file as an HTM file. To do this click on File and then Save As.

When the Save As window opens, you need to find the folder created for saving your HTML documents. Once you have done this, enter the filename **Mywebpage.htm** and then click on Save.

Important note: Always remember to add the .htm file extension to the filename, as if you do not do this then the browser you use to open the web page will not recognise it as a web page.

Check that your file has been saved correctly like this:

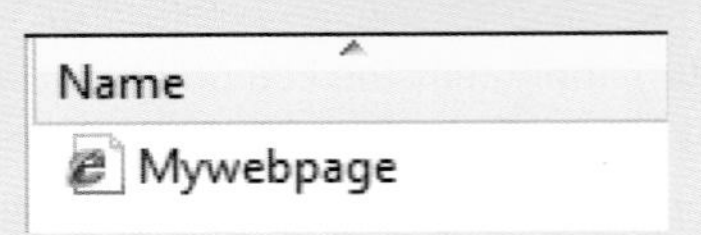

Notice the browser icon is shown next to the filename. If you do not see this icon (or the icon for whichever is your default browser) then you have probably missed out the .htm file extension and you will have saved the file as a text file by mistake.

3 To open the HTML file you have just saved it is necessary to use a web browser program.

Load the web browser program you usually use (e.g. Internet Explorer, Firefox, Chrome).

4 Click on File and then Open.

You will now need to find the HTML folder.

Once you have found the folder click on the file Mywebpage.

Your web page will now load and you can see the results:

This is my main heading of size h1

Text here in size h2 is smaller

Text here in size h3 is smaller again

This is the start of a new paragraph

This is the start of another new paragraph

You can now close this web page.

There are two steps to create an external stylesheet:

1 Create the external stylesheet in .css format.
2 Link the stylesheet to the HTML pages you want it to apply to.

Font families

There is a problem with some fonts because not all web browsers can display them. This means when specifying fonts you need to produce a prioritised list of fonts. This means you do not just give one font; instead, two or more choices are given. What then happens is that the web browser looks along the list of fonts starting from the first one that is specified and if it cannot be used it then moves onto the next font in the list and so on.

There are two ways of specifying a font:

- by the name of the font such as Arial, Times New Roman, Calibri, etc.
- by whether the font is serif or sans serif.

This stylesheet specifies the font families for the h1 headings:

```
h1 {font-family: Calibri, Arial, sans-serif)
```

This means that if Calibri can be used by the browser, it will be used and if not, Arial will be used and if this is not available then any sans serif font will be used.

It is always best to end a font family with either sans-serif or serif as these will always be available no matter which web browser is used to display the web page.

If you use the name of a multi-word font such as Times New Roman, the name of the font must be placed inside inverted commas like this:

```
body {font-family: 'Times New Roman', Century, serif}
```

Here is a list of properties and the values they can have:

Property	Value
Font-style	normal
Font-style	italic
Font-weight	normal
Font-weight	bold
Font-size	12pt, 22pt, etc.
Text-align	left
Text-align	right
Text-align	center
Text-align	justify

Background colour

You can change the background colour in the stylesheet like this:

```
body {background: red}
```

or by using a code for red (#FF0000)

```
body {background: #FF0000}
```

Note the background colour of the body is being defined here.

There are a number of colour names you can use shown here:

black (#000000)	silver (#C0C0C0)	gray (#808080)	white (#FFFFFF)
maroon (#800000)	red (#FF0000)	purple (#800080)	fuchsia (#FF00FF)
green (#008000)	lime (#00FF00)	olive (#808000)	yellow (#FFFF00)
navy (#000080)	blue (#0000FF)	teal (#008080)	aqua (#00FFFF)

Note: the American spelling of the colour 'gray'.

If you want to use a colour that is not one of the sixteen named colours, you can refer to a colour palette and then use the hexadecimal code for the colour.

Palette of colours

To get a palette of colours with their codes use the following website:

http://www.webmonkey.com/2010/02/color_charts/

Activity 21.2

Creating a stylesheet

In this activity you will learn how to:

- create a stylesheet in .css file format
- link a web page to a stylesheet.

In this activity you will be creating a stylesheet which will then be used by a section of HTML code to set the styles for a web page.

For this activity you will be creating two files:

A stylesheet using the .css file format

A web page using the .htm file format

Both of these files can be created and saved using a text editor such as Notepad.

1 From your operating system screen, click on the Start button then click on All Programs and then Accessories and then on Notepad.
2 Notepad is an editor and we can use it to assemble the formatting instructions that make up the stylesheet.

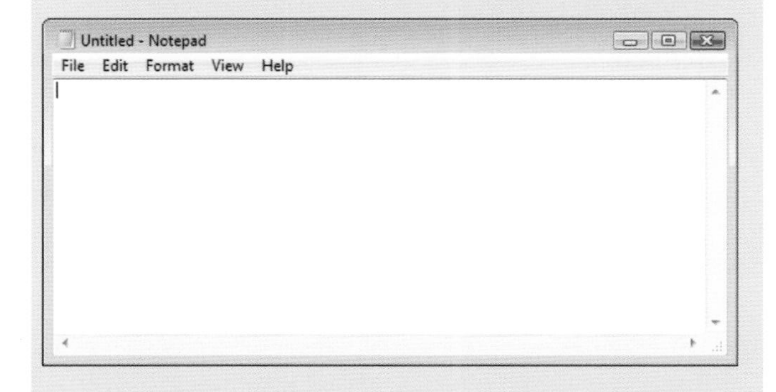

Type the following text into the editor exactly as it is shown here:

```
body {background: #FEF76E; font-family: Arial, Verdana, sans serif }
h1 {font-family: Arial, Verdana, sans serif; font-size: 32pt; font-color: navy; font-weight: bold}
```

Important note: you must spell 'colour' as 'color' in both stylesheets and HTML code.

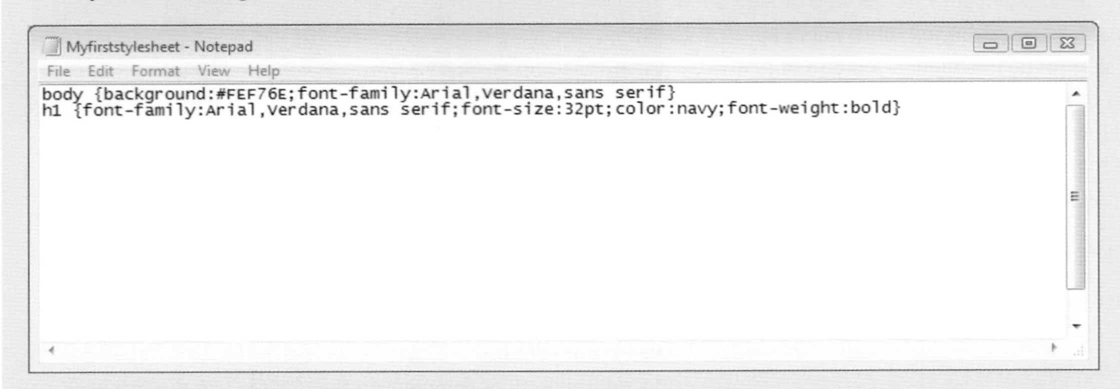

The text file needs to be saved in css format.

To do this click on File and then Save As

Call the file **Myfirststylesheet.css**.

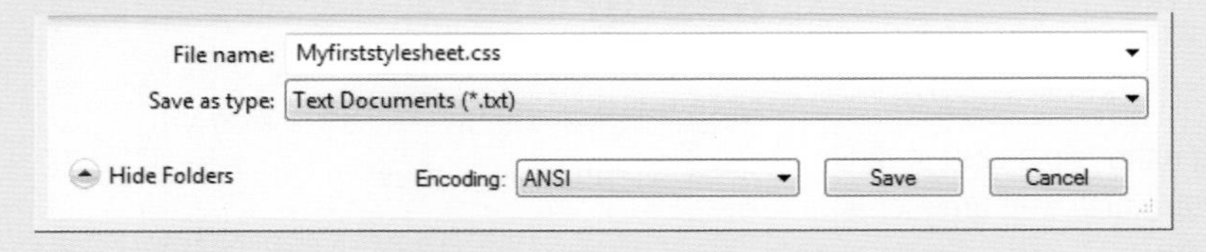

Click on Save.

3 With the stylesheet still being shown in Notepad, click on File and then New and type in the following HTML exactly as it is shown here:

```
<html>
  <head>
    <title>My first stylesheet</title>
    <link rel="stylesheet" type ="text/css"
    href="Myfirststylesheet.css">
  </head>
  <body>
    <h1>This heading should be point size 32 and colour
    navy and in bold</h1>
    <p>This is my first stylesheet so I hope it works</p>
  </body>
</html>
```

Notice the following step which is used to inform the computer that this page is to be linked to a stylesheet. Notice that the filename of the stylesheet is included so that the computer can find the stylesheet that is to be used with this web page.

```
<link rel="stylesheet" type="text/css"
href="Myfirststylesheet.css">
```

⇨

There are two important points about this line of HTML:

- It must be placed in the Head section of the HTML.
- The only part of the line you will need to change when you create your own HTML with stylesheets is the last part shown in blue here. All you need to change is the name of the file you are using for the cascading stylesheet.

Click on File and then Save As.

Check that you are saving the file in the folder you are using to hold your HTML and CSS files.

Type the following name for the file **Web_page_with_a_stylesheet.htm**.

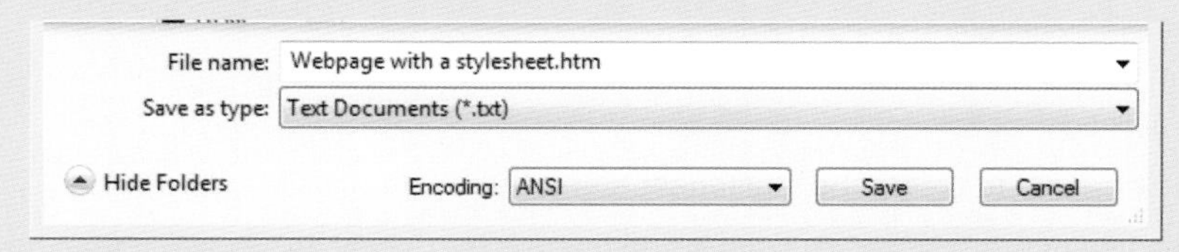

Click on Save.

You have now completed the HTML code of the web page including the link to the stylesheet.

4 Load your web browser software and click on File and then Open.

You will now need to locate the folder where your HTML and CSS files are stored.

You will see the HTML file for the web page displayed like this:

Click on this to open it.

The web page now opens in your browser like this:

This heading should be point size 32 and colour navy and in bold

This is my first style sheet so I hope it works.

You have now created a stylesheet and attached it successfully to the HTML code for a web page.

Tags to create lists

There are tags to create bulleted lists called unordered lists, and tags to create numbered lists, which are called ordered lists.

The HTML to create a bulleted (unordered) list is as follows:

```
<ul>
<li> Saudi Arabia
<li> UAE
<li> Egypt
<li> Kuwait
<li> Argentina
</ul>
```

- Saudi Arabia
- UAE
- Egypt
- Kuwait
- Argentina

The HTML to create a numbered (ordered) list is as follows:

```
<ol>
<li> Saudi Arabia
<li> UAE
<li> Egypt
<li> Kuwait
<li> Argentina
</ol>
```

1 Saudi Arabia
2 UAE
3 Egypt
4 Kuwait
5 Argentina

Activity 21.3

Creating a stylesheet according to instructions

In this activity you will learn how to:

- change the appearance of a bullet in a list
- use a stylesheet with common styles such as h1, h2, h3, p, and li.

In the examination you may be asked to create a stylesheet for use within all the pages of a website to aid consistency.

The stylesheet for a particular website must contain the following styles:

h1 – red, sans-serif font, 34 point, centre-aligned, bold
h2 – dark blue, serif font, 20 point, left-aligned
h3 – green, sans-serif font, 12 point, left-aligned
li – blue, sans-serif, 10 point, bullet points, left-aligned
p – black, serif font, 10 point, left-aligned

1 Load Notepad and key in the following text to create the stylesheet shown here.

Tip: Once you have completed one line of the stylesheet, copy it and paste it for the second line. You then have to make only minor changes rather than type in the whole line, which will save time.

Style_sheet - Notepad
File Edit Format View Help

```
h1{font-family:Arial,sans-serif;color:red;font-size:34pt;text-align:center;font-weight:bold}
h2{font-family:"Times New Roman",Serif;color:navy;font-size:20pt;text-align:left}
h3{font-family:Arial,sans-Serif;color:green;font-size:12pt;text-align:left}
li{font-family:Arial,Sans serif;color:blue;font-size:10pt;text-align:left;list-style-type:circle}
p{font-family:"Times New Roman",Serif;color:black;font-size:10pt;text-align:left}
```

Look carefully at each line of the stylesheet and notice how the lines match the required styles.

Notice the way the font for each style is specified. First a font name is given (e.g. Arial or Times New Roman). If the font name is more than one word, the font name must be enclosed between quote marks like this 'Times New Roman'. After the font name and a comma, the words sans-serif or serif are used. This is so that if the required font is not available with the browser, then an alternative serif or sans-serif font can be used.

Notice that the colours of the fonts are specified like this:

color: navy

There are 16 colours that have names and these are; black, silver, gray, white, maroon, red, purple, fushia, green, lime, olive, yellow, navy, blue, teal, and aqua.

There are many other colours that you can use, but they are accessed using hexadecimal codes and further information about the colours and their codes can be found at: http://www.w3schools.com/html/html_colors.asp

In the style for the list, notice that the type of bullet used for the list can be specified like this:

list-style-type: circle

This means a circle is used for the bullet. Here is a summary of the types of bullet you can have:

List-style-type	Bullet appearance	Example
disk	Solid circle	• UAE • Egypt
circle	Circle	○ UAE ○ Egypt
square	Square	✶ UAE ✶ Egypt
decimal	Decimal number	1. UAE 2. Egypt

2 Save this stylesheet in your HTML folder using the filename **Style_sheet.css**.

⇨

3 You are now going to create the web page that makes use of the stylesheet you have just created.

At the top of the Notepad click on File and then New.

Type in the following HTML:

```
<html>
  <head>
  <title>Task setting up a stylesheet</title>
  <link rel="stylesheet" type="text/css" href="style_sheet.css"/>
  </head>
  <body>
    <h1>This is the h1 heading</h1>
    <h2>This is the h2 heading</h2>
    <h3>This is h3 heading</h3>
    <p>This is the text used for the majority of paragraphs on each page</p>
    <ul>
    <li>This is point one in the list
    <li>This is point two in the list
    <li>This is point three in the list
    <li>This is point four in the list
    </ul>
  </body>
</html>
```

Notice the fourth line where the stylesheet is linked to this web page.

Click on File and then Save As... and call the filename **Second_web_page_with_a_stylesheet.htm**.

4 Load your web browser software and click on File and then Open and locate your HTML folder and open the file **Second_web_page_with_a_stylesheet.htm**.

The web page now opens in your browser like this:

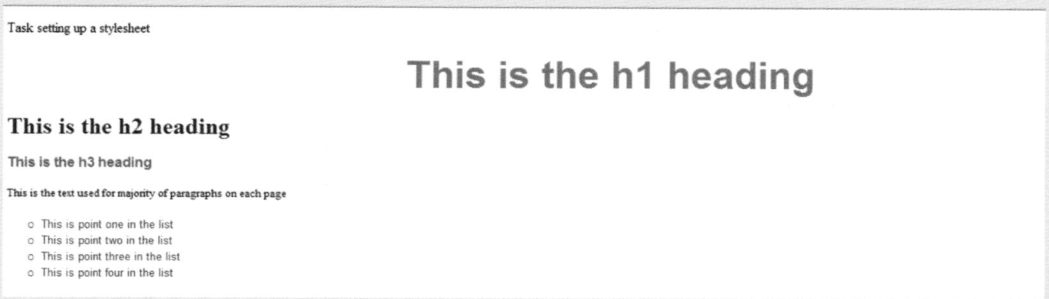

Task setting up a stylesheet

This is the h1 heading

This is the h2 heading

This is the h3 heading

This is the text used for majority of paragraphs on each page

- This is point one in the list
- This is point two in the list
- This is point three in the list
- This is point four in the list

Activity 21.4

Practising creating a stylesheet

In this activity you will learn how to create a stylesheet.

In this activity you are required to create a stylesheet according to instructions. You may find it quicker to amend the stylesheet created in the last activity and then save the stylesheet using a different filename. You will then use the HTML file created in the previous activity to link to this new file. You will have to amend the HTML slightly as the filename for the stylesheet will have changed.

1 Create a stylesheet for use by a web page. The styles for this stylesheet are as follows:

   h1 – black, serif font, 28 pt, centre-aligned, bold

   h2 – red, serif font, 20 pt, centre-aligned, italic

   h3 – green, sans-serif font, 12 pt, right-aligned

   li – black, sans-serif, 10 pt, square bullet points, left-aligned

   p – blue, serif font, 10 pt, left-aligned

Save this stylesheet using the filename **Style_sheet_act4.css**.

2 Load the web page created in the last activity called **Second_web_page_with_a_stylesheet.htm** into Notepad. With the HTML displayed in Notepad, change one of the instructions so that the web page now uses the stylesheet with filename **Style_sheet_act4.css**.

   Save this HTML using the filename **Third_web_page_with_a_stylesheet.htm**.

3 Load your browser and load the file **Third_web_page_with_a_stylesheet.htm**.

   Check carefully that the stylesheet is producing the correct styles for the web page.

Hyperlinks (often just called links)

Hyperlinks are an important part of websites as they allow users to move from one web page to another or within the same web page. Links are text or images that a user can click on in order to jump to a new web page, which can be on the same website or a completely different one. They can also be used to move between sections on the same web page which is especially useful if the web page is long. When the cursor is moved over a link, the cursor arrow turns into a small hand.

Creating external links to a website using the URL

You can create HTML code that will provide a link to an external website using the URL (i.e. web address) of the external site. External links are used to provide links to web pages and websites that are outside your directory where you store your website.

For example, if we wanted to provide a link to the Oxford University Press website the HTML for this is:

```
<a href="http://www.oup.com/">Oxford University Press website</a>
```

The text shown in blue is the URL of the website you want to link to and the text to supply the link (i.e. Oxford University Press) also appears in the tag.

The link on the web page appears as follows:
Oxford University Press website

Important note

To create a link to an external website you have to use http:// as the starting part of the URL and not just www.

It is also important to check external links regularly as websites sometimes change their URL, thus the links to them will no longer work.

Creating a link in a new window

Unless specified, a link will open in the same window.

To create the HTML to open a link in a new window you add target="_blank" to the anchor tag as shown here:

```
<a href="http://www.oup.com/"target="_blank">Oxford University Press website</a>
```

This results in the link appearing like this on the web page: Oxford University Press website

When the user clicks on the link they are taken to the website with URL **http://www.oup.com/** opened in a new window.

Creating links to other locally stored web pages

Websites usually consist of many web pages and users will want to be free to move between these pages using links. These links are called relative links and are used to enable users to move to other pages that are saved in the same directory on the computer or server.

The HTML you need to enter in the editor for this feature looks like this:

```
<a href="webpage1.html">Introduction to HTML</a>
```

The text shown in blue is the name of the file for the web page you are linking to and after it is the text that appears on the web page and acts as the link. When viewed in the browser the link appears as follows:

Introduction to HTML

Creating links to send mail to a specified email address

Visitors to your website may want to send you a message and you can create a link that will allow them to do this using the **mailto** tag. When the user of your web page clicks on the text underlined acting as the link, a dialog box addressed to you will appear into which they can type their message.

The HTML you need to enter in the editor for this feature looks like this:

```
<a href=mailto:emailaddress@aol.com>Name to go here</a>
```

The text shown in blue is your email address and after it is the name that appears on the web page that acts as the link. When viewed in the browser, the link appears as follows: Name to go here

Creating anchors and links to anchors on the same web page

Anchors act as points of reference on a web page. They enable a user to move to part of the same web page, usually using an item in a menu as a link. It makes it easier for a user because they do not have to waste time looking at part of the web page they are not interested in.

Suppose there is a menu like this at the start of the web page and then material about each of these in turn is discussed in sections on the same web page:

1. What is copyright?
2. When can you use copyrighted material?
3. Myths about copyright

To create the above links to the anchors in the main body of text in the document, you can use the following HTML:

```
<a href="#copyright">1. What is copyright?</a>
<a href="#use">2. When can you use copyrighted material?</a>
<a href="#myths">3. Myths about copyright</a>
```

The above links will link to the anchors that are positioned in the text to which they refer. So, for example, the link called 'copyright' will need to have an anchor called 'copyright' at the start of the 'What is copyright?' section. Usually you would place the anchor near the heading for the section.

The HTML used to provide an anchor for the first item in the list could be:

```
<a name="copyright"></a>
```

The HTML for the second item in the list could be:

```
<a name="use"></a>
```

The HTML for the third item in the list could be:

```
<a name="myths"></a>
```

Using tables to organise a web page

Planning a web page is difficult but it can be made easier by using a table. The table produces a page layout into which text, images and other elements can be placed. Tables are also used in web pages for presenting content in a similar way to the way tables are used in ordinary documents.

In the examination you may be asked to create a table consisting of a certain number of columns and rows. In many cases there will be a diagram to show what the table should look like on the web page.

There are three basic tags used to create a table:

```
<table> this is the tag to start a table
    <tr> this is the tag to start a row of a table
        <td> this is the tag to start a data cell
        </td>
    </tr>
</table>
```

Here is a section of HTML used to create a table:

```
<table>
    <tr>
        <td>Cell 1</td>
        <td>Cell 2</td>
    </tr>
    <tr>
        <td>Cell 3</td>
        <td>Cell 4</td>
    </tr>
</table>
```

The table for this section looks like this:

Cell 1	Cell 2
Cell 3	Cell 4

Notice that there is a <tr> and </tr> for each row in the table. Also, there is a <td> and </td> for each column in the row (or each cell, as the intersection of a column and a row is called a cell).

Notice that there is no border to the table. To include a border, a border attribute has to be used as you will see later.

Borders around tables

The tables so far did not have borders around them. To place a border around the table you use the following:

```
<table border="2">
```

Activity 21.5

Creating anchor links

In this activity you will learn how to create anchor links in a web page.

Sometimes anchor points take the user to another point on a long web page and, to avoid the user having to use the scroll bar to scroll the page back to the beginning, it is best to supply them with a link. This is done by inserting an anchor at the top of the page and then placing a link further down the page. You can see the link at the bottom of the page in the following screenshot:

Myths about copyright

Here are some myths about copyright:

There is no copyright symbol, so it is ok to use

Wrong. As soon as a piece of work is produced it is copyrighted automatically with or without the copyright symbol.

I have acknowledged the source of the material so it is ok to copy it

Not true. You still should ask permission.

It is on a website, so it is ok to copy it

Wrong. Someone has spent time and money producing the item you are copying. Copying it is wrong and illegal

How do I know whether work I wish to copy is copyright free?

You have to assume that no work is copyright free unless it specifically says so. Even then you need to check the conditions carefully.

How do I copyright the original work that I produce?

Copyright material is sometimes marked with a © next to it or near it. If you stress that the material is copyright, it is a good idea to include this symbol in the following way:

© Author or owner of copyright. Date of publication

© Stephen Doyle. 2012

It is not a legal requirement to put the symbol in, because any original work is automatically given copyright.

Copyright when taking your own photographs

If possible, it is often quicker to go out and take your own photographs because there are no copyright problems. You own the copyright on photographs you take yourself. There are cases where you would have to use other photographs taken by others. An example of this would be if you wanted a photograph of a distant landmark such as the Great Wall of China or someone famous such as the President of the USA. There are many photograph libraries on the Internet where you can obtain photographs and use them on your website for a small fee.

Go back to the top

1. Load Notepad and open the HTML file called **Anchors**.
2. Look carefully at the HTML and notice the construction of the anchor link which takes a user from the bottom of the page to the top.
3. Use this document to create appropriate anchor links of your own.

Activity 21.6

Creating the code for a table

In this activity you will learn how to create a basic table.

Write the section of HTML that can be used to produce a table with three columns and four rows. Each cell in the table should hold the following text.

Cell 1	Cell 2	Cell 3
Cell 4	Cell 5	Cell 6
Cell 7	Cell 8	Cell 9

This needs to be added after the <table> and before the instructions which specify the rows and cells.

The number '2' is used to specify the thickness of the border.

Here is a table that contains a border of thickness 2 pixels:

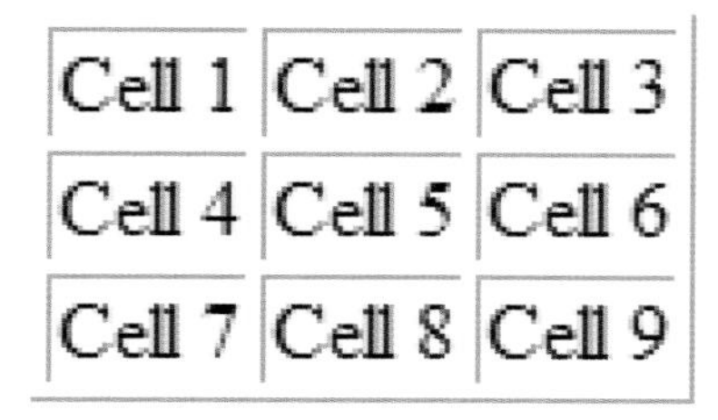

Cell 1	Cell 2	Cell 3
Cell 4	Cell 5	Cell 6
Cell 7	Cell 8	Cell 9

Changing the width of a table

By adding the following, when the table is drawn it will have a border of 1 pixel and the whole table will occupy a width of 50% of the entire screen width.

```
<table border ="1" width="50%">
```

In the browser the table will appear like this:

table4

Cell 1	Cell 2	Cell 3
Cell 4	Cell 5	Cell 6
Cell 7	Cell 8	Cell 9

The table width can also be specified by the number of pixels:

```
<table border ="1" width="500">
```

Changing the height of a table

The following HTML will set the height of a table to 250 pixels and the width of the table to 300 pixels. The border command ensures that a border is drawn around the table:

```
<table height="250" width="300" border="2">
```

Other attributes of tables

Aligning the text horizontally within a cell

The contents of cells can be aligned horizontally left, centred or right using the following:

```
<td align="left"></td>
<td align="center"></td>
<td align="right"></td>
```

These are used like this with text:

```
<td align="right">This text will be aligned to the right</td>
```

Using merged cells

If you want to merge cells in a table together there are two attributes you can use depending on whether you want to merge cells in columns or rows. These attributes are:

- Colspan
- Rowspan

Colspan

Colspan is an attribute using in the <td> tag and you can use it to specify how many columns the cell should span. Look at the following section of HTML:

```
<table border ="2">
    <tr>
        <td colspan="3">Cell 1</td>
    </tr>
    <tr>
        <td>Cell 2</td>
        <td>Cell 3</td>
        <td>Cell 4</td>
    </tr>
</table>
```

You can tell that this table consists of two rows as there are two <tr> tags. You can also tell that the second row is divided into 3 cells as there are three <td> tags (one for each cell). There is a border around the table of 2 pixels.

Colspan is short for column span.

```
<td colspan="3">Cell 1</td>
```

The above HTML tells us the first row spans 3 columns and contains the text 'Cell 1'.

This section of code produces the following table:

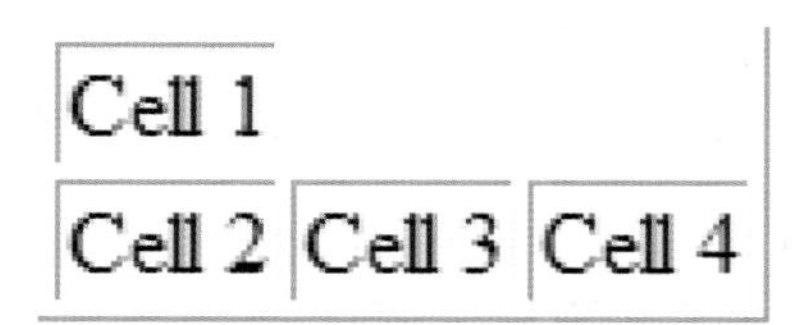

<table>
<tr><td colspan="3">Cell 1</td></tr>
<tr><td>Cell 2</td><td>Cell 3</td><td>Cell 4</td></tr>
</table>

Rowspan

Rowspan specifies over how many rows a cell should span.

Aligning a table horizontally on the screen

You may be asked in the examination to align the table centrally in a horizontal direction on the screen. To do this, you replace the <table> command at the start of the table with:

```
<table align ="center">
```

Cell spacing and cell padding

To add spacing between the cells in a table you use the cellspacing attribute. To add spaces between the content of a cell and the cell itself you use the cellpadding attribute. You will see how both of these attributes can be used in the following section.

Changing the width of a cell

The width of a cell can be set like this:

```
<td width="20%">Cell 1</td>
```

This changes the width of the cell to 20% of the table width.

The width of a cell can also be specified in pixels like this:

```
<td width="200">Cell 1</td>
```

The cell here has a width of 200 pixels.

Activity 21.7

Changing the width and height of a table and the width of cells in the table

In this activity you will learn how to:

- change the width and height of a table
- change the width of cells in a table.

1 Load your HTML editor (e.g. Notepad) and enter the following HTML code:

```
<table height="250" width="300" border="2">
 <tr>
      <td width="200">Cell 1</td>
      <td width="50">Cell 2</td>
      <td width="50">Cell 3</td>
 </tr>
 <tr>
      <td width="200">Cell 4</td>
      <td width="50">Cell 5</td>
      <td width="50">Cell 6</td>
 </tr>
</table>
```

Notice how the height and width of the table in pixels are specified in the first line.

Notice also how the width of each cell is specified in pixels.

2 Save this HTML in your HTML folder using the filename **Table_Activity_15.7**.

Open this file in your browser to check it correctly produces the basic table.

Check your table looks like this:

Cell 1	Cell 2	Cell 3
Cell 4	Cell 5	Cell 6

Table headers

Table headers can be used to specify the header of a column. In the following table 'Country', 'Capital' and 'Currency' have been used as the headers.

Country	Capital	Currency
Greece	Athens	Euro

Here is the HTML used to create the above table. Notice that the table width is 60% of the screen width and the cell spacing and cell padding have both been set to 5 pixels.

```
<table width="60%" border="1" cellspacing="5"cellpadding="5">
    <tr>
        <th>Country</th>
        <th>Capital</th>
        <th>Currency</th>
    </tr>
    <tr>
        <td>Greece</td>
        <td>Athens</td>
        <td>Euro</td>
    </tr>
</table>
```

Activity 21.8

Border, cell spacing and cell padding

In this activity you will learn how to insert a table using a table header, table rows, table data, cell spacing, and cell padding.

1 Load your HTML editor (e.g. Notepad) and enter the code to produce the Country, Capital and Currency table shown above but alter the first line to be:

```
<table width="60%" border="1" cellspacing="10" cellpadding="50">
```

Your table should look the same as this:

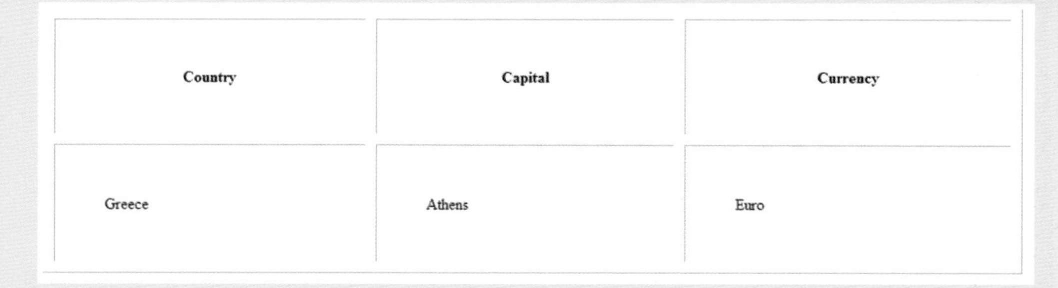

Country	Capital	Currency
Greece	Athens	Euro

2 Now amend the HTML in the following way:

Set the width of the table to 80%

Set the table border to 1

Set the cellspacing to 1

Set the cellpadding to 10

Save the file using a suitable filename and then load it into your browser.

Your table should appear like this:

Country	Capital	Currency
Greece	Athens	Euro

3 Add another row to the table containing the text: France, Paris and Euro.

4 Save the file using a different filename to the name given in step 2 and then load it into your browser.

Activity 21.9

Producing tables

In this activity you will learn how to use tables to organise a web page.

Being able to produce tables according to a design is needed for the examination. Here are some tables that you need to work through and understand. Study the code carefully and then type them in and save each using the filenames **Example_Table1.htm, Example_Table2.htm, etc.**

Table 1

This table has a horizontal width of 60%. Notice that the width of a cell can be specified and this width is a percentage of the table size and not the screen size.

```
<table border="1" cellpadding="5" cellspacing="5" width="60%">
    <tr>
      <td width="20%">Cell 1</td><td>Cell 2</td>
    </tr>
    <tr>
      <td width="20%">Cell 3</td><td>Cell 4</td>
    </tr>
</table>
```

Cell 1	Cell 2
Cell 3	Cell 4

Table 2

This table uses a header that spans two columns:

```
<table border="1" cellpadding="5" cellspacing="5" width="50%">
    <tr>
      <th colspan="2">Table header</th>
    </tr>
    <tr>
      <td width="30%">Cell 1</td><td>Cell 2</td>
    </tr>
</table>
```

Table header	
Cell 1	Cell 2

Table 3

Notice that in the first row of the table, the first cell occupies two rows and the second cell occupies the normal one row. In the second row the cell is placed in the space left to the right of cell 1.

```
<table border="1" cellpadding="5" cellspacing="5" width="50%" >
    <tr>
      <td rowspan="2">Cell 1</td><td>Cell 2</td>
    </tr>
    <tr>
      <td>Cell 3</td>
    </tr>
</table>
```

⇨

<table>
<tr><td rowspan="2">Cell 1</td><td>Cell 2</td></tr>
<tr><td>Cell 3</td></tr>
</table>

Table 4

The first row of this table is used as a header and spans two columns. The second row contains Cell 1 which spans two rows and Cell 2 which spans a single row. The third row only contains Cell 3. The fourth row contains Cell 4 which spans two columns.

```
<table border="1" cellpadding="5" cellspacing="5" width="50%" >
  <tr>
    <th colspan="2">Table Header</th>
  </tr>
  <tr>
    <td rowspan="2">Cell 1</td><td>Cell 2</td>
  </tr>
  <tr>
    <td>Cell 3</td>
  </tr>
  <tr>
    <td colspan="2">Cell 4</td>
  </tr>
</table>
```

<table>
<tr><th colspan="2">Table Header</th></tr>
<tr><td rowspan="2">Cell 1</td><td>Cell 2</td></tr>
<tr><td>Cell 3</td></tr>
<tr><td colspan="2">Cell 4</td></tr>
</table>

Activity 21.10

Producing tables to a design

In this activity you will learn how to use tables to organise a web page.

In the examination you may be given a diagram of a table and then have to produce the section of HTML needed to create it.

Produce the HTML code for each of the following tables.

Table 1

This table is to have a width of 50% of the screen, a table border of 1, cell spacing of 5 and cell padding of 20.

The table is to appear like this:

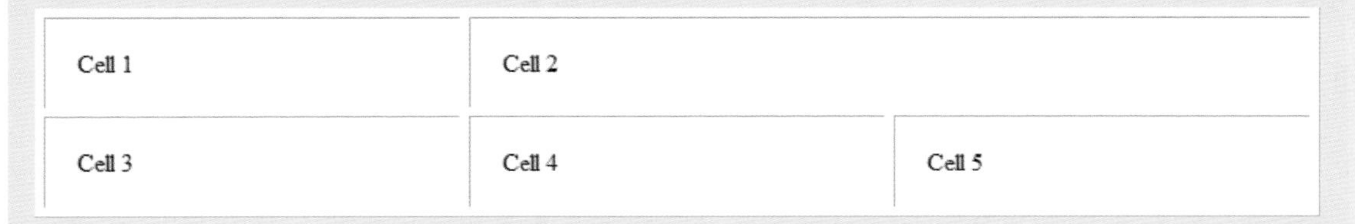

<table>
<tr><td>Cell 1</td><td colspan="2">Cell 2</td></tr>
<tr><td>Cell 3</td><td>Cell 4</td><td>Cell 5</td></tr>
</table>

⇨

Table 2

This table is to have a width of 800 pixels, a table border of 1, cell spacing of 5 and cell padding of 20. Cell 2 is to span two rows.

The table is to appear like this:

<table>
<tr><td>Cell 1</td><td rowspan="2">Cell 2</td></tr>
<tr><td>Cell 3</td></tr>
<tr><td>Cell 4</td><td>Cell 5</td></tr>
</table>

Table 3

This table is to have a width of 500 pixels and a table border of 1. Cell 1 is to span three rows.

The table is to appear like this:

<table>
<tr><td rowspan="3">Cell 1</td><td>Cell 2</td></tr>
<tr><td>Cell 3</td></tr>
<tr><td>Cell 4</td></tr>
</table>

Changing the background colour of a table, row or cell

The background colour of a table, row or cell can be changed. You can either use the name of the colour or pick the colour from a colour chart containing the hexadecimal code for the colour.

This section of HTML is used to change the background colour of a row to red:

```
<table border="1" cellpadding="10" cellspacing="5">
    <tr bgcolor="#FF0000">
        <td>Cell 1</td><td>Cell 2</td>
    </tr>
    <tr>
        <td>Cell 3</td><td>Cell 4</td>
    </tr>
</table>
```

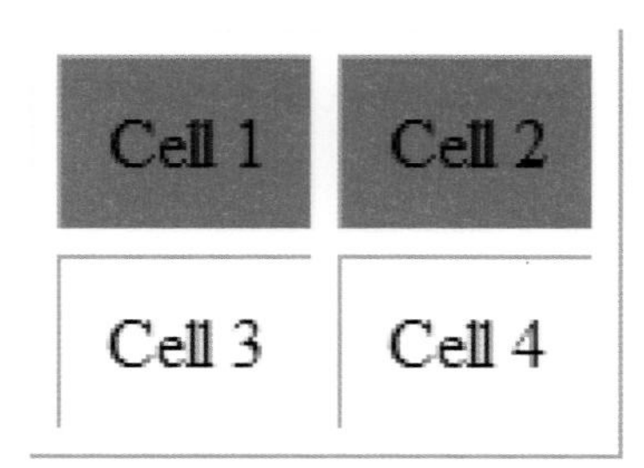

Cell 1	Cell 2
Cell 3	Cell 4

Activity 21.11

Changing background colours in tables

In this activity you will learn how to change the background colour of a row, cell or table.

1. Enter the HTML to draw the table shown on the left into Notepad.
2. Save the file using the filename **bgcolour1.htm** and load the file into the browser to check that it produces the table shown on the left.
3. Go back to the HTML and select the text bgcolor="#FF0000" by highlighting it and cut it and paste it into the position shown here:

```
<table border="1" cellpadding="10" cellspacing="5">
    <tr>
      <td bgcolor="#FF0000">Cell 1</td><td>Cell 2</
td>
    </tr>
    <tr>
      <td>Cell 3</td><td>Cell 4</td>
    </tr>
</table>
```

The background colour is now applied to a single cell (i.e. Cell 1).

Save the file using the filename **bgcolour2.htm**.

Load it into the browser to check it is producing the correct result.

4 Go back to the HTML and then make your own adjustment to make the whole table have a red background.

Save the file using the filename **bgcolour3.htm** and check that it works correctly.

Using images on web pages

There are many different file formats for images and not all of them work with all browsers. It is therefore best to use images in the following formats:

- .jpg (digital cameras store images in this format)
- .gif (often used for simple line diagrams and clip art).

In order to use an image on a web page, the image file must be stored in the same folder as the web page. This means the images you intend to use must be copied to the same folder.

Adding an image to a web page

Depending on the file format for the image, this is done using either:

```
<img src="filenameofyourimage.jpg"/>
```

or

```
<img src="filenameofyourimage.gif"/>
```

Finding out file format and size of an image

It is important to know the file format and size of an image. It is easy to do this by displaying the image and then right-clicking on the mouse so the following menu appears:

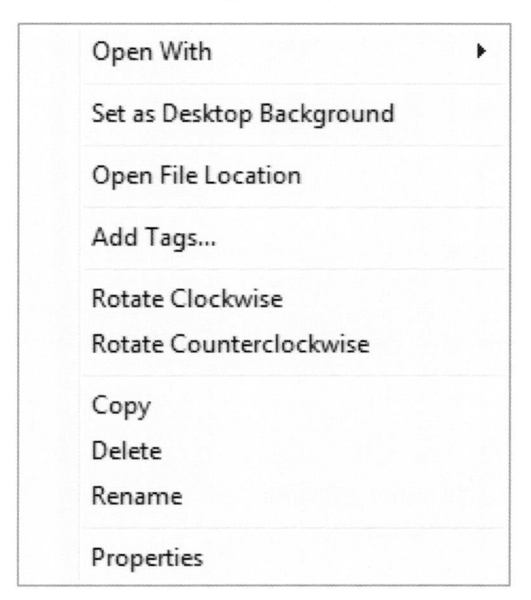

Click on Properties and the following information about the image will be displayed:

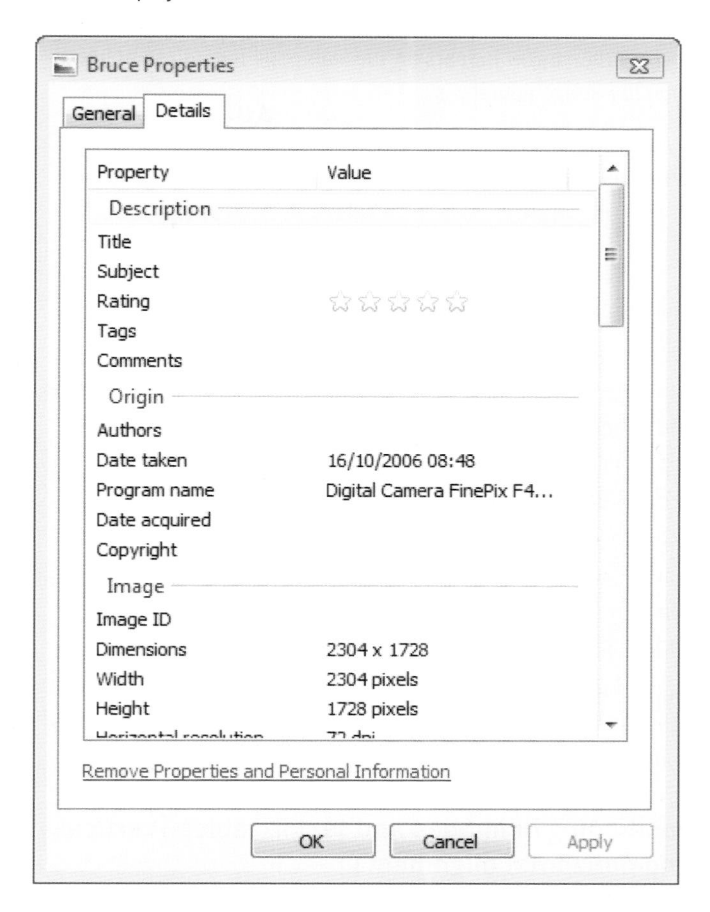

We require the dimensions of the image. This image is 2304 pixels wide and 1728 pixels high. If you scroll further down you will see the file format and also the size of the file.

Images taken with a digital camera are far too large to be used on a web page. They need to be resized and saved using a different filename before being used on a web page. To resize an image you need to use a photo-editing package such as Photoshop or Fireworks. Here the image size has been altered to 200 pixels wide and 150 pixels high.

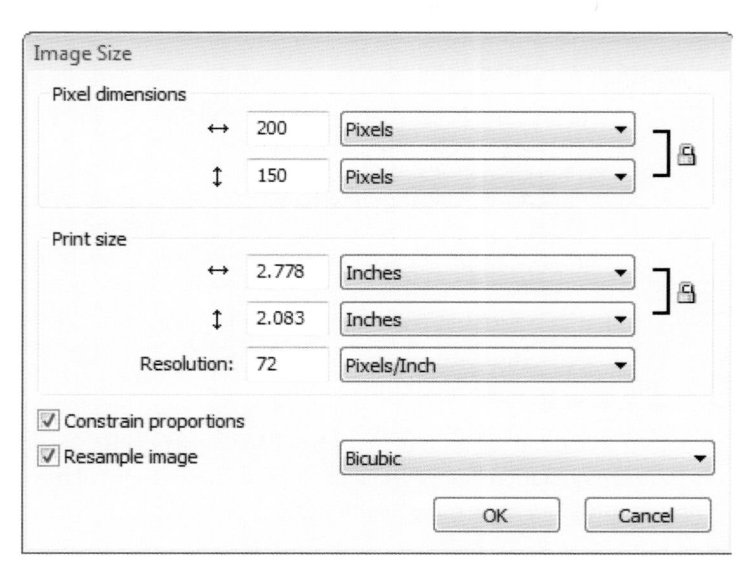

Using tags to adjust image size

The height and width attributes are used to set the height and width of an image on a web page. Using these you can specify how big the image appears on the web page regardless of how large the image actually is.

If you do not set a height and width for the image, it will appear in the actual size it was saved.

```
<img src="filenameofyourimage.jpg" width="100"
height="100"/>
```

This displays the image whose filename is 'filenameofyourimage' using a width of 100 pixels and a height of 100 pixels.

Although you may think that this reduces the file size of the image and hence the time taken to load the image onto the web page, this is not the case, as the image is still loaded from the disk and then adjusted to fit the size specified.

It is therefore better to reduce the image size using an image editing program, as this will reduce the file size when the adjusted image is saved. This will make the file size for the image smaller and hence faster to load on the web page.

It is still useful to use width and height attributes because it tells the browser software how much space is needed in the final layout and this makes the web page download faster.

Positioning of images and text in tables (vertical and horizontal alignment)

Once a table structure has been defined, content such as text and images can be added to the cells. How an image appears in a cell depends on the alignment used for that cell. Both text and images can be aligned vertically in a cell using the following:

```
<td valign="top"><This text will appear aligned to the top of
the cell></td>
```

The above valign will set the vertical alignment of the image to the top of the cell.

```
<td valign="top"><img src="picture1.jpg"></td>
```

The above valign will set the vertical alignment of the image in the cell to the top.

You can have the following vertical alignment:

```
valign ="top"
valign ="middle"
valign ="bottom"
```

For horizontal alignment of text and images, the following is used:

```
<td align="center"><This text will be aligned centrally></td>
```

The above HTML aligns the text centrally in the cell.

```
<td align="right"><img src="picture1.jpg"></td>
```

The above HTML aligns the text to the right in the cell.

You can have the following vertical alignment:

```
align ="left"
align ="center"
align ="right"
```

Activity 21.12

Inserting an image into a cell in a table

In this activity you will learn how to:

- insert an image into a cell in a table
- align an image horizontally and vertically in a cell.

1. Before starting this activity you will need to check that the image file called **Oxford_banner.jpg** appears in the same folder that you use to keep your HTML files. If the file is not in the folder, you need to download it into the folder.
2. Load Notepad and open the file called **Table_with_image.htm**.

   The following HTML appears:

```
<table border="1" cellspacing="5" cellpadding="5" width="50%">
        <tr>
                <td><img src="Oxford_banner.jpg" width="600" height="200"></td><td>Cell 2</td>
        </tr>
        <tr>
                <td>Cell 3</td><td>Cell 4</td>
        </tr>
</table>
```

Notice the way the tag for the image has been inserted into the first cell of the table. The filename of the image to be inserted is between quotation marks and also the width and height of the image have been specified in pixels.

```
<img src="Oxford_banner.jpg" width="600" height="200">
```

⇨

3 Load the file **Table_with_image.htm** into the browser and you will see the image appear in the table like this:

Cell 2

Cell 3

Cell 4

4 There are three images you now need to place in the other cells. Each image is to have a width of 200 pixels and a height of 100 pixels. The image files are to be placed as follows:

In Cell 2: oup1

In Cell 3: oup2

In Cell 4: oup3

Adjust the HTML by replacing the text Cell 2, Cell 3 and Cell 4 with the above images. Check that when viewed in the Browser the web page looks like this:

⇨

5 You are now going to alter the alignment of the content of the cells so that the image in the top right cell is aligned to the top and the image in the bottom left cell is aligned to the right.

Change the tag for the top right data cell to:

```
<td valign="top"><img src="oup1.jpg" width="200" height="100"></td>
```

Change the tag for the bottom left data cell to:

```
<td align="right"><img src="oup2.jpg" width="200" height="100"></td>
```

Save the file using the file name **Table_with_image1.htm**

Load the file into your browser and check it appears as in the screenshot above.

Adjusting the alignment of the image on a web page

You can adjust the alignment of the image on a web page. This section of HTML aligns the image to the middle of the page as well as setting the height and width of the image.

```
<img src="filenameofyourimage.jpg" align="middle"
width="100" height="100"/>
```

Here is a list showing the alignments you can have:

```
align="middle"
align="left"
align="right"
align="top"
align="bottom"
```

All these alignments need to be inserted into the tag for the image similar to that shown above.

For example:

```
<img src="image.jpg" align="left"/>
```

aligns the image from image.jpg to the left of the page.

Using software to resize an image

It is often necessary to resize an image so that it can be put into a certain space on a web page. There are two ways this can be done:

Use image editing software (e.g. Adobe Photoshop, Microsoft Photo Editor, Macromedia Fireworks, etc.)

Specify the required height and width of the image in pixels in the image tag like this:

```
<img src="oup1.jpg" width="200" height="100">
```

Maintaining the aspect ratio

Maintaining the aspect ratio means keeping the ratio of the width to the height of an image the same when the image is resized. This can be done in the image tag as follows:

```
<img src="oup1.jpg" width="200">
```

The image in the image tag is resized so that it has a width of 200 pixels. Notice that the height is not specified in the tag as it will be automatically set to maintain the aspect ratio.

Distorting an image

Sometimes an image needs to be stretched in one direction only. For example, the shape of the original image is not quite the same shape as the gap on the web page it needs to fill. Stretching an image in one direction means that the aspect ratio is not being maintained.

The top image is the original and the bottom image is distorted because it has been widened while keeping the height the same as the first image. You can see that the image has been stretched in the horizontal direction only.

Aspect ratio

The aspect ratio of an image is the ratio of the width of the image to its height. It is expressed as two numbers separated by a colon. Suppose the aspect ratio of an original image is 2:1. You can distort the image using a tag by specifying the width and height like this:

```
<img src="oup1.jpg" width="200" height ="120">
```

Notice that the ratio of the width to the height is no longer 2:1 so the aspect ratio has changed meaning that the image has been distorted slightly.

Using images as links

Links can be made using images. When the user clicks on the image used as the link they will be taken to a new web page or website. It is important to make it clear to a user that the image is acting as a link and that they have to click on the image to follow the link. This can be done by putting a short message next to the image.

```
<html>
    <h1>International resources</h1>
    <p align="center"><a href="index.htm"><img src="oup1.jpg"
    width="200" height="200" border="3" alt="Click here for
    the link to international resources"</a></p>
    <p align="center">Click on the image to follow the link</p>
</html>
```

Look carefully at the above section of HTML. The text 'International resources' will appear at the top of the page as a heading.

A new paragraph will be started and an image will be centred in the horizontal direction and the image with the filename 'oup1.jpg' will be sized and placed with a border in this position.

This image also acts as a link to the locally stored web page with the file name 'index.htm' and when the user puts the cursor over the image, the text 'Click here for the link to international resources' will appear.

Look at the following line in the above code:

```
<p align="center">Click on the image to follow the link</p>
```

This will display the message 'Click on the image to follow the link' which is positioned centrally and under the image like this:

Click on the image to follow the link

The need for low resolution images for data transfer

When an image is used on a website, the user of the website has to wait while the image is obtained from the server, transferred over the internet and assembled on the web page using the browser software. The time they have to wait for the image file to be transferred is called the data transfer time (also called the download time). If the user has a high speed broadband internet connection this time will be low. If they have a low speed connection there will be a time delay and this is annoying. The time for the image to appear depends on the file size of the image. It is therefore advisable to keep image files as small as possible.

For the above reason, images in websites are normally used at low resolutions as low resolution images have a smaller file size.

Amending the file type

There are many file types for images but only three that are used with websites:

- GIF file type images consist of only 256 colours. This file type is good at compressing images so that they load quickly.
- JPEG/JPG file type images comprise millions of colours and this makes them the main file type for photographic images.
- PNG file type images are able to use millions of colours and are good at compressing images.

All graphics packages such as Fireworks or Photoshop are able to change different file types to those shown above. All you need to do is to select Save As... and select the file type you want to save in.

Printing websites in browser view

Printing websites in browser view means printing the web pages exactly as they appear when viewed in the browser you use. In the examination, you will usually be asked to add your name and candidate number on the web pages so that when they are printed, your work can be identified.

Printing evidence of HTML code or stylesheets

If you are asked to supply a copy of the HTML code or stylesheet or both, you will be asked to supply a printout containing details such as your name, candidate number and centre number. As always, you must follow the exact instructions given about the details required in the examination paper.

To produce a printout containing these details it is best to take a screenshot of the HTML or stylesheet using the Prt Scr key on the keyboard. This will take a copy of the screen and store it in the clipboard. The screenshot can then be pasted into a new word-processing document where it can be cropped to remove any unwanted parts and resized so that all information is clearly seen. The details such as name, candidate number and centre number can then be added before saving the document and then printing it out.

You can also print the HTML code directly from some editors. Print screens may not show enough for a long/wide string of code.

Practical exam tip All documents and printouts must have your name and some other details printed (not written) on them.

You might also have to print the CSS file - again putting your name on it. You can use Print screen and copy the screen to word-processing software such as MS Word and then type the details such as your name and other details before printing out the document.

Glossary

3D printer printer that can print in three-dimensions by repeatedly building up layers of material.

Abnormal data data that is unacceptable and that should be rejected by a validation check; for example, entering text into a numeric field or inputting data which is outside the range specified.

Absolute reference a reference to a cell used in a formula where, when the formula is copied to a new address, the cell address does not change.

Address book the names and email addresses of all the people whom you are likely to send email stored as a file.

Alphanumeric data sometimes called text and it includes letters, digits and punctuation marks.

Analogue a continuously changing quantity that needs to be converted to digital values before it can be processed by a computer.

Analogue-to-digital converter (ADC) a device that changes continuously changing quantities (such as temperature) into digital quantities.

ANPR (automatic number plate recognition) a method using OCR on images obtained from video cameras to read car registration plates.

Anti-spyware software used to detect and remove spyware which may have been put on your computer without your knowledge or permission.

Anti-virus software software that is used to detect and destroy computer viruses.

Applet a program designed to be executed from within another application. Unlike an application, applets cannot be executed directly from the operating system.

Applications software software designed to do a particular job such as word-processing or database software.

Artificial intelligence (AI) creating computer programs or computer systems that behave in a similar way to the human brain by learning from experience, etc.

Aspect ratio the ratio of the width of an image to its height.

ATM (automatic teller machine) another name for a cashpoint/cash dispenser.

Attachment a file which is attached to an email which the recipient can open and view the contents provided they have suitable software to open the file.

Audio-conferencing conducting a meeting held by people in different places using devices that allow voice to be sent and received.

Backing storage storage which is not classed as ROM or RAM. It is used to hold programs and data. Backing storage devices include magnetic hard drives, optical drives (CD or DVD), flash/pen drives, etc.

Backup keeping copies of software and data so that they can be recovered should there be corruption or loss of some or all of the ICT system.

Backup file copy of a file which is used in the event of the original file being corrupted (damaged) or lost.

Bandwidth a measure of the amount of data that can be transferred per second over the internet or other network.

Bar code a series of lines of differing thickness which are used to represent a number which is usually written below the bar code. Can be read by a scanner to input the number accurately.

Batch processing type of processing where all the inputs needed are collected over a period of time and then batched together, inputted and processed in one go. For example, questionnaires from a survey are collected over a few weeks and then batched together and processed in one go.

bcc (blind carbon copy) this is useful when you want to send an email to one person and others but you do not want the others to see each other's email addresses.

Biometric a unique property of the human body such as fingerprints or retinal patterns which can be used to identify a person and allow them access to a computer system.

BIOS (basic input/output system) stored in ROM and holds instructions used to 'boot' (i.e. start) the computer up when first switched on.

Bit a binary digit 0 or 1.

Blog a website that allows comments to be posted; usually in reverse chronological order.

Blogger a person who maintains a blog.

Bluetooth a method used to wirelessly transfer data over short distances from fixed and mobile devices. The range of Bluetooth depends on the power of the signal and can typically be from 5m to 100m.

Blu-ray optical disk that has a much higher storage capacity than a DVD. Blu-ray disks have capacities of 25 Gb, 50 Gb, and 100 Gb. These high capacity Blu-ray disks are used to store high definition video. They are used for storing films/movies with a 25 Gb Blu-ray disk being able to store 2 hours of HDTV or 13 hours of standard definition TV. It is possible to play back video on a Blu-ray disk while simultaneously recording HD video.

Bookmark storage area where the URL (i.e. the web address) of a website can be stored so that it can be accessed later using a link.

Boolean data data that can exist only in two states; for example True or False.

Bridge a hardware device used to connect two local area networks to each other. The purpose of a bridge is to decide whether a message needs to be transferred between the two networks or just confined to one of them. This reduces network traffic.

Browser (also called web browser) software program you use to access the internet. Microsoft Internet Explorer is an example of a web browser.

Bullet point a block or paragraph of text that has a symbol placed in front to make the section of text stand out.

CAD (computer-aided design) software software used to produce technical drawings, plans, designs, maps etc.

cc (carbon copy) is used when you want to send an email to one person but you also want others to see the email you are sending. To do this you enter the email address of the main person you are sending it to and in the box marked cc you enter all the email addresses, separated by commas, of all the people you wish to receive a copy.

CD-R (CD recordable) optical storage where data is stored as an optical pattern. The user can record their data onto the disk once only.

CD-ROM (CD read only memory) optical storage where data is stored as an optical pattern. Once data has been written onto CD-ROM it cannot be erased. It is mainly used for the distribution of software.

CD-RW (CD rewritable) optical storage that allows data to be stored on the disk over and over again, just like a hard disk. This is needed if the data stored on the disk needs to be updated. You can treat a CD-RW like a hard drive but the transfer rate is less and the time taken to locate a file is greater. The media is not as robust as a hard drive.

Cell an area on a spreadsheet produced by the intersection of a column and a row, in which data can be placed.

Changeover the process by which an older ICT system is replaced with a newer one.

Character any symbol (letter, number, punctuation mark etc.) that you can type from the keyboard.

Check digit a decimal number (or alphanumeric character) added to a number for the purpose of detecting the sorts of errors humans normally make on data entry.

Chip and PIN Chip readers are the devices into which you place a credit/debit card to read the data which is encrypted in the chip on the card. The PIN pad is the small numeric keypad where the personal identification number (PIN) is entered and the holder can be verified as the true owner of the card.

CLI (command line interface) type of user interface where a user has to type in instructions in a certain format to accomplish a task.

Clipboard temporary storage area used for copying or cutting data to and then pasting it somewhere else.

Cloud computing internet-based computing where programs and data are stored on the internet rather than on the user's own computer.

Command line interface (CLI) type of user interface where a user has to type in instructions in a certain format to accomplish a task.

Compression storing data in a format that requires less space. A compressed file takes less time to be transferred across a network.

Computer Misuse Act a law which makes illegal a number of activities such as deliberately planting viruses, hacking, using ICT equipment for fraud, etc.

Content the actual text, images, etc.

Control system system used to control a process automatically by making use of data from sensors as the input to the system.

Copyright, Designs and Patents Act a law making it a criminal offence to copy or steal software or use the software in a way that is not allowed according to the software licence.

CPU (central processing unit) the computer's brain. It interprets and executes the commands given to it by the hardware and software.

Cropping only using part of an image.

CSS (cascading style sheet) file used to format the contents of a web page. The file contains properties on how to display HTML elements. HTML is a special code used for making web pages. For example, a user can define the size, colour, font, line spacing, indentation, borders, and location of HTML elements. CSS files are used to create a similar look and feel across all the web-pages in a website.

CSV comma separated variables. A way of holding data in a file so that it can be transferred into databases or spreadsheets.

Data raw facts and figures, e.g. readings from sensors, survey facts, etc.

Data capture term for the various methods by which data can be entered into the computer so that it can be processed.

Data logger a device which collects readings from one or more sensors. The time interval between each reading can be varied (called the logging rate) and the total time over which the data is logged (called the logging period) can also be varied.

Data logging the process of using an ICT system to collect data from sensors at a certain rate over a certain period of time. Remote weather stations use data logging.

Data Protection Act an act that restricts the way personal information is stored and processed on a computer.

Data redundancy where the same data is stored more than once in a table or where the same data is stored in more than one table.

Data type check validation check to ensure the data being entered is the same type as the data type specified for the field.

Database a series of files stored in a computer which can be accessed in a variety of different ways.

Device driver a short specially written program that understands the operation of the device it controls/operates. For example, driver software can operate/control a printer or scanner. Driver software is needed so as to allow the systems or applications software to control the device properly.

Digital camera a camera that takes a picture and stores it digitally so that it can be transferred to and processed by a computer or other device.

Digital certificate An electronic passport that allows a person, computer or organisation to exchange information securely using the internet. The digital certificate provides identifying information and is issued by a trusted body and is used with websites to ensure that a website you are visiting is genuine.

Digital-to-analogue converter (DAC) a device that changes digital quantities into analogue ones.

Direct/random access data is accessed immediately from the storage media. This is the method used with storage media such as magnetic hard disks and optical media such as CD and DVD.

Dot matrix printer a printer which uses numerous tiny dots to make up each printed character. It works by hitting tiny pins against an inked ribbon to make the dots on the page, so this makes this type of printer noisy.

Double entry of data entering the details into an ICT system twice; only if the two sets of data are identical will they be accepted for processing. It is a method of verification. This method is often used when you set up a new password.

Download to copy files from a distant computer to the one you are working on.

Drag and drop allows you to select objects (icons, folders, files etc.) and drag them so that you can perform certain operations on them such as drag to the recycle bin to discard, add a file to a folder, copy files to a folder and so on.

DVD-R (DVD recordable) a type of optical storage. DVD-R allows data to be stored on a DVD only once.

DVD-ROM (digital versatile disk read only memory) DVD-ROM is optical storage and offers much higher storage capacity compared to CD. It is used for the distribution of movies where you can only read the data off the disk. A DVD-ROM drive can also be used for the reading of data off a CD. DVD is mainly used for the distribution of films and multimedia encyclopaedias.

EFTPOS (Electronic funds transfer at point of sale) where electronic funds transfer takes place at a point of sale terminal. This means that money is transferred from the bank or credit card company to the store when you pay for goods at a store.

Encryption the process of coding files before they are sent over a network to protect them from hackers. Also the process of coding files stored on a computer/storage device so that if the computer/storage device is stolen, the files cannot be read. Only the person who has a special key can see the information in its original form.

EPOS (Electronic point of sale) a computerised till which can be used for stock control.

Ergonomics an applied science concerned with designing and arranging things people use so that the people and things interact most efficiently and safely.

e-safety using the internet in a safe and responsible way.

Evaluation the act of reviewing what has been achieved, how it was achieved and how well the solution works.

Expert system an ICT system that mimics the decision-making ability of a human expert.

External hardware devices those hardware devices situated outside the computer casing.

Extreme data is data on the borderline of what the system will accept. For example, if a range check specifies that a number from 16 to 21 inclusive is entered, the extreme data would be 16 and 21.

Favourites storage area where the URL (i.e. the web address) of a website can be stored so that it can be accessed later using a link.

Fax a machine capable of sending and receiving text and pictures along telephone lines.

Field a space in an information handling system/database used for inputting data. For instance, you could have fields for surname, date of birth, etc.

File a collection of related data.

File attachment (sometimes called an attachment) a file that is attached to an email and can be sent to another person or a group of people.

File compression using special software to reduce file size before sending them over the internet or to reduce their size so that they take up less space on the storage media.

File Transfer Protocol (FTP) a common protocol for the movement of files across the internet. Users can access a server that contains the file they want and then download it to their computer using FTP. Alternatively, they can upload a file to the server from their computer using FTP.

Firewall a piece of software, hardware or both that is able to protect a network from hackers.

Flash/pen drives portable storage media which offer cheap and large storage capacities and are ideal media for photographs, music and other data files. They consist of printed circuit boards enclosed in a plastic case.

Flat file method used for storage of data in a database where all the data is held in a single table.

Font a set of letters and characters in a particular design.

Footer text placed at the bottom of a document.

Format checks checks performed on codes to make sure that they conform to the correct combinations of characters.

Generic file formats file formats that are able to be used by different software no matter who the manufacturer of the software is.

GIF (Graphics Interchange Format) file type used for images. Images in this format are reduced to a maximum of 256 colours. Images in this format are compressed so this means that they load quickly. Used for simple line diagrams or clip art.

GIGO abbreviation for garbage in garbage out. It means that if you put rubbish into the computer then you get rubbish out.

GIS (Geographic information system) an ICT system that is used to capture, manage, analyse and display geographically referenced information.

GPS (Global Positioning System) system which uses the signals from several satellites to obtain the exact position of any object (e.g. aircraft, ship, car, etc.) on the Earth's surface. Many cars are equipped with satellite navigation systems which use GPS so that the driver can locate their position on a map on a small screen inside the car.

Graph plotter an output device which draws by moving a pen. Useful for scale drawings and is used mainly with CAD packages.

Geographic information system (GIS) an ICT system that is used to capture, manage, analyse and display geographically referenced information.

Graphics tablet an input device which consists of shapes and commands on a tablet which can be selected by the user by touching. The use of a graphics tablet means that more space is left on the screen for a plan or diagram.

GUI (graphical user interface) interface that allows users to communicate with the computer using windows, icons, menus and pointers.

Hackers people who break into a computer/computer network illegally.

Hacking process of trying to break into a secure computer system.

Hard copy printed output from a computer which may be taken away and studied.

Hardware the physical components of a computer system.

Header text placed at the top of a document.

Hot spot an image or piece of text used as a link. When you click on the image or text, you are taken to another part of the same page, a different page or a different site, or it may open a new file or a new window.

htm a popular file format used for storing web pages.

HTTP (HyperText Transfer Protocol) a protocol that defines the process of identifying, requesting and transferring multimedia web pages over the internet.

HTTPS (HyperText Transfer Protocol secure variant) a protocol that defines the process of identifying, requesting and transferring multimedia web pages over the internet, except unlike HTTP it uses encrypted connections to hide passwords, bank information and other sensitive material from the open network.

Hub a hub contains multiple ports (i.e. connection points). When a packet of data arrives at one port, it is copied to the other ports so that all network devices of the LAN can see all packets. Every device on the network will receive the packet of data which it will inspect to see if it is relevant or not.

Hyperlink a feature of a website that allows a user to jump to another web page, to jump to part of the same web page or to send an email message.

Hypertext Mark-Up Language (HTML) a code used to create documents on the world wide web. You use it to specify the structure and layout of a web document.

Identity theft using someone's banking/credit card/personal details in order to commit fraud.

Inference engine one of four parts of an expert system. It uses the input data along with the rules in the rules base and the knowledge in the knowledge base to arrive at conclusions/decisions/answers which are output using the user interface.

Ink-jet printer printer that works by spraying ink through nozzles onto the paper.

Input device the hardware device used to feed the input data into an ICT system such as a keyboard or a scanner.

Instant messaging (IM) a method of two people using real-time text to conduct a conversation using the internet.

Integer a whole number which can be positive, negative or zero.

Interactive where there is a constant dialogue between the user and the computer.

Internal hardware devices those hardware devices situated inside the computer casing.

Internal memory memory inside the computer casing (e.g. ROM, RAM and internal hard disk).

Internet a huge group of networks joined together. The largest network in the world.

Internet service provider (ISP) a company that provides an internet connection.

Intranet a private network used with an organisation that makes uses of internet technology (e.g. web pages and web browsers) used for sharing internal information.

IP (Internet Protocol) address a number which uniquely identifies the physical computer linked to the internet.

Joystick input device used instead of the cursor keys or mouse as a way of producing movement on the screen.

JPEG a file format used for still images which uses millions of colours and compression, which makes it an ideal file format for photographic images on web pages.

K Kilobyte or 1024 bytes. Often abbreviated as KB. A measure of the storage capacity of disks and memory.

Key field this is a field that is unique for a particular record in a database.

Key logging the process of someone recording/monitoring the keys you press when you are using your computer using a key logger which can either be hardware or software. As it can record keystrokes someone can record your passwords and banking details.

Knowledge base one of four parts of an expert system. A huge organised set of knowledge about a particular subject. It contains facts and also judgemental knowledge, which gives it the ability to make a good guess, like a human expert.

LAN (local area network) a network of computers on one site.

Landscape page orientation where the width is greater than the height.

Laser printer printer which uses a laser beam to form characters on the paper.

LCD (liquid crystal display) technology used with thin flat screens.

Length check validation check to make sure that the data being entered has the correct number of characters in it.

Light pen input device used to draw

directly on a computer screen or used to make selections on the screen.

Linkers programs that are usually part of the compiler which take care of the linking between the code the programmer writes and other resources and libraries that make up the whole program file that can be executed (i.e. run).

Login accessing an ICT system usually by entering a user-ID/username and/or a password.

Magnetic stripe stripe on a plastic card where data is encoded in a magnetic pattern on the stripe and can be read by swiping the card using a magnetic stripe reader.

Magnetic stripe reader hardware device that reads the data contained in magnetic stripes such as those on the back of debit/credit cards.

Mail merge combining a list of names and addresses with a standard letter so that a series of letters is produced with each letter being addressed to a different person.

Main internal memory memory which is either ROM (read only memory) or RAM (random access memory).

Master slides (also called slide masters) used to help ensure consistency from slide to slide in a presentation. They are also used to place objects and set styles on each slide. Using master slides you can format titles, backgrounds, colour schemes, dates, slide numbers, etc.

Megabyte a unit of file or memory size that is 1024 kilobytes.

Megapixel one million pixels (i.e. dots of light).

Memory cards thin cards you see in digital cameras used to store photographs and can be used for other data.

Memory stick/flash drive/pen drive solid state memory used for backup and is usually connected to the computer using a USB port.

MICR (magnetic ink character recognition) system used for cheque clearing which is able to read data printed onto cheques in magnetic ink.

Microprocessor the brain of the computer consisting of millions of tiny circuits on a silicon chip. It processes the input data to produce information.

MIDI (musical instrument digital interface) enables a computer and musical instrument to communicate with each other.

Monitor another name for a VDU or computer screen.

Motherboard the main printed circuit board in a computer that connects the central processing unit, memory and connectors to the external hardware devices such as keyboard, screen, mouse, etc.

MP3 music file format that uses compression to reduce the file size considerably, which is why the MP3 file format is popular with portable music players such as iPods and mobile phones.

Multimedia making use of many media such as text, image, sound, animation and video.

Multimedia projector output device used to project the screen display from a computer onto a much larger screen that can be viewed by a large audience.

Network group of ICT devices (computers, printers, scanners etc.) which are able to communicate with each other.

Network interface card (NIC) circuit board which connects to the motherboard of the computer. It prepares the data ready for sending over a network. They include connectors that allow connection to network cables.

Networking software systems software which allows computers connected together to function as a network.

Normal data data that is acceptable for processing and will pass the validation checks.

OCR (optical character recognition) a combination of software and a scanner which is able to read characters into the computer.

OMR (optical mark reader/recognition) reader that detects marks on a piece of paper. Shaded areas are detected and the computer can read the information contained in them.

Online processing the system is automatically updated when a change (called a transaction) is made. This means that the system always contains up-to-date information. Online processing is used with booking systems to ensure seats are not double booked.

Online shopping shopping over the internet as opposed to using traditional methods such as buying goods or services from shops or trading using the telephone.

Operating system software that controls the hardware of a computer and is used to run the applications software. Operating systems control the handling of input, output, interrupts, etc.

Optical character recognition (OCR) input method using a scanner as the input device along with special software which looks at the shape of each character so that it can be recognised separately.

Optical disk a plastic disk used for removable storage. Includes CD and DVD.

Optical mark recognition (OMR) the process of reading marks (usually shaded boxes) made on a specially prepared document. The marks are read using an optical mark reader.

Output the results from processing data.

Password a series of characters chosen by the user that are used to check the identity of the user when they require access to an ICT system.

pdf (portable document format) used with a particular piece of applications software for viewing documents. This file format is used by the software Adobe Acrobat. This file format is popular because the software used to view the files is widely available and free.

Personal data data about a living identifiable person which is specific to that person.

Pharming malicious programming code is stored on a computer. Any users who try to access a website which has been stored on the computer will be re-directed automatically by the malicious code to a bogus website and not the website they wanted. The fake or bogus website is often used to obtain passwords or banking details so that these can be used fraudulently.

Phishing fraudulently trying to get people to reveal usernames, passwords, credit card details, account numbers, etc., by pretending to be from a bank, building society, or credit card company, etc. Emails are sent asking recipients to reveal their details.

PIN (personal identification number) secret number that needs to be keyed in to gain access to an ATM or to pay for goods/services using a credit/debit card.

Piracy the process of illegally copying software.

Pixel a single point in a graphics element or the smallest dot of light that can appear on a computer screen.

Plotter a device which draws by moving a pen. Useful for printing scale drawings, designs and maps.

Podcast a digital radio broadcast created using a microphone, computer and audio editing software. The resulting file is saved in MP3 format and then uploaded onto an internet server. It can then be downloaded using a facility called RSS onto an MP3 player for storing and then listening.

Point a length which is 1/72 inch. Font size is measured in points. For example, font size of 12 pts means 12/72=1/6 inch which is the height the characters will be.

Portrait page orientation where the height is greater than the width.

Presence checks validation checks used to ensure that data has been entered into a field.

Print preview feature that comes with most software used to produce documents. It allows users to view the page or pages of a document to see exactly how they will be printed. If necessary, the documents can be corrected. Print preview saves paper and ink.

Process any operation that transfers data into information.

Processing performing something on the input data such as performing calculations, making decisions or arranging the data into a meaningful order.

Processor often called the CPU and is the brain of the computer consisting of millions of tiny circuits on a silicon chip. It processes the input data to produce information.

Program the set of step-by-step instructions that tell the computer hardware what to do.

Proofreading carefully reading what has been typed in and comparing it with what is on the data source (order forms, application forms, invoices etc.) for any errors, or just reading what has been typed in to check that it makes sense and contains no errors.

Protocol a set of rules governing the format of the data and the signals to start, control and end the transfer of data across a network.

Proxy-server a server which can be hardware or software that takes requests from users for access to other servers and either forwards them onto the other servers or denies access to the servers.

Query a request for specific information from a database.

RAM (random access memory) type of main internal memory on a chip which is temporary/volatile because it loses its contents when the power is removed. It is used to hold the operating system and the software currently in use and the files being currently worked on. The contents of RAM are constantly changing.

Range check data validation technique which checks that the data input to a computer is within a certain range.

Read only a user can only read the contents of the file. They cannot alter or delete the data.

Read/write a user can read the data held in the file and can alter the data.

Real-time a real-time system accepts data and processes it immediately. The results have a direct effect on the next set of available data.

Real-time processing type of processing where data received by the system is processed immediately without any delay. Used mainly for control systems, e.g. autopilot systems in aircraft.

Record the information about an item or person. A row in a table.

Relational database a database where the data is held in two or more tables with relationships (links) established between them. The software is used to set up and hold the data as well as to extract and manipulate the stored data.

Relative reference when a cell is used in a formula and the formula is copied to a new address, the cell address changes to take account of the formula's new position.

Report the output from a database in which the results are presented in a way that is controlled by the user.

Resolution the sharpness or clarity of an image.

RFID (radio frequency identification) reader reads data from a small chip/tag at a distance.

ROM (read only memory) type of internal memory on a chip which is permanent/non-volatile and cannot have its contents changed by the user. It is used to hold the boot routines used to start the computer off when the power is switched on.

Router hardware device which is able to make the decision about the path that an individual packet of data should take so that it arrives in the shortest possible time. It is used to enable several computers to share the same connection to the internet.

RSI (repetitive strain injury) a painful muscular condition caused by repeatedly using certain muscles in the same way.

rtf (rich text format) file format that saves text with a limited amount of formatting. Rich text format files use the file extension '.rtf'.

Rules base one of four parts of an expert system and made up of a series of IF, AND, THEN statements to closely follow human-like reasoning.

Sans serif a set of typefaces or fonts that do not use the small lines at the end of characters which are called serifs.

Scanner input device that can be used to capture an image; useful for digitising old non-digital photographs, paper documents or pictures in books.

Screenshot copy of what is seen on a computer screen. It can be obtained by pressing the Prt Scrn button and then a copy of the screen will be placed in the clipboard. The copy of the screen can then be pasted.

Search engine program which searches for required information on the internet.

Secondary storage storage other than ROM or RAM and is non-volatile, which means it holds its contents when the power is removed. It is used to hold software/files that are not instantly needed by the computer. Also used for backup copies.

Secure Socket Layer (SSL) see SSL.

Sensors devices which measure physical quantities such as temperature, pressure, humidity, etc.

Serial/sequential access data is accessed from the storage media by starting at the beginning of the media until the required data is found. It is the type of access used with magnetic tape and it is a very slow form of access when looking for particular data on a tape.

Serif a small decorative line added to the basic form of a character (letter, number, punctuation mark, etc.).

Smartphone a mobile phone with an advanced operating system which combines the features of a mobile phone with other mobile devices such as web browser, media player, Wi-Fi, camera, GPS navigation, etc.

Smishing using text messaging to send fraudulent messages that try to steal your credit card, banking or other personal details.

SMS (short messaging service) service that provides texting.

Social networking site a website used to communicate with friends, family and to make new friends and contacts.

Software programs controlling the operation of a computer or for the processing of electronic data.

Software licence document (digital or paper) which sets out the terms by which the software can be used. It will refer to the number of computers on which it can be run simultaneously.

Solid state backing storage the smallest form of memory and is used as removable storage. Because there are no moving parts and no removable media to damage, this type of storage is very robust. The data stored on solid state backing storage is rewritable and does not need electricity to keep the data. Solid state backing storage includes memory sticks/pen drives and flash memory cards.

Sound card expansion card consisting of circuitry that allows a computer to send audio signals to audio devices such as speakers or headphones.

Spam unsolicited bulk email (i.e. email from people you do not know, sent to everyone in the hope that a small percentage may purchase the goods or services on offer).

Spellchecker program usually found with a word-processor and most packages which make use of text which checks the spelling in a document and suggests correctly spelled words.

Spyware software that is put onto a computer without the owner's knowledge and consent with the purpose of monitoring the user's use of the internet. For example, it can monitor keystrokes so it can be used to record usernames and passwords. This information can then be used to commit fraud.

SSL (Secure Socket Layer) standard used for security for transactions made using the internet. SSL allows an encrypted link to be set up between two computers connected using the internet and it protects the communication from being intercepted and allows your computer to identify the website it is communicating with.

Standalone computer if a computer is used on its own without any connection (wireless or wired) to a network (including the internet), then it is a standalone computer.

Storage media the collective name for the different types of storage materials such as DVD, magnetic hard disk, solid state memory card, etc.

Style sheet a document which sets out fonts and font sizes for headings and subheadings, etc., in a document. Changes to a heading need only be made in the style sheet and all the changes to headings in the document will be made automatically.

Swipe card plastic card containing data stored in a magnetic stripe on the card.

Switch a device that is able to inspect packets of data so that they are forwarded appropriately to the correct computer. Because a switch only sends a packet of data to the computer it is intended for, it reduces the amount of data on the network, thus speeding the network up.

Systems software programs that control the computer hardware directly by giving the step-by-step instructions that tell the computer hardware what to do.

Tablet a portable hand-held computer device that is primarily controlled via a touch screen.

Tags special markers used in HTML to tell the computer what to do with the text. A tag is needed at the start and end of the block of text to which the tag applies.

Tape magnetic media used to store data.

Templates electronic files which hold standardised document layouts.

Terabyte a unit of file or memory size that is 1024 gigabytes.

TFT (thin film transistor) a thin screen used in laptops/notebooks or in desktops where desk space is limited.

Thesaurus software which suggests words with similar meanings to the word highlighted in a document.

Touchscreen a special type of screen that is sensitive to touch. A selection is made from a menu on the screen by touching part of it.

Tracker ball an input device which is rather like an upside down mouse and is ideal for children or disabled people who find it hard to move a mouse.

Transaction a piece of business, e.g.

an order, purchase, return, delivery, transfer of money, etc.

Transcription error error made when typing data in using a document as the source of the data.

Transposition error error made when characters are swapped around so they are in the wrong order.

txt text files just contain text without any formatting. Text files use the file extension '.txt'.

Update the process of changing information in a file that has become out of date.

URL (Uniform Resource Locator) another way of saying a web address.

USB (Universal Serial Bus) a socket which is used to connect devices to the computer such as webcams, flash drives, portable hard disks, etc.

User a person who uses a computer.

User interface one of four parts of an expert system that uses an interactive screen (which can be a touch screen) to present questions and information to the operator and also receive answers from the operator. Can also be one of the three parts of an expert system.

User log a record of the successful and failed logins and also the resources used by those users who have access to network resources.

Username or User-ID a name or number that is used to identify a certain user of the network or system.

Utility software part of the systems software that performs a specific task such as create a new folder, copy files, etc.

Validation checks these are checks that a developer of a solution sets/creates, using the software, in order to restrict the data that a user can enter so as to reduce errors.

Verification checking that the data being entered into the ICT system perfectly matches the source of the data.

Video cards circuits that generate signals that enable a video output device to display text and graphics.

Video-conferencing ICT system that allows virtual face-to-face meetings to be conducted without the participants being in the same room or even the same geographical area.

Virus a program that copies itself automatically and can cause damage to data or cause the computer to run slowly.

Voice recognition the ability of a computer to 'understand' spoken words by comparing them with stored data.

WAN (wide area network) a network where the terminals/computers are remote from each other and telecommunications are used to communicate between them.

Web logs (blogs) websites that are created by an individual with information about events in their life, videos, photographs, etc.

Webcam a digital video camera that is used to capture moving images and is connected to the internet so the video can be seen by others remotely. They are often included as part of the screen in computers or bought separately and connected to a USB port.

Web-conferencing service allowing conferencing events to be shared with users who are in remote locations using internet technologies. Usually there will be one sender and many recipients.

Web page a document that can be accessed using browser software.

Website a collection of interconnected web pages relating to a topic or organisation.

Wi-Fi a trademark for the certification of products that meet certain standards for transmitting data over wireless networks.

WIMP (windows icons menus pointing devices) the graphical user interface (GUI) way of using a computer rather than typing in commands at the command line.

WLAN a local area network (LAN) where some or all of the links are wireless, making use instead of infrared or microwaves as a carrier for the data rather than wires or cables.

World wide web (www) the way of accessing the information on all the networked computers which make up the internet. WWW makes use of web pages and web browsers to store and access the information.

zip a popular archive format widely used when downloading files from websites using the internet. ZIP files are data containers since they can store one or several files in the compressed form. After you download a ZIP file, you will need to extract its contents in order to use them.

Complete ICT for IGCSE: Index

Q

R

S